Photoprotective Green Pharmacology: Challenges, Sources and Future Applications

Vinod K. Kannaujiya • Rajeshwar P. Sinha
Md. Akhlaqur Rahman • Shanthy Sundaram
Editors

Photoprotective Green Pharmacology: Challenges, Sources and Future Applications

Springer

Editors
Vinod K. Kannaujiya
Department of Botany, MMV
Banaras Hindu University
Varanasi, Uttar Pradesh, India

Rajeshwar P. Sinha
Center of Advanced Study in Botany
Banaras Hindu University
Varanasi, Uttar Pradesh, India

Md. Akhlaqur Rahman
Department of Biotechnology
S. S. Khanna Girls' Degree College
Prayagraj, Uttar Pradesh, India

Shanthy Sundaram
Center of Biotechnology
University of Allahabad
Prayagraj, Uttar Pradesh, India

ISBN 978-981-99-0748-9 ISBN 978-981-99-0749-6 (eBook)
https://doi.org/10.1007/978-981-99-0749-6

This Springer imprint is published by the registered company Springer Nature Singapore Pte Ltd.
The registered company address is: 152 Beach Road, #21-01/04 Gateway East, Singapore 189721, Singapore

Preface

Photobiology is an important field of science that covers various intense forms of energy that directly and indirectly affect the health of living organisms. Ionizing radiations have very high energy, which causes major health problems leading to the damage of nucleic acids and essential biomolecules. However, non-ionizing radiations have low energy, but they are suspected to cause more damage after long exposure. The world is aware of the hazardous effects of radiation coming from nuclear power plants and medical therapeutics. Yet, many other applications of radiation in agriculture, therapeutics research, atomic and non-atomic industries indirectly affect the genetic stability and healthy lifespan of individuals on Earth's surface. Apart from nuclear radiation, solar radiation is also one of the greatest obligate sources of energy and radiation since the beginning of Earth's evolution. Solar electromagnetic radiation consists of mainly ultraviolet radiation that is suspected to cause several detrimental impacts on flora and fauna after exposure for a long duration of time. It has been recorded that certain atmospheric pollutants such as halocarbon and its derivatives constantly deplete the thickness of stratospheric ozone layer that results in increased ultraviolet radiation on Earth's surface.

To overcome the detrimental effects of radiation, several synthetic drugs including amifostine, radiogardase, potassium iodide, sargramostim, calcium trisodium and palifermin are formulated that are known to reduce the effects of radiation in radiotherapy. However, they also have the risk of chronic toxicity after a long duration of time. Thus, it is utmost necessary to search for novel alternatives to overcome the radiation toxicity and save human life. Natural radioprotectors have an inherent property to protect the cell from acute radiation. Thus, they could be considered as one of the best alternatives to harmful synthetic products. Microalgae are an evolutionary ancient source of radioprotective compounds including mycosporine-like amino acids, scytonemin, protein biomolecules, vitamins, amino acids and different fatty acids. Apart from microalgae, marine algae have also many metabolites that efficiently reduce the effects of acute radiation. Several radioprotective compounds have been reported in higher plants, bryophytes, pteridophytes and gymnosperms of plant kingdom. In Ayurveda, several herbal formulations that are enriched with bioactive compounds are developed for reducing the deadly effects of radiation. The combination of dietary modulation with herbal formulations plays a great role in the suppression of life-threatening diseases induced by ionic and non-ionic radiations. Hence, herbal medicine could be a promising source of many radioprotective drugs. Recently, advancement in drug discovery is novel to explore the hidden diversity of radioprotective metabolites in various taxa. It has become a challenging platform to identify chemical entities and their functional behaviour. The exploration of stress tolerance mechanisms, gene expression, evolutionary relationships and biosynthetic routes could be the key elements for the identification and drug discovery of radioprotective natural metabolites.

This book critically explains the source, history and detrimental effects of radiation on living creatures. Certain chapters deal with the status of radioactive pollution, ultraviolet radiation and their impacts on the flora and fauna of Earth's surface. This book primarily focuses on exploring the various classes of radioprotective metabolites derived from cyanobacteria, algae, fungi, bryophyte, pteridophyta, gymnosperm and higher plants. In addition, we also discuss the potential applications, structural variations, medical importance including commercial utilization and practices in modern therapeutic science. This book has been designed to illustrate the stress tolerance strategies, evolutionary and biosynthetic mechanisms of radioprotective compounds. The detailed view of the structural and functional diversity of genes and metabolites will be used for the development of production technology and cost-effective drug formulations of radioprotective compounds. This book will play an indispensably role in the understanding of radioprotective mechanisms and curative measures for various deleterious diseases that lead to development of cancer. It will provide state-of-the-art knowledge about life-threatening issues and therapeutic recovery mechanisms arising from solar and atomic radiations. We have made an attempt to combine the entire knowledge of the detrimental effects of radiation and radioprotective compounds with future perspectives in a single unit. Thus, the composition of the book is unique and rarely available in the market.

Varanasi, Uttar Pradesh, India — Vinod K. Kannaujiya
Varanasi, Uttar Pradesh, India — Rajeshwar P. Sinha
Prayagraj, Uttar Pradesh, India — Md. Akhlaqur Rahman
Prayagraj, Uttar Pradesh, India — Shanthy Sundaram

Contents

Editors and Contributors

About the Editors

Vinod K. Kannaujiya has obtained his Ph.D. in Botany from Banaras Hindu University, Varanasi, India. He is presently working as Assistant Professor in the Department of Botany, MMV, Banaras Hindu University, Varanasi, India. His main research interest is in photobiology, molecular biochemistry, biotechnology, bioenergetics and nanotechnology of cyanobacteria. He is a life member of several Indian scientific societies. He has published more than 45 research papers, book chapters, reviews, conference proceedings and authored/edited two books.

Rajeshwar P. Sinha, DAAD Fellowship Awardee and Fellow, Society for Applied Biotechnology, India, is a Professor of Molecular Biology, Department of Botany, Banaras Hindu University (BHU), Varanasi, India. He is also an Adjunct Professor at the University Center for Research & Development (UCRD), Chandigarh University, Chandigarh, India. He received his Ph.D. (Biotechnology) form BHU, Varanasi, and visited countries such as Argentina, Austria, Belgium, Canada, China, Germany, Greece, France, Italy, Japan, Luxembourg, Norway, Poland, the Republic of Korea, Spain, Switzerland, the Netherlands, the UK and the USA in the field of academics/research. He is working on the physiological, biochemical, molecular, nanobiotechnological and computational biology aspects of cyanobacteria. His primary research focus is on UV-B radiation impacts on DNA damage and repair, phycobiliproteins, mycosporine-like amino acids, scytonemin, etc. He has published over 490 research papers/reviews/book chapters/conference proceedings and edited/authored 12 books. He is a lifetime member of several national and international scientific societies and an editorial board member of various national and international journals. He has over 14490 citations with an h-index of 56 and i10-index of 141. He is among the top 02% most cited scientists in the world.

Md. Akhlaqur Rahman has completed his B.Sc. Biotechnology (Hon's), M.Sc. (Microbiology) and D. Phil. in Biotechnology from the University of Allahabad, Prayagraj, India. He is presently working as Assistant Professor in the Department of Biotechnology, S. S. Khanna Girls' Degree College, Prayagraj, India. His area of research interest includes microbiology, stress physiology, bioinformatics and bioactive compounds from microalgae and medicinal plants. He has worked as a visiting researcher at Solution Research Laboratory, Department of Nanoparticle Translational Research, Tokyo Medical University, Tokyo, Japan. He has published over 25 research papers including reviews, book chapters and conference proceedings.

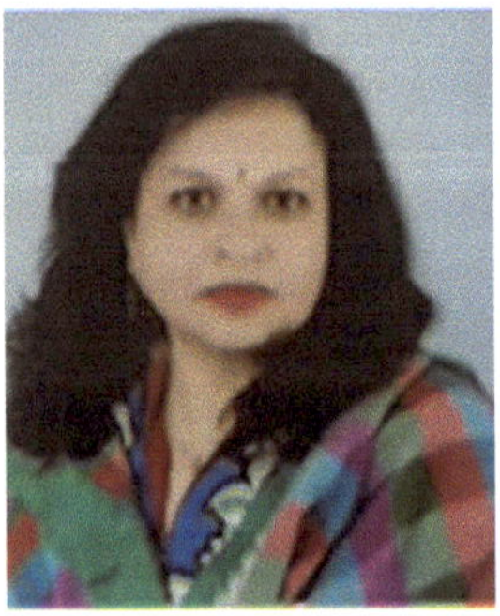

Shanthy Sundaram is Gold medallist in Microbiology from Bombay University, Bombay, India. She has completed her Ph.D. in Microbiology from the Department of Microbiology, Barkatullah University, Bhopal. She is presently working as Professor and Coordinator, Centre of Biotechnology, University of Allahabad. Her areas of research interest are microbiology, plant tissue culture, nanotechnology and algal biotechnology. She has been awarded with different prestigious fellowships and awards such as British Council postdoctoral Fellowship, University of Warwick, Coventry, UK, and UGC Academic Staff Fellowship, University of Warwick, Coventry, UK. She has worked as Visiting Professor, SUN University, Naples, Italy. She is a member of different national and international scientific societies such as member of International Union of Material Research Society, USA, American Society of Microbiology, USA, and Biotechnological Society of India. She has completed 08 research projects funded by DST, DRDO, UGC, INDO-GERMAN and INDO-JAPAN. Prof. Sundaram has published more than 150 research publications, reviews and book chapters in national and international journals. Prof. Sundaram has published authored book in Springer-Nature, Singapore, and edited book in Elsevier.

Contributors

Sarita Agrawal Department of Zoology, S.S. Khanna Girls' Degree College, Prayagraj, India

Iffat Zareen Ahmad Natural Products Laboratory, Department of Bioengineering and Biosciences, Integral University, Dasauli, Lucknow, Uttar Pradesh, India

Atifa Haseeb Ansari Department of Zoology, S.S. Khanna Girls' Degree College, Prayagraj (A Constituent College of University of Allahabad, Prayagraj), Prayagraj, Uttar Pradesh, India

Krishna Kumar Choudhary Department of Botany, MMV, Banaras Hindu University, Varanasi, India

Sonam Dwivedi Natural Products Laboratory, Department of Bioengineering and Biosciences, Integral University, Dasauli, Lucknow, Uttar Pradesh, India

Amit Gupta Laboratory of Photobiology and Molecular Microbiology, Centre of Advanced Study in Botany, Institute of Science, Banaras Hindu University, Varanasi, India

Meenu Jain Viral Research & Diagnostic Lab (VRDL), GR Medical College, Gwalior, Madhya Pradesh, India

Jyoti Jaiswal Laboratory of Photobiology and Molecular Microbiology, Centre of Advanced Study in Botany, Institute of Science, Banaras Hindu University, Varanasi, India

Urmila Jarouliya School of Studies in Biochemistry, Jiwaji University, Gwalior, Madhya Pradesh, India

Vinod K. Kannaujiya Department of Botany, MMV, Banaras Hindu University, Varanasi, India

Neha Kumari Laboratory of Photobiology and Molecular Microbiology, Centre of Advanced Study in Botany, Institute of Science, Banaras Hindu University, Varanasi, India

Sabitri Kumari Laboratory of Morphogenesis, Centre of Advanced Study in Botany, Institute of Science, Banaras Hindu University, Varanasi, Uttar Pradesh, India

Shubhra Malviya Department of Zoology, S.S. Khanna Girls' Degree College, Prayagraj, India

Sonal Mishra Laboratory of Photobiology and Molecular Microbiology, Centre of Advanced Study in Botany, Institute of Science, Banaras Hindu University, Varanasi, India

Saumi Pandey Department of Botany, MMV, Banaras Hindu University, Varanasi, India

Shashi Pandey-Rai Laboratory of Morphogenesis, Centre of Advanced Study in Botany, Institute of Science, Banaras Hindu University, Varanasi, Uttar Pradesh, India

Jainendra Pathak Department of Botany, Pt. Jawaharlal Nehru College (Affiliated to Bundelkhand University, Jhansi), Banda, India

Nidhi Rai Laboratory of Morphogenesis, Centre of Advanced Study in Botany, Institute of Science, Banaras Hindu University, Varanasi, Uttar Pradesh, India

Rajneesh Laboratory of Photobiology and Molecular Microbiology, Centre of Advanced Study in Botany, Institute of Science, Banaras Hindu University, Varanasi, India

Pajeb Saha Laboratory of Morphogenesis, Centre of Advanced Study in Botany, Institute of Science, Banaras Hindu University, Varanasi, Uttar Pradesh, India

Niharika Sahu Laboratory of Photobiology and Molecular Microbiology, Centre of Advanced Study in Botany, Institute of Science, Banaras Hindu University, Varanasi, India

Ashish P. Singh Laboratory of Photobiology and Molecular Microbiology, Centre of Advanced Study in Botany, Institute of Science, Banaras Hindu University, Varanasi, India

Durgesh Singh Department of Zoology, S.S. Khanna Girls' Degree College, Prayagraj (A Constituent College of University of Allahabad, Prayagraj), Prayagraj, Uttar Pradesh, India

Pooja Singh Department of Botany, MMV, Banaras Hindu University, Varanasi, India

Prashant R. Singh Laboratory of Photobiology and Molecular Microbiology, Centre of Advanced Study in Botany, Institute of Science, Banaras Hindu University, Varanasi, India

Sippy Singh Department of Zoology, S.S. Khanna Girls' Degree College, Prayagraj (A Constituent College of University of Allahabad, Prayagraj), Prayagraj, Uttar Pradesh, India

Sneha Singh Laboratory of Morphogenesis, Centre of Advanced Study in Botany, Institute of Science, Banaras Hindu University, Varanasi, Uttar Pradesh, India

Varsha K. Singh Laboratory of Photobiology and Molecular Microbiology, Centre of Advanced Study in Botany, Institute of Science, Banaras Hindu University, Varanasi, India

Rajeshwar P. Sinha Laboratory of Photobiology and Molecular Microbiology, Centre of Advanced Study in Botany, Institute of Science, Banaras Hindu University, Varanasi, India

University Center for Research & Development (UCRD), Chandigarh University, Chandigarh, India

Niharika Srivastava School of Engineering and Sciences, GD Goenka University, Gurugram, Haryana, India

Chapter 1
Photobiology: Historical Background, Sources, and Complications

Sarita Agrawal and Shubhra Malviya

Abstract The modern understanding of ionizing radiation started in 1895 with Wilhelm Röntgen. In the process of conducting various experiments on applying currents to different vacuum tubes, he discovered that despite covering one in a screen to block light, there seemed to be rays penetrating through to react with a barium solution on a screen he'd placed nearby. After several experiments, including taking the first photo (of his wife's hand and skeletal structure) with the new rays, he temporarily named them X-rays as a designation of something unknown, and the name stuck. Living organisms are continuously exposed to ionizing radiations from natural sources. In addition, exposures occur as a result of human activities and medical practices. Radiations are broadly categorized into natural and humanmade sources. More than 90% of radiation exposure to humans occurs from natural sources—e.g., cosmic rays; terrestrial sources that come from radionuclides in the Earth's crust, the air, food, and water; and the human body itself. Humanmade radiation exposure to populations occurs mainly from medical uses of radiation and radioisotopes in healthcare, occupational sources in the generation of electricity from nuclear power reactors, industrial uses of nuclear techniques, and (mostly in the past) testing nuclear weapons. The use of ionizing radiation in medical diagnosis and therapy is widespread and continues to increase with new, useful applications in healthcare. Diagnostic radiation exposures can be significantly reduced by following adequate safety measures and optimizing nuclear-based procedures and practices. To better understand the biological effects of ionizing radiation, it can be divided into directly ionizing and indirectly ionizing. Electromagnetic radiation includes radio waves, microwaves, visible light, ultraviolet radiation, X-rays, and γ-rays. These waves are essentially characterized by their energy, which inversely varies with wavelength. They can be thought of as moving packets of energy (quanta), and in this form, they are called photons.

Keywords Radiation · Radiation sources · Radiation effects · Types of radiation

S. Agrawal (✉) · S. Malviya
Department of Zoology, S.S. Khanna Girls' Degree College, Prayagraj, India

V. K. Kannaujiya et al. (eds.), *Photoprotective Green Pharmacology: Challenges, Sources and Future Applications*,
https://doi.org/10.1007/978-981-99-0749-6_1

1.1 Introduction

Photobiology deals with radioactive substances and ultraviolet radiation in the environment. We now address the effects of radiation on individuals, populations, communities, and ecosystems. Radiation biology deals with the fate of radioactive substances released into the environment and how ecological communities and populations control the distribution of radioactivity. The rapid development of nuclear energy for peaceful and other purposes has spread potentially hazardous materials into the environment. For example, materials such as radioactive strontium and radioactive iodine fall from the sky after nuclear explosions. Such materials enter soils and natural systems as radioactive waste. Many organisms take up these substances and concentrate them in their tissues. In high concentrations, such substances have biological radiation effects, including mutations, cancers, and death.

When reactants are exposed to ionizing radiations such as X-rays, protons, and neutrons, the molecules of the reactants are ionized, and this ionization gives rise to chemical changes through secondary processes. Thus, radiation biology is the study of the action of ionizing radiation on living organisms. The action is veritably complex, involving drugs, chemistry, and biology. Introductory principles are used in radiation treatment to ideally treat cancer with minimal damage to normal tissues. Not all living cells are equally sensitive to radiation. Those cells that are actively proliferating are more sensitive than those that are not proliferating. This is because dividing cells require DNA information to be collected in order for the cell's offspring to survive. Direct interactions between the radiation and the active cell may occur. It leads to cell death or mutation, while a direct interaction between radiation and the DNA of a dormant cell has less of an effect. As a result, living cells can be classified according to their rate of reproduction, which also indicates their relative sensitivity to radiation. This means that different cell systems have different sensitivities. Lymphocytes (white blood cells) and cells that produce blood are constantly replenished and are therefore the most sensitive. The reproductive cells and the digestive system do not quickly regenerate and are therefore less sensitive. Neuromuscular cells are the slowest to regenerate and the cells least sensitive to radiation.

Cells in the human body have a tremendous ability to repair damage, but not damage from all radiation. The effects are irreversible. In many cases, cells can repair any damage and function normally. If the damage is severe enough, though, the affected cell will die. In some cases, the cell is damaged but is able to reproduce. However, the daughter cells may lack some important parts that sustain its life, in which case it may die. Another possible consequence of radiation exposure is that the cell is affected in such a way that it does not die but instead mutates. The mutant cell multiplies and thus perpetuates the mutation. This could mark the onset of a malignant tumor. The most exposed human organ is the skin, and skin cells can be damaged by the effects of ultraviolet (UV) radiation (Gonzalez et al. 2008). When skin is exposed to UV radiation, some part of this radiation is absorbed, while another part is reflected. The absorbed part reaches various layers (Gallagher and

Lee 2006). When DNA, proteins, RNA, etc. absorb UV radiation, it causes secondary interactions and some photochemical reactions. Because of these reactions, harmful effects occur on the human body and give rise to various diseases, such as skin damage, allergies, and premature aging.

The aims of this chapter are to outline the historical background of radiation biology, to give detailed information on its sources and to elucidate the complications occurring mainly because of this radiation.

1.2 Historical Background

Research about X Rays started after its discovery by Conrad Rontgen on 8th November 1895 (Rontgen 1895; Farmelo 1995), which was crucial to the field of medical diagnostics (Wojcik and Ringdahl 2019). Probing the human body via X-rays already had widespread use. At that time, it was assumed that no harm was associated with radiation exposure. Although radiobiological studies started immediately after the finding of X-rays, access was tightly controlled until the end of World War II. Quantitative radiobiological investigations started after 1945, when the nuclear weapons race and nuclear energy programs started. Radiation biology does not report which radiation threats can be directly derived, but it provides the data necessary for determining the nature of the threats. In 1896, the first case of X-ray injury to human tissue was reported. A US electrical engineer, Elihu Thomson, purposely exposed his one finger to X-rays and gave actual findings on the burns generated. A fluorescent X-ray lamp was developed by Thomas Alva Edison in that same year, and he observed the deleterious effects of new rays when he saw that his assistant's hair had fallen out and that that assistant's scalp had become inflamed and ulcerated. Later on, in 1904, his assistant had grown severe ulcers on both hands and arms, which soon became cancerous and caused his early death.

After the first few years, skin burns and tissue necrosis became apparent, and radiation was determined to be the energy source responsible for inducing biochemical reactions in the body. Over the next few years, many researchers and physicians developed radiation injuries in the form of skin burns and cancer after their exposures to X-rays. This understanding generated new branches of science, namely radioactivity and radiation biology. French physicist Henri Becquerel discovered radioactivity in 1906. This was accidentally discovered while he was carrying radioactive materials in his pocket and burned himself. The co-discoverer of radium, Pierre Curie, also developed a similar burn. Henri Becquerel and Marie and Pierre Curie found that the water in many areas was radioactive (Macklis 1996). In early 1925, a number of women became sick with lesions on their jawbones and mouths and developed anemia because they applied luminescent radium-containing paints to clocks and instrument dials; they eventually developed bone cancer.

In 1933, Ernest O. Lawrence and his coworkers built the first full-scale cyclotron at the University of California at Berkeley. The type of particle in this device was a source of neutrons. Lawrence et al. found that such radiations were more effective in killing rats than X-rays were when they used the cyclotron to expose the rats to fast neutron products. Moreover, in 1942, in Chicago, the first nuclear reactor was built, and by that time, further knowledge about neutrons and its effect on living beings had been acquired. The nuclear reactor generated an enormous number of neutrons and became a prime source of energy for the world. It also produces other forms of radiation. Health physics was arising because of the extensive use of nuclear reactors, which feature high-energy particle accelerators known to be sources of ionizing radiation. Moreover, thanks to the introduction of spaceflight in the late 1950s, positive sorts of radiation from houses and their effects on human fitness have attracted attention. The particles in the Van Allen radiation belts (two doughnut-shaped zones of high-energy particles trapped in the Earth's magnetic field) include protons and heavier ions ejected in photovoltaic flares, and comparable particles close to the pinnacle of the environment are especially important.

Although injuries to the pores, skin burns, and tissue necrosis became obvious within the first few years after the discovery of radiation and radioactivity (Codman 1902; Pitkin 1904; Martland 1929; Martland and Humphries 1929; Lambert 2001), Low levels of radiation exposure were beleived to cause less injuries (Walker 2000). Moreover, it was once widely believed that the causes of such injuries were not radiation itself but rather phenomena such as static electricity or injured people's sensitivity (Lambert 2001). Also, when the concept of genes was esoteric (Summers 2011), it was once expected that radiation would affect DNA, which we now refer to as stochastic (Hulse et al. 1982). There used to be no need for research aimed at understanding the organic mechanisms of radioactive motion in terms of radiation protection.

Radiation biology used to be a dynamic department of science during the first 40 years of the twentieth century (Spear 1953), but the incentive for pursing experiments used to be the popular fascination with radiation and its capacity to discern electricity in a well-defined, spatiotemporal placement at the microscale (Sloan and Fogel 2011). The scenario dramatically changed after World War II, following improvements in and the testing of nuclear weapons and the growing public awareness of humanmade radiation. Ever since, the intent for carrying out radiation biology research must be tied determining the qualitative and quantitative effects of radiation, particularly following exposure to low doses (Salomaa et al. 2017).

The Manhattan Project, undertaken at some stage in World War II to advance the first atomic bomb, led immediately to the second lengthy time period during which we learned about the effects of long-term radiation exposure, specifically from the survivors of the bombs dropped on Hiroshima and Nagasaki, Japan. The bombings altogether killed more than 150,000 people (some estimates have asserted that the total is closer to 245,000 or more) and additionally left more than 600,000 survivors (hibakusha, which translates to "explosion-affected people"), many of whom have been studied in the following years. Since the introduction and detonation of the atomic bombs ushered in the Atomic Age, much has changed in our understanding and implementation of radiation and radioactive materials. Throughout the Cold War, experimentation was conducted on radioactive material to assess reactors and their associated sites, aiming to harness the destructive power of radioactive materials for use in nuclear weapons but also in fields such as medicine, radiography, and others.

The desire for increased awareness of the human health effects of the radioisotopes from nuclear fallout led to funding for radiobiological studies. In contrast to studies during pre–World War II technology, these studies investigated clinical uses for radiation. Interest in radiation biology grew to assess the effects of radiation on human health risks. A wide variety of large-scale animal research was initiated to analyze responses to a range of radiations and radionuclides (Haley et al. 2011). The focus used to be on the outcomes of doses, dose rates, and fractionation to determine tumorigenesis and genetic changes. The results, including those from prewar research on the genetic consequences of radiation in *Drosophila*, showed scientists, notably Herman Muller, that there is no protective dose of radiation and that the testing of nuclear weapons must stop.

In the United States, this claim was once rejected under the authority of personnel from a governmental agency, the Atomic Energy Commission (AEC), which was set up to improve nuclear technology; these personnel were satisfied that nuclear testing was safe. In order to clear up this issue, an impartial investigation into the biological effects of atomic radiation (BEAR) was initiated under the auspices of the National Academy of Sciences (NAS). In the United Kingdom, a comparable investigation was commenced under the auspices of the British Medical Research Council. Despite some disagreements on how to evaluate genetic risks, each study reached a comparable conclusion: The genetic outcomes after radiation exposure increase linearly with dosage, and no threshold dosage exists. Reviews of these investigations were published in June 1956. The BEAR committee was later renamed the National Academy of Sciences Advisory Committee on the Biological Effects of Ionizing Radiation (BEIR). In 1972, it published seven reviews on the effects of ionizing radiation on organic substances and human health. In 1956, the National Academy of Sciences Committee presented a file declaring that there is no safe threshold for radiation exposure. The document condemned both the immoderate use of X-rays in clinical and dental practices and exposing pregnant people to radiation. Former AEC official John C. Bugher announced at an American Public Health Association assembly that atomic electricity would incur increased health threats from the large quantities of radioactive chemical substances emitted into the environment during electricity generation.

Debate started among scientists and politicians on the risks of radiation to human populations. However, some of the expertise on this subject comes from experiments on human exposure to radiation, which were often carried out between 1944 and 1974 in the United States and at various nuclear testing sites. Medical patients, prisoners, troops, and teenagers were deliberately exposed to radiation and isotopes, usually without their informed consent. Human experiments are no longer important to understanding the effects of radiation, but they were in line with "moral" priorities (or the lack of such) at the time. Also, the technique of obtaining scientific information from the life span study on Hiroshima and Nagasaki survivors at some point during the first 20 years after the bombings that was carried out by the Atomic Bomb Casualty Commission (ABCC) was once criticized for treating survivors as mere guinea pigs (Beatty 1993).

Since the 1970s, international residents from rich countries have started to become more and more conscious of the human health and environmental dangers from present-day scientific experiments, such as nuclear testing (Beck 1986; Harremoes 2001). The subject was once empowered by the nuclear accidents at Three Mile Island, Chernobyl, and Fukushima Daiichi. This, along with the developing use of radiation in clinical diagnostics, initiated countrywide and worldwide efforts to guide and coordinate radiobiological investigations into determining the human health outcomes of exposure to low doses of radiation. Such investigations have been carried out by a number of nations (Fukunaga et al. 2017; Brooks 2018; Cho et al. 2019; Wang et al. 2018). In the European Union (EU), radiation safety assessments are currently being reorganized thanks to funding from EU member states in an effort to improve the level of available funding for determining the human health outcomes after low doses of radiation (Ruhm et al. 2015, 2016, 2018).

The records of this effort are concisely described below. Any discussion of "low-dose radiation" on human health must define what *low* means. Although *low* is often a relative term, principal national and global advisory bodies such as the have defined low doses of radiation as those at or below 100 mGy. This definition is supposed to roughly describe the dose under which there is scientific uncertainty about the associated effects of radiation on human health. On one hand, for the public, a low dose frequently corresponds to one equal to the level of radiation that humans have historically been normally exposed to; that is, it should vary from 1 to 5 mGy. On the other hand, for healthcare experts who treat cancer patients and aim to destroy most cancers cells with radiation, a low-dose method uses a higher number of Gy.

Although investigations into the human health effects of radiation commenced quickly after the discovery of radiation in 1895, the biology of radiation safety commenced only after the nuclear bombings on Japan in 1945. The biology of radiation safety does now not supply direct facts on radiation dangers to humans. Such outcomes come from epidemiological studies. However, this discipline contributes to radiological safety for humans by supplying qualitative and quantitative data on the effect of radiation on cells and organisms, such as the dose-response curve, dose charge results, and mechanisms of sensitivity to radiation.

We conclude this section with a citation from the first UNSCEAR file, which even today remains applicable to radiation safety: "radiobiology is no longer a science in itself; it is however a utilized science and it rests totally on our know-how of the extraordinary standards of biology, which can't be studied independently of one another" (1958, annex G). Some parts of research develop faster than others, but in the long run, all these parts need to be researched holistically. The research is now pushing ahead to learn more about genetics and carcinogenesis. Our lack of knowledge of biology (taken in its widest possible sense), though, is the principal limitation on our perception of sources of radiation.

Radioactivity may be a part of the Earth in that it has persisted throughout the planet's lifetime. Current radioactive sources include the Earth's crust; the floors and walls of our homes, schools, and offices; and the food we tend to consume. There are also radioactive gases in the air we tend to breathe. Our own bodies' muscles, bones, and tissues generate radiation. Humankind has continuously been exposed to natural radiation emanating from the planet, just as it has from those outside the planet. The radiation we tend to be exposed to from space comes from cosmic rays. We conjointly receive exposure from human-caused radiation, such as X-rays, other diagnostic testing, and treatments for cancer. Fallout from nuclear explosives testing, and small quantities of radioactive materials discharged to the atmosphere from coal and nuclear energy plants are also sources of radiation exposure.

Radioactivity refers to the disintegration of atoms. An atom is characterized by the number of protons within its nucleus. Some natural parts are unstable, and their nuclei disintegrate or decay, at which point energy is released in the form of radiation. The rate of decay over time is expressed in units known as becquerels: 1 becquerel equals 1 decay per second (dps). The radionuclides decay at a characteristic rate that remains constant no matter the external influence, such as temperature or pressure. The time that it takes for half of a radionuclide to disintegrate or decay is called its half-life. This differs for every radioelement, starting from fractions of a second to billions of years. For example, the half-life of iodine-131 is 8 days, whereas the half-life of uranium, which is in varying amounts all over the planet, is 4.5 billion years. Metal 40, the most prominent source of radiation in our bodies, has a half-life of 1.42 billion years.

We are endlessly exposed to radiation from several sources. All species on Earth existed and evolved in environments exposed to natural background radiation. More recently, humans and various other organisms have additionally been exposed to humanmade sources that have developed over the past century. More than 80% of our exposure comes from natural sources, and only 20% comes from humanmade sources—mainly radiation applications used in medical treatments.

In this book, radiation exposure is classified according to its sources, emphasizing the total amount that the public is exposed to. For containment purposes (e.g., radiation protection), radiation exposure is handled separately for different groups. Therefore, additional data on patients whose regional unit was exposed to the medical use of radiation and on people exposed to radiation in the workplace are presented here. All living creatures since the beginning of time have been and are still exposed to radiation.

1.3 Sources of Radiation

Sources of radiation can be classified into two types: natural and anthropogenic.

1.3.1 Natural Sources of Radiation

Radiation has been emitting forever and is all around us. Life evolved on a world containing low levels of radiation, and our bodies adapted to it. The United Nations Scientific Committee on the Effects of Atomic Radiation (UNSCEAR) has identified the following main sources of the propagation of natural radiation.

1.3.1.1 Solar Radiation

Solar rays returning from the Sun retain a small number of gamma rays, cosmic rays, and particles. Star storms intensify these rays; however, the Earth's atmosphere shields us from the most dangerous radiations. Star radiations consisting of cosmic rays are extremely energetic particles (109 MeV) that reach the Earth's surface from other galaxies. Analyses on cosmic rays are lacking. Indian contributions to ionizing radiation analysis have come from pioneers such as D.M. Bose, H.J. Bhabha, V.A. Sarabhai, and P.S. Gill et al., who have addressed its scientific problems.

Earth's outer atmosphere is often bombarded with radiation. Typically, radiation consists of fast-moving particles and come from a number of sources, such as the Sun and other celestial objects and events. Cosmic rays are composed mainly of protons, but they can also contain particles or wave energy. Some of this radiation penetrates the Earth's atmosphere and is absorbed by humans, resulting in natural radiation exposure. Doses caused by natural sources of radiation vary depending on the source and its location. Areas at higher altitudes receive a lot of radiation, but radiation levels vary by altitude and latitude in North America. Space exploration presents innumerable challenges to the ingenuity of human race. Vast distances separate our planet from those nearby and those on the far side our solar system, and they necessitate advances in engineering to alleviate several of the adverse effects of prolonged spaceflight on human physiology. Although several threats to the success of such extraterrestrial missions have been popularized in the media and recreation industries, one area that has not received such attention is the human health risks posed by radiation exposure. When the National Aeronautics and Space Administration (NASA) launches a mission to Mars, astronauts are inevitably exposed to low fluences of extremely energetic and totally ionized nuclei that make up the spectrum of galactic cosmic rays (GCRs) (Abu Bakar et al. 2018; Wojcik and Ringdahl 2019; Beatty 1993). Charged particles in these GCRs are parts of radiation that are deflected from the surface of the planet by its protecting magnetic field. Thanks to their high energy, multicharged particles can penetrate the hull of a ballistic capsule and the tissues of the human body, carving out particle trajectories of dense ionizations (Beatty 1993). Inside the human body, the ionizations ensuing from these interactions injure a range of essential molecular targets, causing advanced lesions that compromise cellular repair processes and delay the recovery of irradiated tissues. Recovery from radiation injury is complicated and delayed by secondary ionizations caused by delta rays that emanate from primary particle trajectories, significantly increasing the extent of cellular damage (Beck 1986; Brent 1999).

Usually, two strategies are used to research the cosmic radiation spectrum at low and high energies, where the former refer to phenomena nearer to Earth, such as solar activity, and the latter refer to those with more-distant origins (galactic or extragalactic). No matter their origin, cosmic rays have effects on a wide range of human activities. The cosmic radiation spectrum extends over 14 orders of magnitude in energy and 12 in intensity.

There are mainly two methods of detecting ionizing radiation. First is the direct detection of the first cosmic rays at high altitudes, which incorporates experiments on stratospheric balloons, satellites, or orbiting stations. Second is the indirect detection of secondary particles, namely the in-depth air showers created by primary cosmic rays' reaching Earth's atmosphere. Cosmic rays are often divided into two types: (1) galactic cosmic rays (GCRs) and gathering cosmic rays (i.e., high-energy particles coming from outside our solar system) and (2) solar energy particles (high-energy particles—mostly protons—emitted by the Sun, mainly from stellar eruptions). However, the term *cosmic radiation* is usually used to refer exclusively to those from extrasolar sources.

A primary cosmic particle collides with an atmospheric molecule, creating an air shower. Cosmic rays originate from primary cosmic rays, which were originally created as a result of numerous astronomical processes. Primary cosmic rays are composed mostly of protons and alpha particles (99%) but also a small number of heavier nuclei (≈1%) and a particularly small proportion of positrons and antiprotons (Gonzalez et al. 2008). Secondary cosmic rays, caused by the decay of primary cosmic rays upon impacting an environment, contain photons, leptons, and hadrons such as electrons, positrons, muons, and pions. The last three of these were first detected in cosmic rays. Once in Earth's atmosphere, secondary particles form air showers or emit radiation within the atmosphere.

Cosmic rays constitute one of the most vital sources of energy transformation in the universe. They teach us about our galaxy, our universe, and the sources of our energies. Even a century after their discovery, we have no definitive models on the origins, method of acceleration, or propagation processes of this radiation. The main reason is that there are still vital discrepancies among the results obtained by different experiments conducted at ground level.

1.3.1.2 Terrestrial Radiation

The temperature of the world is maintained by a balance between incoming radiation and the emissions of the Earth's radiation into space. The former come in wavelengths between 0.3 and 3.0 μm, whereas the latter come in wavelengths between 5.0 and 50 μm. Substances whose unit area is close to one could also be opaque, although at these long wavelength ranges, the substances could go through several minima or maxima of opacity and transmittance. Regardless of this opacity, which is the main problem, solar radiation is prevented from passing to the surface of the Earth mainly by the water vapor in Earth's atmosphere and the scattering of air molecules. Losses of direct sunlight owing to the scattering of air molecules are for the most part compensated for by indirect radiation. On the other hand, the direct rays are lost because they are absorbed by the vapor of the Earth's atmosphere, which would otherwise heat up the planet even more if they reached the surface. The Earth is maintained at its current temperature by a balance between incoming radiation and emitted terrestrial radiation. The average intensity of the radiation is considerably higher when outside the atmosphere of the Earth,. However, it seems to vary from year to year and also be irregular, arriving in short intervals of days, weeks, or months. These variations, which have been observed for about 15 years by the Smithsonian Astronomical Observatory at its station on Mt. Wilson, were later compared with terrestrial temperatures and atmospheric pressures at various stations across the planet by many meteorologists, and the consequences of changes in the stars appear to be important in inducing changes in Earth's temperature.

The study of the outgoing radiation from the world has lately been carried out by many observers. Investigations about the results of water vapor in long atmospherically columns on decreasing the rays of long wavelengths such as those that Earth emits have only extended to wavelengths of 17 μm because no appropriate optical media on the market can yet measure longer wavelengths. It seems that iodide is such a medium, and it may soon produce giant crystals of the salt appropriate for creating prisms to study longer wavelengths of terrestrial radiation.

The subject of the darkness of the Earth is crucial. Late estimations show that a cloud's surface reflects about 78% of the Sun-originating radiation that hits it. Given that approximately 67% of the Earth's surface is shaded by cloud cover according to studies by NASA, mists clearly play a vital role in determining Earth's temperature. Clouds are successful at reducing not just incoming radiation from the Sun but also friendly radiation from the Earth. If the aforementioned autographic device were available, it should be used at various stations across the planet and, if possible, on the oceans.

Very little is known about the impact of ozone on the Earth's level of radiation. A prominent band of radiation occurs at a frequency of around 10 μm where water fumes are present. It is somewhat difficult to determine how best to measure the impact of ozone. Wavelengths of 10 μm are undeniably challenging to work with in the Sun-based range. All earthbound environmental elements emit radiation at this frequency. It is as if the source of this range were mixed with stray light from each, hampering the precision of measurements. In addition, the more extraordinary beams within the Sun-based range, coming from a more-limited range of frequencies, are experienced as stray light, in many cases stronger than the beams within the local range. Rock salt should be used for visual assessments, but its hygroscopicity is not a minor problem. Finally, estimations of energy should be quantitative.

1.3.1.3 Inhalation

Radioactive material is also found throughout nature. It is in dirt, water, and vegetation. Low levels of uranium and thorium and their decays can be found everywhere. Some of these materials are ingested through food and water, while others, such as radon, are inhaled. The portion from terrestrial sources also fluctuates depending on the region of the planet. Some areas have higher concentrations of impurities containing uranium and thorium than other areas. Significant isotopes of concern in terrestrial radiation include uranium and uranium decays, such as thorium, radium, and radon. Radioactive aerosol emerges from different sources, such as atomic accidents, regular rot processes, and the decommissioning of atomic reactors (Beatty 1993; Abu Bakar et al. 2018; Wojcik and Ringdahl 2019). For example, profoundly radioactive miniature particles were released into the atmosphere around the Chernobyl and Fukushima Daiichi accidents (Beck 1986; Abu Bakar et al. 2018). The spray particles let out of accidents and normal rot cycles might stay suspended in the air for prolonged periods, fuse with soil particles that can return to the air, or be breathed in by human and nonhuman animals. Whenever spray particles are breathed in, a small portion of the breathed-in particles are stored in the respiratory system, while the rest are breathed out (Codman 1902; Farmelo 1995; Fukunaga et al. 2017; Brent 1999; Cho et al. 2019). Estimations of the number of particles inhaled are subject to numerous obstacles, such as the math of the respiratory routes, the molecule size of the breathed-in spray, and each person's breathing condition. Breathed-in radioactive sprayers can expose internal organs to radiation for prolonged periods and may induce a plethora of morphological changes, such as hereditary transformations and carcinogenesis.

1.3.1.4 Environmental Radiation

The radioisotopes of a naturally occurring radioelement release huge amounts of radiation in the form of alpha, beta, and gamma rays. In addition to radioisotopes, other radiations come from the air, soil, rocks, and groundwater. Radioactive elements found primarily in the lithosphere include radioisotopes of uranium, thorium, potassium, and carbon (C-14). These radiations mix and interact with natural particles in the atmosphere, increasing the extent of radioactive pollution. In seawater, sediments have a higher concentration of radioactive isotopes. Strontium (Sr-90), a radioisotope of Sr that resembles calcium (Ca), can be deposited into the bones and calcareous masses of aquatic organisms. In some planktonic protozoa, such as Acantharea, whose skeleton is composed of celestine ($SrSO_4$), Sr-90 can be deposited into bodily tissues from the surrounding water. If a radionuclide is mixed into particles in seawater, it can be immediately adsorbed by filamentous algae, marine crustacean shells, and sediments (Real et al. 2004). Although the radioactive elements present in the air occur in minute quantities—i.e., mrad per year—they have seriously adverse effects on living biota.

Uranium and thorium occur throughout nature and are contained in ores, rocks, soils, rivers, and seawaters and in animal and plant organisms. The half-lives of many natural uranium and thorium isotopes are so long that they have been present in the Earth's crust since their formation. The isotopes of uranium and thorium are the parents of three radioactive families: ^{238}U, ^{235}U, and ^{232}Th. All the other natural radioactive elements occur in nature as the products of the radioactive decay of uranium and thorium. In old minerals and ores that have not been exposed to the actions of chemical reagents, a radioactive equilibrium is established, at which the ratio of the radioisotopes of various elements obeys the law of radioactive equilibrium. However, because of the breakdown and weathering of rocks, radioactive elements migrate from one place to another, thus disturbing the radioactive equilibrium. Radioactive elements broken down from their parent elements uranium and thorium rapidly undergo disintegration; as a result, short-lived elements rapidly disappear, and only ^{230}Th, ^{231}Pa, and ^{236}Ra survive. Radium has been found in soils and in seawaters and rivers. Because of the widespread presence of radium in nature, water basins and the atmosphere contain its decay products, such as radon (emanations), thoron, and actinon. In a suspended state in the air and in a dissolved state in water, the decay products of emanations include isotopes of thallium, lead, bismuth, polonium, and astatine (Choppin and Rydberg 1980). In soils, natural radioactive elements are passed on to plants and from plants to animals. For example, uranium and radium contents in plants are 10–5 to 10–8 and 10–11 to 1012 percent, respectively. The radium content in animals is 10–12 to 1013 %. Radioactive series and thus uranium and thorium ores contain isotopes of elements whose atomic numbers are 85 and 87. In addition to uranium, thorium, and their decay products, radioactive isotopes of K, Ca, Rb, Sn, etc. are also found in nature. The natural radioactive isotopes of nonradioactive chemical elements include K_{19}^{40}, Ca_{20}^{40}, V_{23}^{50}, Rb_{37}^{87}, Zr_{40}^{96}, and Cd_{48}^{113}. This clearly indicates that many chemical elements exhibit radioactivity, and the average content of these elements in the Earth's crust is about 0.1%. Plants and animals, therefore, also contain radioactive isotopes of other chemical elements (along with small amounts of uranium, thorium, and radium), such as radioisotopes of the aforementioned elements. They contain considerable amounts of K_{19}^{40}. In addition to the process of radioactive decay, which leads to the formation of radioactive nuclides (i.e., members of uranium, actino-uranium, and thorium), numerous nuclear reactions also occur in nature and give rise to radioactive isotopes.

Our planet, heavenly bodies, and cosmic space may be regarded as a kind of specialized laboratory, where natural nuclear processes take place and give rise to the formation of radioactive isotopes. Large amounts of radioactive carbon are formed in the atmosphere of the Earth. The action of cosmic rays on oxygen and nitrogen in the atmosphere results in the splitting of nuclei of these elements, and fast neutrons also appear. The neutrons interact with nuclei of nitrogen atoms, and radiocarbon is formed according to the reaction $N_7^{14}(n,p)C_6^{14}$. The so-formed recoil atoms of carbon-14 interact with oxygen to form CO_2 containing radioactive carbon (C-14), the half-life of which is 5760 years. Because the intensity of cosmic rays has been assumed to be constant for centuries, radioactive carbon dioxide is continuously being formed at the same rate in the Earth's atmosphere. Because the decay of radioactive carbon also takes place at a constant rate, the atmosphere invariably contains a definite fraction of C-14, which is assimilated by plants. This indicates why the content of radioactive carbon in the tissues of living plants is constant. Animals and human beings also contain radioactive carbon, which comes from consuming plants in their diets. The interaction of nitrogen with neutrons from the cosmic rays gives rise to a radioactive isotope of hydrogen, ^{3}H (tritium), the half-life of which is 12.3 years.

$$N_7^{14} + n_0^1 \rightarrow C_6^{12} + H_1^3$$

The reactions ^{2}H (n, ϒ) ^{3}H and ^{3}He (n, p) ^{3}H also take place in the upper layer of the atmosphere. Tritium leaches into water and is absorbed by plants and animals just like radiocarbon-14.

A number of nuclear reactions also take place in the Earth's crust under the influence of α-particles emitted from the decay of naturally occurring radioactive elements. For example, ^{22}Na is the radioactive isotope of sodium that is formed from fluorine according to the following reaction:

$$F_9^{19} + He_2^4 \rightarrow Na_{11}^{22} + n_0^1$$

Although many nuclear reactions take place in the Earth's crust, very few radioactive isotopes accumulate because of these reactions. Neutrons are formed as a result of the interaction between alpha particles and the nuclei of many elements, such as oxygen, fluorine, chlorine, and bromine. Examples include the reactions ^{18}O (α, n) ^{21}N and ^{9}Be (α, n) ^{12}C. Thus α-emitting elements such as uranium and thorium are the natural sources of neutrons. The spontaneous fission of uranium also releases neutrons into nature, and the half-life of ^{235}U undergoing spontaneous fission has been found to be 10^{15} years.

Almost all chemical elements can interact with neutrons, and as a result, radioactive isotopes, some of which have small half-lives and hence do not accumulate in the Earth's crust, are formed. Radioactive nuclides also result from nuclear reactions between chemical elements and neutrons.

Hence, the nuclides of ^{239}Pu, ^{237}Np, and ^{233}U also accumulate in the Earth's crust, in addition to ^{14}C and ^{3}H. Uranium fission products, most of which are the radioactive isotopes of chemical elements with $Z = 30–65$, are formed because of the fission of uranium in the Earth's crust. There are only 14 radioactive nuclides that have comparatively long half-lives, out of more than 200 radioactive nuclides formed from the fission of uranium. The most important of these are Sr-90 ($T_{1/2} = 0.28$ years), Tc-99 ($T_{1/2}$ = 2.15×10^5 = years), Cs-137 ($T_{1/2}$ = 30 years), and Pm-147 ($T_{1/2} = 2.26$ years) (Kaur et al. 2021).

1.3.1.5 Radionuclides in the Earth's Crust

Radioactive minerals such as uranium (U-238), thorium (Th-232), and potassium (K-40), which are widely distributed in the Earth's crust, lead to a phenomenon known as terrestrial radioactivity. The concentration of these radioisotopes in soils determines the intensity of nuclear radiation at a specific location.

Radioactive potassium (K-40) contains 0.012% natural potassium, while rubidium (Rb-87) contains 28% natural rubidium. Because K-40 contributes radioactivity to all potassium-containing systems in soil, it is reported that for each milligram of K-40, there will be two radioactive disintegrations per minute. Thus, K-40 is considered to be responsible for 20–80% of all radioactivity in soil. The water in the hydrosphere is also radioactive in proportion to its potassium content, including its contribution from biological materials. Rubidium (Rb-87) occurs less abundantly in the Earth's crust, so it is relatively less distributed in the environment. Water becomes contaminated with numerous radionuclides when it runs through soils and rocks. Radon (Rn-222) is reported to pass through the ground and contaminate groundwater to a large extent. Rn-222 and its immediate daughter "nuclide radium-A through radium-C" are the common radioactive isotopes in radioactive springs.

Uranium mining produces a gaseous emission of radioradon; as it decays, it releases long-lived polonium and radiolead, which enter soil and groundwater. Uranium and thorium also naturally decay to produce several radioisotopes that differ in their properties, types, and levels of energy of radiation. Crops grown in such soil contain radioactive elements such as C-14, K-40, Rn-222, and Th-232, which are absorbed into the human body along with food. On average, a person receives about 1 rad per year from terrestrial radiation, and it can be up to 2000 mrads per year in areas containing uranium-bearing rocks.

Evidence indicates that natural background radiation at 110 mrem per year has several detrimental effects on human health (Table 1.1).

Table 1.1 Radionuclides of metabolically useful elements

Nuclide	Half-life period
Calcium (Ca-45)	160 days
Iodine (I-131)	8 days
Phosphorus (P-32)	14 days
Iron (Fe-59)	45 days
Manganese (Mn-54)	300 days
Carbon (Co-14)	5568 days
Cobalt (Co-60)	5.25 years
Sodium (Na-22)	15.1 years
Sodium (Na-24)	87.1 years
Hydrogen (H-3)	12.4 years

1.3.1.6 Internal Radiation

In general, radiation comes from our bodies, especially during the breakdown of potassium in our muscles. In addition to potassium, radioactive elements such as uranium, thorium, strontium, and carbon (C-14) exist in trace amounts in the human body. Radioactive materials that emit alpha or beta particles are known as internal emitters because these rays seriously affect living tissue when absorbed or ingested. Radioactive substances that emit gamma rays are called external emitters because they are highly penetrating and can produce their effects without being absorbed into the body. For humans, the value of internal radiation ranges from 25 to 75 mrad per year. Some metabolically important radionuclides that release radiation into the body include calcium, cobalt, iodine, phosphorus, carbon, iron, manganese, and hydrogen, among others. These radioelements are essential components, leaving our body in traces and are not environmentally harmful. In general, emanations that hit our body from the outside represent an external source, while emanations that result from eaten food containing traces of radionuclides represent an internal source (Kaur et al. 2021).

1.3.2 Anthropogenic Sources of Radiation

Recently, humanmade sources have begun to add large doses of radiation to the existing natural radioactive pollution to which our bodies have grown accustomed, having several adverse effects. The major present-day artificial sources of radiation include the following:

(a) Medical X-rays (diagnostic and radiotherapeutic)
(b) Radioisotopes
(c) Nuclear tests
(d) Radioactive fallout
(e) Nuclear reactors
(f) Nuclear power plants
(g) Nuclear installations
(h) Radioactive ore processing
(i) The industrial, medical, and research uses of radioactive materials
(j) Radiation pollution from electric fields

1.3.2.1 Medical X-Rays

Medical X-rays account for about 18% of the artificial radiation used in radiotherapy for diagnostic purposes. These rays are highly penetrating, like gamma rays. X-ray exposure accumulates in the body and creates chronic defects in the internal organs. Recently, a United Nations committee announced that radiation from medical sources is responsible for diseases resulting from genetic damage.

1.3.2.2 Radioisotopes

Radioisotopes administered to patients during radiation therapy are now emerging as dangerous sources of nuclear contamination. The ability of radiation to kill diseased cells while mainly leaving normal cells unaffected has made it an indispensable tool in the diagnosis and treatment of some deadly diseases, such as cancer. However, the indiscriminate use of radionuclides, their excessive doses to patients, and their improper handling have led to dangerous nuclear pollution.

1.3.2.3 Nuclear Tests

Nuclear explosion tests have threatened the entire globe because natural background radiation in the environment has increased enormously because of these tests. During atmospheric nuclear explosion tests, large quantities of long-lived radionuclides are released into the atmosphere, which are distributed all over the world. Generally, the tests including nuclear fission and fusion processes use uranium (^{235}U) and plutonium (^{239}Pu) as fission materials and lighter nuclei such as hydrogen, lithium, or beryllium as fusion elements.

When fissionable materials are exposed to neutrons, a nonmoderate chain reaction takes place, and a huge amount of energy is released as heat, gamma rays, and neutrons. A very large neutron flux makes the surrounding environment radioactive. Most of the heat comes from the kinetic energy of fission products. The actual mass of the fission products obtained in a nuclear reactor is generally equal to the mass of the fuel employed. One study has revealed that radionuclides formed in explosion tests include fission fragments such as strontium (Sr-90), cesium (Cs-137), barium (Ba-141), and iodine (I-131), along with unused explosives and activation products. The activation products are produced by the neutron bombardment of elements in water or soil.

Wagner claimed that when an atomic bomb is detonated, about 50% of the energy released goes into the explosion, 35% is dissipated as heat, and 15% is released as radioactivity. The radioactive dust that falls to the ground after an atomic explosion is known as fallout. It hangs above the Earth's surface up to a height of about 7–8 km, which can be dispersed by air currents around the world. During a nuclear explosion, radioactive products evaporate into hot gases thanks to the greater force of the explosion and the very high temperature. These radioactive materials are ejected high into the air as extremely fine particles representing atmospheric contamination from radioactive fallout. These radionuclides usually settle with rain and mix with soil, water, and vegetation. When radioactive materials are adsorbed into soil particles, they can easily transfer to plants and enter the food chain. Once established in the human body through food, they radiate internally for almost the entire life of the individual. When a stream receives radioactivity through rains from fission fragments, the aquatic flora and fauna absorb and concentrate this radioactivity. When these aquatic flora and fauna are used by humankind as food, the radioisotopes accumulate in dangerous amounts in our bodies. Thus, radioactive pollution hazards not only occur at the test sites but also spread to the remote corners of the Earth.

The cloud of radioactive material produced by nuclear explosions is known as radioactive fallout. It sweeps a path across the entire planet and hence contaminates water bodies and soils. The cloud produced during a nuclear explosion contains a mixture of gases, molten nuclear fuel, and some partially melted radioisotopes. As the fireball cools, these materials condense to form the debris that drops down to Earth in the form of radioactive fallout. The radiation is emitted from radioisotopes in the form of either high-energy particles (α-, β-, or neutron particles) or very short-wavelength electromagnetic waves (gamma rays). Nuclear fallout contains about 200 radioisotopes, including the most dangerous Sr-89, Sr-90, Cs-137, and C-14. A nuclide with a short half-life causes less damage than a nuclide with a longer half-life, and the latter is stored in the body in a greater concentration. Similarly, a more metabolically significant isotope can cause more damage than an isotope that is not actively involved in the body's basic metabolic reactions and that can be immediately eliminated from the body (Kaur 2020). The products of radioactive fallout from nuclear explosions are distributed over large areas of the Earth's surface and are concentrated in agricultural products (e.g., hay and cereals), and after finding their way into waterbodies, they are taken up by fish and accumulate in their bodies. Contaminated vegetables are eaten by dairy cattle, from which the radioactive fallout finds its way into humans. Some of the radioactive isotopes are concentrated in various parts of the human body, where they remain for a long time. For example, Sr-90 is concentrated in human bones.

Biological organisms, including humans, are subjected to contamination by either the consumption or the inhalation of radioactive contamination. An indirect path of nuclide contamination occurs through the food chain. Radioiodine, I-131, has a short half-life, namely 8 days, and effectively makes its way into the food chain. Radioactive isotopes seep into soil, groundwater, and surface water. Water consumed by plants acts as a medium for radioactive isotopes. This contamination passes on to plants and animals through nourishment from soil and water. Such contaminated plants are eaten by cattle, and the isotope is excreted into the milk that humans consume. Radioisotopes may also enter into the human body via the direct consumption of vegetables and fruits, which then accumulate in our bloodstreams or bodies' organs, damaging these systems. The main hazard from I-131 is its accumulation in the thyroid gland, which plays an important role in regulating metabolic activities.

Isotopes of strontium—i.e., Sr-90 (half-life = 28 years) and Sr-89 (half-life = 54 days)—pose potential dangers because they are easily absorbed by living organisms, the air, water, solid grass, and vegetables. Of all the radioactive elements in the air, Sr-90 accounts for about 5%. It reaches the human body primarily through dairy products and vegetables. Sr-90 contamination in plants occurs both from surface deposition and by uptake from soil. After human consume them, this contamination reaches human bodies and accumulates in our bones, causing cancer. The migration of Sr-89 from soil to human bodies is to some extent connected to the simultaneous migration of Ca. Because Sr can replace Ca, it can be easily absorbed by infants, thus, harming them considerably more than it harms adults. It has been reported that Sr-90 has traversed the globe and mixed into the milk that children drink (Kaur 2020).

Another radioactive fallout isotope, Cs-137, has a half-life of 30 years and is a gamma emitter. It behaves in biological systems very similarly to potassium. Cesium tends to be retained in soil and is therefore minutely transported through plant roots. The surface contamination of plants is the main source through which Cs-137 reaches the human body, through the consumption of either food or milk. Its effects are less serious than those of Sr-90 because Cs-137 is more quickly excreted than Sr-90 is. About 50% of Cs is eliminated from the human body within 70–140 days.

The most common radionuclide, C-14, has a half-life of 5670 years. This carbon is an essential constituent of living matter in that CO_2 is used during plant photosynthesis, which releases oxygen. C-14 reaches human bodies either directly from the consumption of plants or indirectly from the consumption of animals. The direct inhalation of contaminated air constitutes another source of C-14. Even in trace amounts, once introduced into the human body, it can have deleterious effects because of its long half-life. In addition to these radionuclides in radioactive fallout, cerium (Ce-144), zinc (Zn-65), iron (Fe-59), and cobalt (Co-60) may also accumulate in the bodies of marine organisms and can also enter into the food chain (Kaur 2020).

Large quantities of radioactive isotopes are likely to accumulate in various parts of the human body. Even trace amounts of radioactive exposure to the human body or its individual organs is intolerable. Today, we know very little about the effects of radiation on heredity, but they may pose serious biological hazards. As a result of nuclear explosions, the overall radioactive dose from radioactive materials in soil and the human body is small at present compared with the overall radioactive dose from cosmic rays and natural radioactive elements. The further accumulation of the products of nuclear explosions, however, will pose unexpected threats to human and animal life.

1.3.2.4 Radioactive Fallout

Atomic explosions produce not only local ionizing radiation but also radioactive isotopes that enter the atmosphere and continue to fall over wide geographic areas for long periods of time. This is known as nuclear fallout or radioactive fallout. This fallout is very dangerous to life because it also produces ionizing radiation. The term *radioactive fallout* was coined in 1945 to denote radioactive debris that settles on the Earth's surface after a nuclear explosion. Radioactive fallout consists mainly of radioactive particles ejected into the air during an explosion test. It also includes their decay products and the radioactive dust that rises from the bomb's crater.

However, the fallout from a neutron bomb is comparatively smaller than that of a hydrogen bomb or an atom bomb.

During a nuclear explosion, an enormous amount of heat is liberated within a fraction of a microsecond. The temperature increases extensively to millions of degrees, enough to vaporize the surrounding materials. The explosion produces a fire ball that rapidly expands toward the sky, creating numerous hazards. If the explosion is on or near the ground, it carves out a larger crater. The radioactive dust from this crater rises into the air, mixing with radioactive pollutants, and it becomes radioactive and produces a mushroom cloud after the nuclear blast. Radioisotopes suspended in air may come down to soil and water in the form of radioactive rain. The radioactive fallout that causes radiation pollution comes in mainly two types:

1. Early fallout: If the nuclear explosion is at a very low altitude, it will consume a large amount of soil and water and seriously adversely affect all nearby living organisms. The fireball also condenses into heavy particles that fall back to Earth in a short time. Radioactive pollutants are carried by the wind in different directions, which have several harmful effects. They cause serious damage even in places far away from the explosion site.
2. Delayed fallout: If the nuclear explosion occurs at a high altitude, it will consume little soil and water. The delayed fallout may be tropospheric or stratospheric, polluting the environment with radioactive materials. The fission products may be ejected into the air and spread over several kilometers. The debris usually drifts from the equator toward the poles, causing deleterious effects on living creatures. The radioactive fallout from testing nuclear weapons is radioactive dust containing radioisotopes that fall to the Earth after nuclear explosions. In the atmosphere, these materials mix and interact with the natural particulate matter and the matter released by humankind from industries and nuclear power plants, and they then continue to gradually fall over broad geographic areas for a long time. They are capable of producing ionizing radiation, and in this way, they pollute the air. The amount of radioactive fallout produced depends on the following:

   (a) The type of bomb, whether fission or fusion
   (b) The size of the bomb
   (c) The amount of environmental material that is mixed up in the explosion

About 80% of the fallout from small and big atomic weapons is deposited on the ground, and about 5% passes into the lower atmosphere, whose debris settles on land within a few weeks. The remaining 15%, which is made up of small particles, ascends to the upper atmosphere, is widely dispersed in the air, and may come down in rain over long distances and enter the food chain. This indicates that the total amount of radioactivity decreases in accordance with its distance from the site of the explosion (Kaur et al. 2021).

1.3.2.5 Nuclear Reactors

The leakage of nuclear radiations from nuclear reactors, nuclear research laboratories, medicines, and industries is also on the increase with the increase in the number of such facilities. Even with the proper handling of radioactive substances and the techniques of shielding screens that prevent radiation, some radioactive emanations and neutrons still leak out of research laboratories and reactor cores. In nuclear reactors, structural materials and components become radioactive when exposed to radiation and generate active corrosion products. Such activation products are formed by fast neutron (n, p) or thermal neutron (n, r) interactions.

The processed nuclear fuel is introduced into the reactor, the operation of which is a major contributor of radiation pollution. The sequential processes are as follows:

(a) Fission process
(b) Activation process
(c) Thermal process

Some nuclear reactors use U-238 ore as the basic raw material. This has to be processed by purification and enrichment with U-235, and it is then manufactured into fuel elements. This fabrication results in the production of solid, liquid, and gaseous wastes as radioactive pollutants. Even the fuels of nuclear reactors, also called radioactive wastes, contain vast amounts of long-lived radionuclides.

The next operation takes place in the reactor itself, where the fission of nuclear fuel releases enough heat energy to drive steam turbines for electrical generators. The fission products of heavy metals include all the elements ranging from zinc (Zn-30) to gadolinium (Gd-64). These elements contain excess neutrons as soon as they are produced. They rapidly convert into stable elements when radioactive decay emits electrons or beta and gamma radiations. During the activation process, the stable elements are activated by neutron flux.

Nuclear reactors produce waste, and the following types are generated by reactors:

(a) Fission products remaining in both the primary and secondary fuels
(b) Extraneous activation products in the coolant
(c) The gaseous wastes of several nuclides containing C-14, H-3, Xe-133, I-131, Kr-85, Ar-41, Fe-54, and I-129, among others
(d) Liquid wastes containing H-3, Co-58, Fe-55, Co-59, and other corrosion products
(e) The highly active radioactive gaseous wastes—namely H-3, I-133, and long-lived nuclides—produced during chemical treatments that separate the reusable components of U-235, U-238, and Pu-239 from the waste products of fission

The transportation, proper disposal, and storage of these wastes are major problems of the current nuclear industry. This radioactive waste contains unwanted excess fission products and hazardous activation products. Because enormous quantities of radioactive waste are produced, they pose extremely critical public health risks. When the fuel element is dissolved in water, all the heavy elements (from

Zn-30 to Gd-64) and fission products are separated from confinement, which are then transferred into the liquid. The inert gases and volatile halogens are released as vapors. In the atmosphere, these radioactive pollutants become potential contaminants of the environment and are partially scrubbed from the air by rain. Thus, these wastes constitute ever-increasing sources of nuclear radiation from reactors that are installed to meet our rising energy demands. No one can properly estimate its disastrous effects and what will happen in the event of a nuclear accident if the reactor core melts down.

1.3.2.6 Nuclear Power Plants

Nuclear energy promised limitless amounts of clean electric power with the commencement of the first power plant. As of 1985, there were about 600 nuclear power reactors in developed countries. The United States alone has 100 licensed nuclear plants. About 53 power plants were cancelled between 1980 and 1984 because of the enormous radiation danger. According to studies it has been suggested that nuclear energy is clean and provides pollution-free power with no greenhouse gas emissions as cooling towers in nuclear plants only emit water vapor and do not release any pollutant or radioactive substance into the atmosphere. Thus nuclear energy is indeed one of the cleanest sources. However, radioactive nuclear waste contains highly poisonous chemicals like plutonium and the uranium pellets used as fuel. Nuclear electric power first became available in the early 1950s. At that time, few realized the potential risk from nuclear power plants. The first mishap occurred within a few months after the commissioning of power plants. By the end of 1960s, there had been fewer than half a dozen major reactor accidents in the United Kingdom, the United States, Canada, and Switzerland, which dispersed radiation pollution all over the globe. The Three Mile Island power plant leakage in 1979 in the United States and the "meltdown" of the reactor in Chernobyl in the former USSR (CIS) in 1986 are some of the instances of nuclear plant accidents.

Although nuclear power plants are inconvenient to build, once fueled, they can operate for several months. These plants are different from conventional electricity-generating plants. In fueled plants, fossil fuels are burned to produce heat. But the fuel used in nuclear plants, being radioactive, is critically dangerous and the waste materials are equally so. No power plant is entirely contamination-proof. Leakage may occur from several sites, which may be chronically radioactive (Wakil 1989).

The radioactive waste generated by nuclear power plants may come in the following forms:

(a) Low-level radioactive liquid wastes: Radioactive wastes in solution coming from power plants contaminate aquatic life. These radioelements eventually enter human bodies from water supplies, the food chain (including livestock), soil, or vegetation.
(b) Gaseous and particulate radioactive wastes: Stack effluents from atomic power plants contain gaseous and particulate radioisotopes such as H-3, C-14, Kr-85,

and I-129. Some of these radionuclides have critically long half-lives and may be distributed in the environment for several years. When these radioisotopes are inhaled by man, they get concentrated in specific organs posing injurious health effects. According to a recent estimate about 36 Mg/GW/year spent nuclear fuels are discharged from a pressurized water reactor causing deleterious effects on living organisms.

(c) Fission fragments: The largest volume of radioactive wastes comes from reprocessing of irradiated fuel. These radionuclides include Sr-90, I-131, Cs-137, and Co-58, among others. These wastes and traces of induced radionuclides, such as P-32, Fe-59, and Zn-65, are released into the rivers, ditches, waste holding ponds and aquatic environment. Sr-90 released from atomic power plants concentrates in the aquatic food web.

(d) Release of tritium: The heavy water reactors contain high tritium (H-3) inventories because of its production through irradiation of deuterium (D_2) in heavy water (D_2O). Tritium is also released from primary coolants in which lithium hydroxide in added to slow down corrosion in pressurized water reactors (PWRs). Tritium so released from power reactors emits beta (β) radiations like radioactive carbon, strontium, and other radionuclides. Tritium distributes to an altitude of about 10 km in the atmosphere and to a depth of 12 m in the oceans, emanating radiations in the environment.

(e) Heat release: In power plants, atomic pellets of uranium metal are used as fuel in nuclear reactors, and this fuel contains three million times as much potential energy as fossil fuel does. An estimate shows that 1 tonne of uranium produces as much energy as 12 million barrels of oil. This nuclear fuel produces enormous heat. Just 1 g of fissionable material releases 23,000 kilowatt hours (kWh) of heat, causing thermal pollution in the air and water bodies.

Since 1945, when the United States dropped nuclear bombs on and Hiroshima and Nagasaki in Japan, a large number of nuclear and thermonuclear explosions have been carried out in various parts of the world. These products of nuclear explosions have spread through the atmosphere over the Earth's surface. An atom bomb contains U-235 or Pu-239, in which occurs a chain reaction of the fission of the atomic nuclei of uranium or plutonium at the site of the explosion. The action of a hydrogen bomb is based on the thermonuclear reaction between deuterium and tritium:

$$H_1^2 + H_1^3 \rightarrow He_2^4 + n_0^1$$

This reaction takes only 3×10^{-6} s and releases an enormous amount of energy. But for the reaction to be triggered, an excessively high temperature is required to induce fusion. Such a temperature can be provided by an atom bomb. Therefore, a hydrogen bomb, which contains a mixture of deuterium and tritium, is detonated by an atomic plutonium bomb. In the thermonuclear explosion of a hydrogen bomb, the atom bomb is exploded first, followed by a thermonuclear or fusion reaction (Fig. 1.1).

Fig. 1.1 Thermonuclear reaction

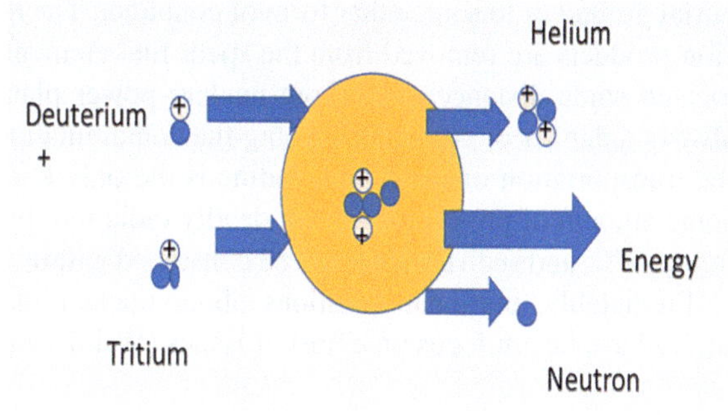

The fission of U-235 or Pu-239 in the explosion of an atom bomb and of a hydrogen bomb (also called a fusion bomb or thermonuclear device) releases an enormous number of neutrons that bombard the surrounding substances and form radioactive isotopes. In this way, a large amount of radioactive carbon-14 and tritium is formed. In addition, large quantities of fission products are also ejected into the atmosphere. The most dangerous of these are Sr-90 and Cs-137.

1.3.2.7 Nuclear Installations

Many advanced countries with massive nuclear power commitments, such as the United Kingdom, the United States, Canada, the former USSR, and France have established several power plants. The advanced scientific and technological developments of nuclear power plants, nuclear reactors, and nuclear testing have recklessly exploited the delicate balance that has existed for millions of years among aquatic flora and fauna and terrestrial flora and fauna (including humans). The result is that hazardous radiations from these energy sources have deteriorated the natural balance in the environment. Now, clean water, clean air, and safe food have become essential commodities.

The dangers posed by nuclear power plant installations from the proliferation of plutonium and other isotopes are astounding. Radioactive materials act as environmental poisons and damage the entire biota. A 1000 MW breeder reactor, for example, will yield 3000 kg of plutonium, and only one invisible speck of plutonium is enough to result in serious bodily harm, along with other irreparable damage. Other potential risks include the release of chronic radiations during normal operations, the release of heat causing thermal pollution, the difficulty of assessing the risks from disastrous reactor accidents, the adequate long-term management of radioactive wastes, safeguarding plutonium from abuse, and the vulnerability of the entire nuclear plant industry to the effects of sabotage and/or war. Thus, public apprehension toward nuclear power installations is not without merit.

In many industries, such as those concerned with nuclear power generation, radioactive isotopes are used as fuel. After the nuclear fuel has been burned up in the nuclear reactor, this spent fuel is transferred to the processing plant and then to a burial ground or to some other form of container. The reprocessing plant, where fission products are removed from the spent fuel elements, and the burial ground are located some distance away from nuclear power plants. For this reason, there is always a danger of accidents causing the contamination of the environment during the transportation of spent fuel. Iodine is the only known substance that can offer some amount of protection against deadly radiation. Iodine pills contain 100 mg of iodine-131, and such a pill should be consumed within an hour of a nuclear accident.

Predictably, pertinent questions about nuclear plant establishment and their safely have begun focus on a host of issues (Wakil 1989).

1.3.2.8 Radioactive Ore Processing

Radioactive ores of uranium, such as pitchblende and thorium, are used in nuclear processes. U_{92}^{235} undergoes natural fission and emits radiations such as alpha, beta, and gamma. The half-life of U-235 is 4.5 × 10^9 years. The disintegrated products that are formed constitute $R_{88}^{226}a$ and $P_{82}^{210}b$, which are extremely toxic in that they emit radiations into nature. U_{92}^{238} and Th_{90}^{232} are employed as artificial fissionable radioactive elements.

The mining and the refining of these materials are the first steps involved in nuclear technology. During mining and washing processes, these radioactive elements release large volumes of mine waters containing a small percentage of ore, while the residue in the form of solid sludges from the metallurgical units contain the daughter elements of uranium and thorium, which have long half-lives. Their daughter products include thorium-230, radium-226, and lead-210 in the natural disintegration of uranium-radium series, while the thorium $\left(Th_{90}^{232}\right)$ series forms fission products as radium-228, radon-222, lead-208, and thorium-228. These radioisotopes persist much longer in the environment. The necessary isotopes of uranium and thorium are separated from other radioisotopes via either the isotopic dilution method or the gaseous diffusion method. The unnecessary radioactive materials can be separated via various chemical techniques to prevent radiation pollution.

During the milling for the recovery of uranium, process effluents are slurried with other residues are released to a tailings pond from which the effluents drain away to join public waters. Of all the radionuclides, radium-226 is most dangerous in aquatic environments because of its longer half-life, its biochemical properties, its highly energetic radiations, and its immediate fission into daughter radionuclides (Kaur 2020).

In addition to radioactive elements, chemical contaminants such as chromium, manganese, sulfate, and nitrate are also present as effluents in water. The chemical treatment of monazite also forms highly toxic effluents. All these processes—i.e., mining, washing, refining, separating, and milling, among others—emit nuclear pollution into the atmosphere. All these treatments during ore processing result in the release of radioactive gases that subsequently adsorb on the particles in the atmosphere. Uranium and thorium ores form dusts in air, which have deleterious effects on living organisms.

1.3.2.9 Industrial, Medical, and Research Uses of Radioactive Materials

Human activities have always affected the environment, whether from agriculture, deforestation, food security, industrialization, or technological developments. The environmental implications of many of these activities are often felt so late that by then certain types of radiations have accumulated in the environment. Thermal power plants, fertilizer companies, and other industries involved in large-scale coal burning are among the main threats of radiation pollution. This finding is based on an investigation into the extent of radiation pollution caused by the combustion of coal in the steam and power units of a fertilizer plant in Sindri, Dhanbad, Bihar, India. It was found that the level of released radioactivity and its exposure to residents from the daily produced ash is within the maximum permissible limit recommended by the International Commission on Radiological Protection. Scientific research has found that radioactive contamination from the former Soviet Union's production of plutonium for nuclear weapons was much higher than previously believed. Since 1948, 8900 becquerels of beta radiation from the radioactive isotopes of strontium-90 and cesium-137 have leaked into the environment from the Maya nuclear complex. Accidents and intentional discharges from Mayak have polluted hundreds of lakes over an area of more than 200 km and is supposed to be the most radioactively contaminated area in the world. Russian scientists have claimed that anyone who comes close to Lake Karachay for a few hours may suffer from radioactive disease. The rapid industrialization of power plants resulted in the devastation of the virgin atmosphere. The leakage of nuclear radiation from reactors, plants, and nuclear research laboratories is increasing. Many wastes from the nuclear industry contain dangerous radionuclides. Some other industries leave their waste products open and allow their effluents to flow into soil or water bodies. These wastewaters also contain small concentrations of radionuclides. The population living nearby is thus permanently exposed to radiation. Because of the recent population boom, the lack of conventional building materials, and the need to provide cheaper houses, industrial wastes are often used in constructing buildings. For example, using energy from power plants and slag from steel mills increases our exposure to nuclear radiation.

Radionuclides administered to patients in medical diagnostics used in radiotherapy and scientific research laboratories have been shown to be major sources of nuclear pollution. X-rays and gamma rays (γ) are the most dangerous. They have extremely high penetration abilities, can destroy internal body tissues, and can quickly cause serious burns. X-rays are also forms of cosmic radiation and can ionize atoms in living tissues. While some radiation has proven to be an indispensable tool in the treatment of deadly diseases and can kill diseased cells, the effects of radiation are dangerous even at the lowest levels. Even then, radiation is often used to destroy malignant tumor cells, to sterilize medical products and food by killing bacteria, to follow the progress of drugs in the body with radioisotopes, and to attack cancer cells and stones from the body's cells.

The nuclear pollution from research laboratories is also bound to increase in the future with the increase in novel technologies. The radiations coming from research work take the form of particles. Some come in the form of alpha and beta rays and others in the form of highly energetic electromagnetic waves—e.g., X-rays and gamma rays. Many of the devices that humans make also emit neutrons (Kaur 2020).

1.3.2.10 Radiation Pollution from Electric Fields

Electrical gadgets and power transmission lines generate electric fields that cause environmental radiation hazard. The modern human is continuously exposed to enough low-frequency electric fields to be harmful. Studies have shown that bees turn violent near high-tension wires. Migratory birds can be disoriented by artificial magnetic and radio waves when sunlight has been completely blocked out.

R. Gavalas Medici and S.R. Day Magdeleno of the Brain Research Institute in Los Angeles, California, USA, have reported that animals acutely respond to radiation, where exposure has elicited changes in behavior. The field is too weak to trigger fine connections between nerve cells. Scientists therefore believe that radiation from electric fields could disrupt molecular aggregations by disrupting ion distributions. Humankind is also affected by a similar danger from radiation. As the world continues to debate the future of nuclear power, Sweden once decided to phase out its large nuclear industry. The government once intended to close at least two of its 12 nuclear reactors by July 2001, and the first was shut down in July 1998. Nuclear power accounts for half of the electricity generated in Sweden and provides cheap energy compared to other sources. It supports important industries such as forestry, mining, steel, and engineering.

1.4 Complications Caused by Radiation

All living organisms are exposed to radiation every day. In addition to diagnostic and therapeutic medical exposures, they are exposed to background radiation from cosmic rays, radioactive wastes, radon decay, nuclear tests, and nuclear accidents. The contribution to the dose of natural radionuclides is now much greater. Radiations have enough energy to affect the atoms in living cells and thereby damage their genetic material. Fortunately, the cells in our bodies are extremely efficient at repairing this damage. However, if the damage is not properly repaired, the cell can die or eventually become cancerous.

Radiation has surrounded us throughout our evolution, so our bodies are designed to deal with the low levels that we are exposed to every day. But too much radiation can damage tissues by changing cell structures and damaging DNA. This can cause serious health problems, including cancer.

1.4.1 *Effects of Radiation*

Exposure to very high levels of radiation, such as those when near a nuclear explosion, can cause acute health effects such as skin burns and acute radiation syndrome ("radiation sickness"). It can also lead to long-term health effects, such as cancer. Our exposure to the low levels of radiation that we encounter in the environment does not cause immediate health effects but rather only slightly contributes to our overall risk of cancer.

The amount of damage that exposure to radiation can cause depends on several factors:

- The type of radiation
- The dose of radiation
- The method of exposure, such as through skin contact, swallowing it or breathing it in, or having rays pass through the body
- The site of radiation concentration in the body and how long it stays there
- The sensitivity of the body to radiation

Fetuses are the most vulnerable to the effects of radiation. Infants, children, older adults, pregnant people, and people with compromised immune systems are more vulnerable to the deleterious health effects of radiation than healthy adults are.

1.4.2 *Types of Effects of Radiation*

When ionizing radiation interacts with cells, it can damage the cells and genetic material—i.e., deoxyribonucleic acid or DNA. If not properly repaired, this damage can result in the death of the cell or potentially harmful changes to the DNA. The health effects from radiation doses can be divided into two categories: deterministic and stochastic.

1.4.3 *Radiation Exposure Pathways*

The type of radiation that a person receives, the way they are exposed (external exposure vs. internal exposure), and the length of their exposure are all important in estimating the health effects of radiation exposure (Choppin and Rydberg 1980).

The risk from exposure to a particular radionuclide depends on the following factors:

- The energy of the radiation that the radionuclide emits.
- The type of radiation that it emits (alpha, beta, gamma, or X-ray).
- The activity of the radionuclide (how often it emits radiation).
- The dose of radiation.

- Whether the exposure is external or internal: External exposure is when the radioactive source is outside of the body. X-rays and gamma rays can pass through the body, depositing energy as they travel. Internal exposure is when radioactive material enters the body via eating, drinking, breathing, or injection (from certain medical procedures). Radionuclides may pose serious health risks if significant quantities are inhaled or ingested.
- The rate at which the body metabolizes and eliminates the radionuclide after ingestion or inhalation.
- Where the radionuclide concentrates in the body and how long it stays there.
- The sensitivity of those affected: Children and fetuses are especially sensitive to radiation exposure. The cells in children and fetuses divide rapidly, providing more opportunities for radiation to disrupt the process and cause cell damage. The US Environmental Protection Agency (EPA) considers differences in sensitivity according to age and sex when revising radiation protection standards.

Cells contain DNA, the blueprint of life. DNA consists of two chains of sugar, phosphate, and four bases. As the genetic information is incorporated in the arrangement of these bases, they are firmly combined to mutually act as a template in order to maintain the arrangement. When DNA is irradiated, it may be partially damaged, depending on the amount of radiation. DNA is damaged not only by radiation but also by carcinogens in food, tobacco, environmental chemicals, reactive oxygen, etc. It is reported that DNA is damaged at 10,000–1,000,000 sites per cell per day. Cells have functions to repair damaged DNA, such as the action of repair enzymes. There are cases where DNA has been completely repaired, partially repaired, or incompletely repaired. When radiation hits a cell, it can damage the DNA inside the cell, but such damage is repaired by the internal systems of the human body. Minor damage can be successfully repaired so that DNA is restored. However, when many parts are damaged, they cannot be fully repaired, and in these cases, the cells die. Even when some cells die, if other cells can replace them, dysfunction does not occur in organs and tissues. However, when too many cells die or degenerate, deterministic effects may appear, such as hair loss, cataracts, skin injury, other acute disorders, or fetal disorders. When a cell whose genes were not completely repaired survives, its genes may mutate and cause stochastic effects such as cancer or hereditary disorders. Damage from low-dose exposure is very rare compared to metabolic DNA damage. However, radiation delivers local energy and causes complicated damage that affects multiple parts of DNA. Approximately 85% of radiation effects are indirect insofar as they are caused by active oxygen created by the radiation, and approximately 15% of these radiation effects cause direct damage.

1.4.4 Lapse of Time After Exposure and Its Effects

Within 1000th of a second after irradiation, DNA breaks and base damage occurs. Within a second of exposure, DNA repair begins, and if this repair fails, cell death and mutation can occur within 1 h to 1 day. It takes some time for such a response at the cellular level to develop into clinical symptoms at the individual level. This period is called the incubation period. The effects that cause symptoms to appear within a few weeks are called acute (early) effects, while the effects that take a relatively long time to develop are called delayed (late) effects. In particular, it takes several years to decades for a person to develop cancer.

1.4.5 Radiosensitivity of Tissues and Organs

Actively dividing less-differentiated cells tend to have higher radiosensitivity (Fig. 1.2). For example, hematopoietic stem cells in bone marrow are differentiated into various blood cells while actively dividing. Immature (undifferentiated) hematopoietic cells that have divided (proliferated) from stem cells are highly sensitive to radiation and more easily die from a small amount of radiation than differentiated cells do (Liu et al. 2021).

As a result, the supply of blood cells is suspended and the number of various types of cells in the blood decreases. In addition, the epithelium of the digestive tract is constantly metabolized and is also highly sensitive to radiation. High-dose radiation exposure may also affect reproductive systems and cause genetic mutations (Cwikel et al. 2020). On the other hand, nerve tissues and muscle tissues, which no longer undergo cell division at the adult stage, are known to be resistant to radiation.

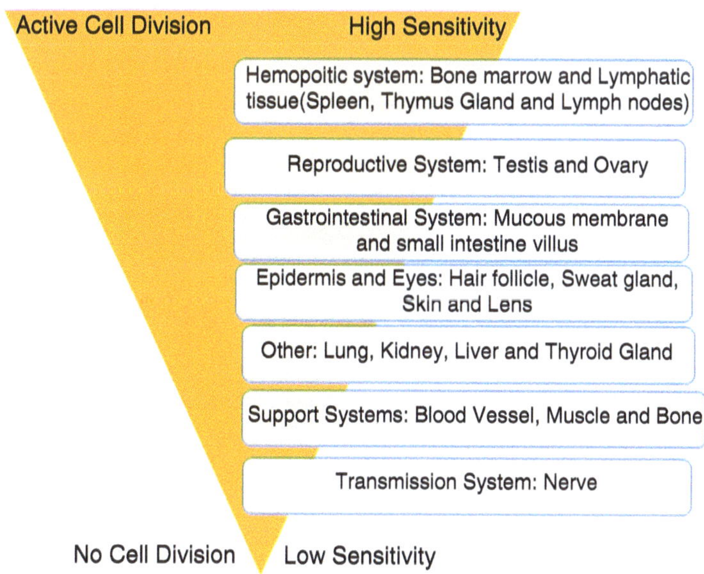

Fig. 1.2 Radiosensitivity of tissues and organs

1.5 Conclusion

The danger of radiation to human health has been well known for a century. It can have biological effects, especially changing the molecules in cells. The severity of radiation damage depends on the sensitivity of the cell and the type and dose of radiation. DNA damage can be repaired, but there are deterministic effects that can be seen in both the short term and the long term. A late somatic effect can manifest as malignancy or other genetic mutations.

References

Abu Bakar NF, Othman SA, Azman NA, Jasrin NS (2018) Effect of ionizing radiation towards human health: a review. Published under licence by IOP Publishing Ltd. IOP Conf Ser Earth Environ Sci 268:012005

Beatty J (1993) Scientific collaboration, internationalism, and diplomacy: the case of the atomic bomb casualty commission. J Hist Biol 26:205

Beck U (1986) Risikogesellschaft. Auf dem Weg in eineandereModerne. Suhrkamp Verlag, Frankfurt am Main

Brent RL (1999) Utilization of developmental basic science principles in the evaluation of reproductive risks from pre- and postconception environmental radiation exposures. Teratology 59:182–204

Brooks AL (2018) Low dose radiation: the history of the U.S. department of energy research program. Washington State University Press, Washington

Cho K, Imaoka T, Klokov D, Paunesku T, Salomaa S, Birschwilks M, Bouffler S, Brooks AL, Hei TK, Iwasaki T (2019) Funding for radiation research: past, present and future. Int J Radiat Biol 2:1–25

Choppin GR, Rydberg J (1980) Nuclear chemistry—principles and applications. Pergamon Press

Codman EA (1902) A study of the cases of accidental X-ray burns hitherto recorded. Reprinted from Philadelphia Med J

Cwikel J, Sergienko R, Gutvirtz G, Abramovitz R, Slusky D, Quastel M, Sheiner E (2020) Reproductive effects of exposure to low-dose ionizing radiation: a long-term follow-up of immigrant women exposed to the Chernobyl accident. J Clin Med 9(6):1786

Farmelo G (1995) The discovery of X-rays. Sci Am 273:68–73

Fukunaga H, Yokoya A, Taki Y, Prise KM (2017) Radiobiological implications of Fukushima nuclear accident for personalized medical approach. Tohoku J Exp Med 242:77–81

Gallagher RP, Lee TK (2006) Adverse effects of ultraviolet radiation: a brief review. Prog Biophys Mol Biol 92:119–131

Gonzalez L, Brown MS, Slate JR (2008) Teachers who left the teaching profession: a qualitative understanding. Qual Rep 13(1):1–11

Haley B, Wang Q, Wanzer B, Vogt S, Finney L, Yang PL, Paunesku T, Woloschak G (2011) Past and future work on radiobiology megastudies: a case study at Argonne National Laboratory. Health Phys 100:613–621

Harremoes P (2001) Late lessons from early warnings: the precautionary principle 1896–2000. European Environmental, Copenhagen

Hulse EV, Path FRC, Mole RH (1982) Reflections on the terms stochastic and non-stochastic as currently used in radiological protection. Br J Radiol 55:321–324

Kaur H (2020) Environmental chemistry. Published May 4th 2011 by Pragati Prakashan (first published January 1st 2010). ISBNn1282803395 (ISBN13: 9781282803398)

Kaur H, Batra SK, Batra K (2021) Taxmann's environmental studies an imperative educational resource to broaden the understanding of the science behind environmental issues | Choice Based Credit System (CBCS). 5th ed
Lambert B (2001) Radiation: early warnings; late effects. In: Harremoes P, Gee D, MacGavin M, Stirling A, Keys J, Wynne B, Guedes VS (eds) Late lessons from early warnings: the precautionary principle. European Environmental Agency, Copenhagen, pp 1896–2000
Liu N, Peng Y, Zhong X, Ma Z, He S, Li Y, Zhang W, Gong Z, Yao Z (2021) Effects of exposure to low-dose ionizing radiation on changing platelets: a prospective cohort study. Environ Health Prev Med 26:14
Macklis RM (1996) Chapter 11. Radiomedical fraud and popular perceptions of radiation. In: Gagliardi R, Wilson FJ (eds) A history of radiological sciences: oncology. Radiology Centennial, Reston, pp 277–292
Martland HS (1929) Radium poisoning. Mon Lab Rev 28:20–95
Martland HS, Humphries RE (1929) Osteogenic sarcoma in dial painters using luminous paint. Arch Pathol 7:406–417
Pitkin JT (1904) Dangers to the X-ray operator. Am X-Ray J 14:9
Real A, Sundell-Bergman S, Knowles JF, Woodhead DS, Zinger I (2004) Effects of ionising radiation exposure on plants, fish and mammals: relevant data for environmental radiation protection. J Radiol Prot 24(4A):A123–A137
Rontgen WC (1895) Uber eineneue Art von Strahlen. Sitzungsberichte der Physikalisch-Medizinischen Gesellschaft zu Wurzburg. 132–149
Ruhm W, Woloschak GE, Shore RE, Azizova TV, Grosche B, Niwa O, Akiba S, Ono T, Suzuki K, Iwasaki T (2015) Dose and doserate effects of ionizing radiation: a discussion in the light of radiological protection. Radiat Environ Biophys 54:379–401
Ruhm W, Fantuzzi E, Harrison R, Schuhmacher H, Vanhavere F, Alves J, Bottollier-Depois JF, Fattibene P, Knezevic Z, Lopez MA (2016) EURADOS strategic research agenda: vision for dosimetry of ionising radiation. Radiat Prot Dosimetry 168(2):223–234
Ruhm W, Friedl AA, Wojcik A (2018) Coordinated radiation protection research in Europe: is it the beginning of a new era? Radiat Environ Biophys 57:1–4
Salomaa S, Jourdain JR, Kreuzer M, Jung T, Repussard J (2017) Multidisciplinary European low dose initiative: an update of the MELODI program. Int J Radiat Biol 93:1035–1039
Sloan PR, Fogel B (2011) Introduction. In: Sloan PR, Fogel B (eds) Creating a physical biology. The three-man paper and early molecular biology. The University of Chicago Press, Chicago
Spear FG (1953) Radiations and living cells. Wiley, New York
Summers WC (2011) Physics and genes. In: Sloan PR, Fogel B (eds) Creating a physical biology. The three-man paper and early molecular biology. The University of Chicago Press, Chicago
Wakil MME (1989) Power plant technology. McGraw Hill Book Co., New York
Walker JS (2000) Permissible dose: a history of radiation protection in the twentieth century. University of California Press, London
Wang Y, Bannister LA, Sebastian S, Le Y, Ismail Y, Didychuk C, Richardson RB (2018) Program at Canadian nuclear laboratories: past, present and future. Int J Radiat Biol 24:1–11
Wojcik A, Ringdahl MA (2019) Radiation protection biology then and now. Int J Radiat Biol 95(7):841–850

Chapter 2
Algal Photoprotective Phytochemicals: Sources and Potential Applications

Neha Kumari, Sonal Mishra, Jyoti Jaiswal, Amit Gupta, Varsha K. Singh, and Rajeshwar P. Sinha

Abstract Ultraviolet (UV; 280–400 nm) radiation, mainly UV-A (315–400 nm) and UV-B (280–315 nm) coming from sunlight are very harmful as they form reactive oxygen species leading to cell damage, thus affecting the growth of photosynthetic organisms. There is a need to investigate alternate sources of UV-shielding materials from nature because synthetic sunscreens have a number of negative consequences on both individuals and the environment. Algae are a natural source of a large number of bioactive phytochemicals such as polyunsaturated fatty acids (PUFAs), pigments, terpenoids, polysaccharides (carrageenan, fucoidan, agar and alginates), phenolic compounds, tocopherol, unsaturated fatty acids, bio-pesticides and algaecides. They are also good biocatalysts and can be used for the enhancement of green energy sources such as biodiesel and the sustainable production of food. Algae can absorb harmful solar radiation due to the presence of photoprotective phytochemicals such as MAAs, sulphated polysaccharides, carotenoids, halogenated compounds, glucosyl, glycerols and polyphenols. Besides UV filters, these metabolites reveal extensive biological activities such as matrix-metalloproteinase inhibitors, antioxidants, antibacterial and anti-inflammatory properties. Thus, such novel bioactive compounds from algae have auspicious applications in therapeutics and also show a beneficiary effect on the skin, mainly overcoming rashes, pigmentation, ageing, and cancer. Therefore, they can be extensively used in skincare, cosmetics, biomedical, nutraceutical and pharmaceutical products. In this chapter, various bioactive substances derived from algae and their promised applications in various fields have been discussed. Using combinations of algal extracts enhanced

N. Kumari · S. Mishra · J. Jaiswal · A. Gupta · V. K. Singh
Laboratory of Photobiology and Molecular Microbiology, Centre of Advanced Study in Botany, Institute of Science, Banaras Hindu University, Varanasi, India

R. P. Sinha (✉)
Laboratory of Photobiology and Molecular Microbiology, Centre of Advanced Study in Botany, Institute of Science, Banaras Hindu University, Varanasi, India

University Center for Research & Development (UCRD), Chandigarh University, Chandigarh, India

V. K. Kannaujiya et al. (eds.), *Photoprotective Green Pharmacology: Challenges, Sources and Future Applications*,
https://doi.org/10.1007/978-981-99-0749-6_2

with photoprotective phytochemicals, a variety of organic UV-screening products could be designed.

Keywords Algae · Natural photoprotective compounds · Anti-ageing · Mycosporine-like amino acids

2.1 Introduction

Worldwide, there are roughly 59% brown algae, 40% red algae and fewer than 1% green algae, depending on the overall number of algae produced (Wang et al. 2015). Algae are rich sources of certain bioactive substances that are absent from other taxonomic groups. They possess a variety of compounds as compared to terrestrial plants (Fu et al. 2017). Algal product phlorotannin extracted from brown algae is absent in terrestrial plants and comprises 25% of the dry weight. Primary metabolites produced by algae include polysaccharides, vitamins, vital amino acids, and unsaturated fatty acids (Thomas and Kim 2013; Fernando et al. 2017) and secondary metabolites such as sulphated polysaccharide, fucoxanthin, fucosterol, polyphenol and fucoidan. Anti-inflammatory, antioxidant, anticancer, antibacterial, and anti-ageing properties are displayed by secondary metabolites (Fig. 2.1) (Saidani et al. 2012; Hwang et al. 2014; Wang et al. 2017a). Because there are natural extracts that are safe for humans, the demand for algae bioactive components in cosmeceuticals is rising quickly. Algae can survive in extreme conditions (cold, ultraviolet

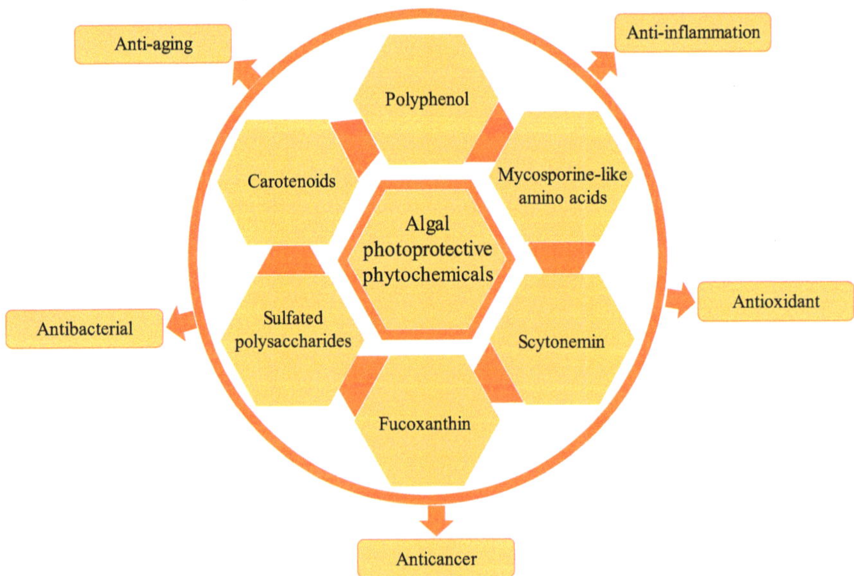

Fig. 2.1 Algal photoprotective phytochemicals and their applications

radiation, heat, desiccation and salinity) (Martins et al. 2014; Wang et al. 2015; Ariede et al. 2017) as they produce various stress-tolerant substances. Aplanspores containing astaxanthin are present to protect against desiccation. Cells are shielded from photo-oxidation by the carotenoid astaxanthin.

Algal carotenoids like *Dunaliella*-derived β-carotene and *Haematococcus*-derived astaxanthin are commercially produced in large-scale processes. Algae receiving high UV radiations (UVRs) contain scytonemins, mycosporine-like amino acids (MAA) and carotenoids are photoprotective substances that serve as antioxidants and shield them. These substances could be utilised in cosmetics to shield cells from harmful UV radiation. The natural products derived from algae could be exploited for environmental, pharmaceutical and cosmeceutical industries. They can be a strong competitor in the market due to low production cost, high biological value and improvement in the cultivation process. This chapter focuses on the production of various photoprotective phytochemicals isolated from algae and their potential uses for enhancing human health in the food, pharmaceutical, and cosmetic industries.

2.2 Photoprotective Phytochemicals Derived from Algae

2.2.1 Phycocolloids

They are a unique class of polysaccharides made by different kinds of seaweed. The sulphated polysaccharides extracted from red algae are agar and carrageenan, whereas alginates are binary polyurodines containing mannuronic acids and guluronic obtained from brown algae (Smit 2004). Phycocolloids act as viscosifier, gelling agent and emulsifier (Cardozo et al. 2007).

2.2.2 Carrageenan

Carrageenan (Fig. 2.2) is a sulphated galactans derived from *Chondrus crispus* (Irish moss), *Eucheuma, Hypnea* and *Gigartina.* Galactose and anhydrogalactose repeating units of the α (1 → 4) and β (1 → 3) types constitute carrageenans. It has thickening and gelling properties. The UV-screening role of carrageenan in UV-B induced human keratinocytes (HaCaT) cells has been reported (Wang et al. 2015). It protects against apoptosis caused by UV-B in HaCaT cells and reduced the generation of reactive oxygen species (ROS). UV-shielding property of carrageenan may be due to immunomodulatory properties and ROS-scavenging activities. It acts as immunomodulatory by activating the production of prostaglandin-E2 (PGE2) and cyclooxygenase-2 (COX-2) gets liberated (Fu et al. 2017). Carrageenan added

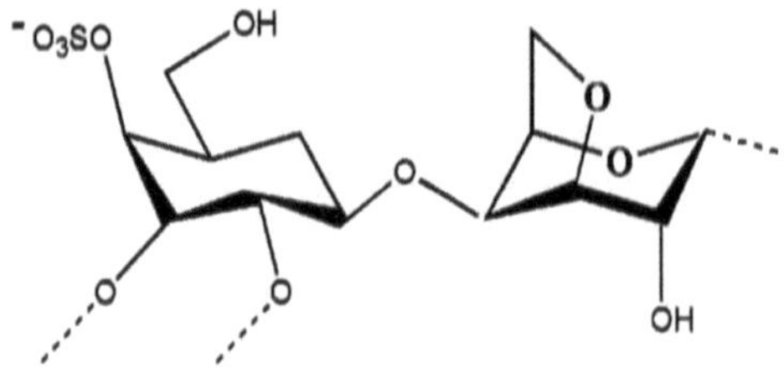

Fig. 2.2 Chemical structure of carrageenan

to a variety of skincare and cosmetic items may reduce the photodamage caused by UVR in comparison to sunscreen alone.

2.2.3 *Fucoidan*

They have α-linked L-fucose backbone with various substitutions. The majority of its occurrences are in brown algae such as *Undaria pinnatifida*, *Fucus vesiculosus*, *Hizikia fusiforme*, *Cladospiphon okamuranus* and *Laminaria japonica.* Fucoidan chemical structure and bioactive properties are different in different algal species (Thomas and Kim 2013). Fucoidan produced by *Costaria costata*, *Fucus evanescens*, *Undaria pinnatifida* and *Ecklonia cava* has shown UV-shielding properties (Wang et al. 2017a; Kim et al. 2018; Su et al. 2020). They protect from UVR by inhibiting the activity of matrix metalloproteinase-1 (MMP-1).

MMP-1 is involved in the collagen damage and photo-ageing caused by UVR on human skin (Fig. 2.3). Fucoidan having a lower molecular weight showed more photoprotection than the fucoidan with higher molecular weight (Kim et al. 2018). This may be due to the fact that fucoidan having lower molecular weight is mostly absorbed before irradiation, and also, they take part in a photoprotective activity instead of UV filtering effects.

2.2.4 *Phlorotannins*

They protect spores and juvenile gametophytes of brown algae against UVR. They can absorb in the UV wavelength range and also contain antioxidant activity. Phlorotannin (Fig. 2.4) contents differ depending on species, habitat and developmental stage and also seasonal and environmental variables. Phlorotannin present in zoospores and their surrounding medium is related to the seasonal maturity of the examined parental algae and with their diverse adaptations to their habitats. *Cystophora congesta* contain phlorotannin such as diphlorethol pentacetate, triphlorethol-A-heptacetate and phloroglucinol triacetate (Hakim and Patel 2020). Four phlorotannins (di- and tri-phlorethols and tri- and 448 tetra-fuhalols) were isolated from the brown macroalga *Halidrys siliquosa.* They contain UV-shielding

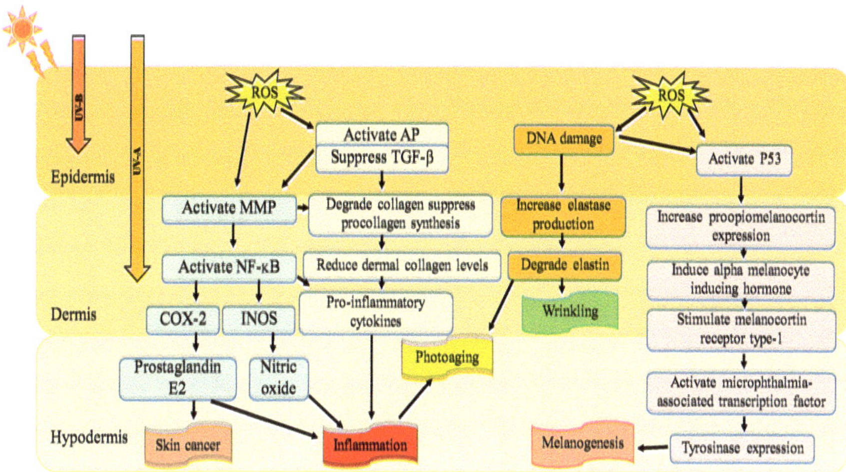

Fig. 2.3 Impact of reactive oxygen species produced by UV radiation (ROS). ROS generation activates the corresponding signalling pathway, causing photo-ageing, skin cancer, inflammation, wrinkles, and melanogenesis. (Modified from Thiyagarasaiyar et al. 2020)

Fig. 2.4 Chemical structure of phlorotannin

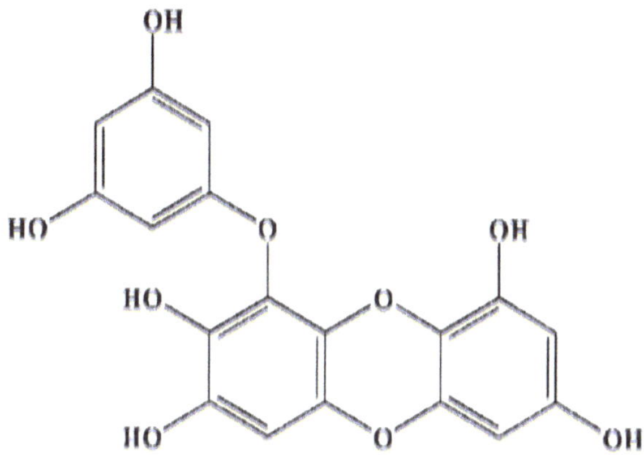

and antioxidant properties. These are the essential elements of algal cell walls. Moreover, they offer chemical protection against herbivores, bacteria, and obstructive organisms. Using MTT, comet assay, microscopic and DCFH-DA analyses, phlorotannins found in *Eckloina cava* were shown to have a photoprotective action against the photo-oxidative stress brought on by UV-B radiation (Heo et al. 2009a).

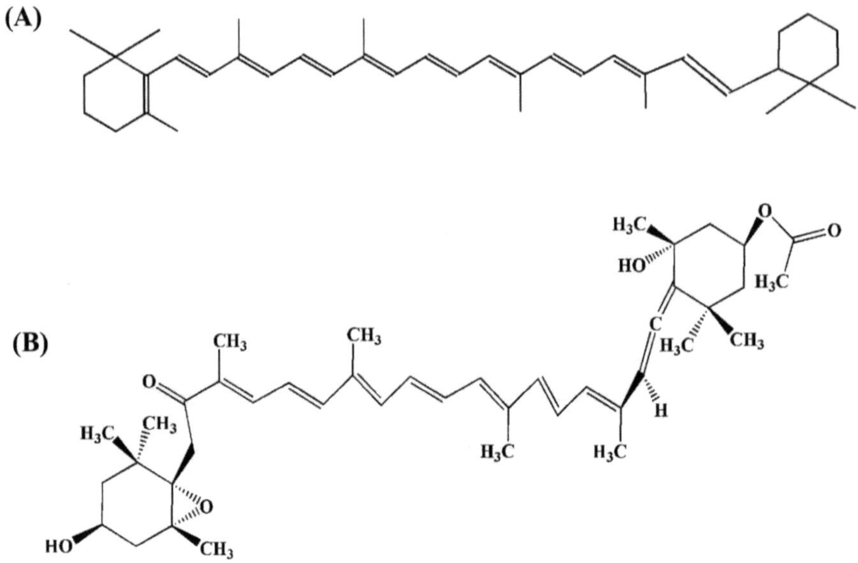

Fig. 2.5 Chemical structure of β-carotene (**a**) and fucoxanthin (**b**)

2.2.5 *Carotenoids*

All types of algae and cyanobacteria have natural compounds called carotenoids. Carotenoids are mainly produced from the Chlorophyceae family. *Dunaliella* contains the maximum amount of β-carotene (Fig. 2.5a) and *Haematococcus pluvialis* contains the maximum amount of astaxanthin. Brown algae contain the xanthins like diatoxanthin, diadinoxanthin and fucoxanthin. Carotenoids can transport energy to chlorophylls by absorbing sun spectrum in the blue-green region (Hashimoto et al. 2015). Carotenoids content increased in UV-B irradiation in brown algae *Pelvetia canaliculata* (Hupel et al. 2011). Fucoxanthin (Fig. 2.5b) obtained from brown algae showed UV-screening property against photo-ageing caused by UVR. They reduce the harmful effects of UVR by converting extra energy into heat. Fucoxanthin activates the filaggrin (filament aggregating protein) promoter activity in sunburn caused by UVR (Matsui et al. 2016).

2.2.6 *Mycosporine-Like Amino Acids*

They are low molecular weight compounds, soluble in water, showing absorption in the range of 310–360 nm. Cyanobacteria, brown algae, red algae and green algae all contain them (Carreto and Carignan 2011). Red algae producing MAAs are *Gelidium* spp., *Palmaria palmata*, *Chondrus crispus*, *Crassiphycus corneus*, *Asparagopsis armata*, *Porphyra* spp., *Grateloupia lanceola*, *Curdiea racovitzae*

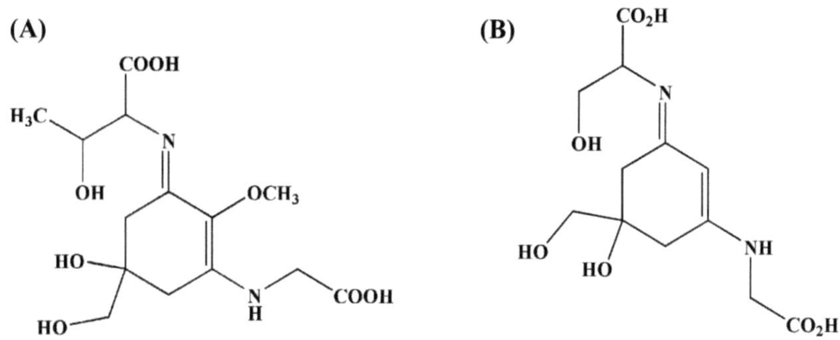

Fig. 2.6 Chemical structure of porphyra-334 (**a**) and shinorine (**b**)

and *Solieria chordalis* (Yuan et al. 2009; Reef et al. 2009; Bedoux et al. 2014). The types of MAAs present in algae depend on the season, depth, climate and environmental variables like temperature, nutrient, pH and salinity (Peinado et al. 2004). MAAs act as UV-screening compounds as well as antioxidants having the ability to scavenge oxygen radicals (Oren and Gunde-Cimerman 2007). A cream with 0.005% Porphyra-334 (Fig. 2.6a) can counteract photodamage of UV-A effectively as a cream containing synthetic UV-A filters (1%) and UV-B filters (4%) (Daniel et al. 2004). It also suppresses the production of ROS and the expression of MMP-1 and MMP-13 is downregulated on human dermal fibroblast following UV-A radiation. Porphyra-334 has enhanced the ability to protect UVR of sunscreen formula (Bhatia et al. 2010).

Shinorine (SH) (Fig. 2.6b) was obtained from *Porphyra rosengurttii*. They have been demonstrated to be extremely photostable, producing no reactive intermediates like radicals when exposed to radiation. The formation of sunburn cells (SBC) protects exposed UVR, and damaged cells are eliminated. UV-exposed skin when treated by combining Porphyra-334 (P-334) + Shinorine (SH) prevented SBC formation. *G. vermiculophylla* contains a high amount of MAAs and absorbs UVR (Barceló-villalobos et al. 2017).

2.2.7 Scytonemin

Scytonemin (Fig. 2.7) is a dimeric compound with a molecular weight of 544 Da, consisting of indolic and phenolic subunits. It is a yellow-brown coloured compound present in the exopolysaccharide sheath of many cyanobacteria such as *Rivularia* sp., *Calothrix* sp., *Scytonema* sp., *Nostoc commune* and *Chlorogloeopsis* sp. (Garcia-Pichel and Castenholz 1991; Sinha et al. 1998; Pathak et al. 2020; Kumari et al. 2021). It is soluble in lipid. It exists in two forms: red-coloured fuscorhodin (reduced) and yellow-coloured fuscochlorin (oxidized) (Wada et al. 2013). UVR is strongly absorbed by scytonemin at 386 nm (Pandey et al. 2020; Kumari

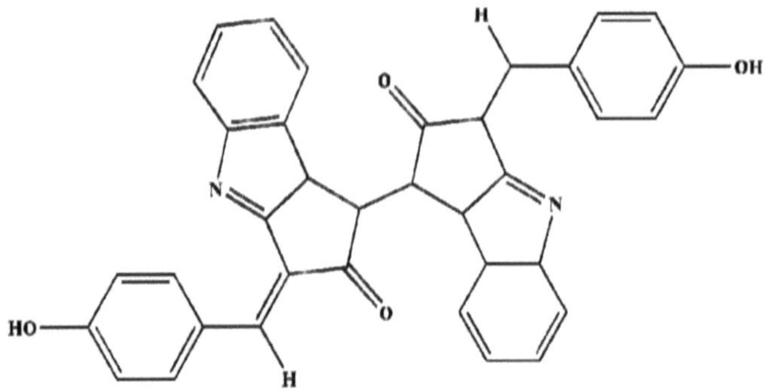

Fig. 2.7 Chemical structure of scytonemin

et al. 2023). Three more forms of scytonemin, that is, tetramethoxyscytonemin, dimethoxyscytonemin, and scytonin pigments are found in *Scytonema* sp. Scytonemin can efficiently decrease the photosynthesis that is inhibited by UV-A radiation and also able to decrease photobleaching of chlorophyll *a* (Gao and Garcia-Pichel 2011). The photoprotective role of scytonemin was reported in *Chlorogloeopsis* sp. (Portwich and Garcia-Pichel 2003). Scytonemin is a very stable compound at high temperatures, strong UV radiation. Due to the high UV-shielding properties, scytonemin can be used as a UV-screening in cosmetics for human beings (Rastogi et al. 2015).

2.2.8 Sporopollenin

Sporopollenin is an inert biopolymer having complex structure among aliphatic, especially, aromatic and isoprenoid components present infrequently in some algal cell walls, spores and in the plant pollens (Osthoff and Wiermann 1987). Some chlorophyte species *Scotiella chlorelloidea, Coelastrum microporum, Characium terrestre, Spongiochloris spongiosa, Scenedesmus* sp., and *Scotiellopsis rubescens* have high sporopollenin content which shows resistance against UV-B radiation (Xiong et al. 1997). It is also found in zygotes of *Dunaliella salina* (Komaristaya and Gorbulin 2006). Sporopollenin increases in response to the UV-B exposure in chlorophytes. UV-B absorbing ability of sporopollenin makes it a harmless, natural and shielding material and it can be used in sunscreens.

2.2.9 Polyphenolic Compounds

Based on the amount of phenol rings and the structural components holding these rings together, they are divided into various classes (Ignat et al. 2011). Phenolic acids, flavonoids, and tannins are the main types of polyphenols. Fucofuroeckol-A (Fig. 2.8a), phloroglucinol (Fig. 2.8b), triphlorethol-A (Fig. 2.8c) and dieckol (Fig. 2.8d) produced from marine brown algae showed protective activity against photodamage induced by UV-B radiation (Kang et al. 2008; Heo et al. 2009a; Ko et al. 2011; Piao et al. 2012; Vo et al. 2018).

In UV-B irradiation, HaCaT cells, phloroglucinol (10 M) scavenged free radicals and prevented macromolecule damage (Kim et al. 2012). Also, it prevented the UV-B-induced overexpression of MMP-1 and the activation of MAPK and AP-1, two proteins that bind to the promoter of MMP-1 (Piao et al. 2012). It can be used in the formation of sun-protective creams and lotions. A free radical or other reactive species can obtain an electron from the hydroxyl group connected to the aromatic ring by using it as an electron donor. Damage caused by ROS on macromolecules is inhibited. As a result, the signal transduction pathways like the MAPK signaling pathway are inhibited. It does not show any toxic effect, so can be used in skincare products. High phenolic content in corallina pilulifera (CPM) reduces UV-induced MMP-2 and MMP-9 expressions in human dermal fibroblast (HDF) cells (Ryu et al. 2009).

2.2.10 Polyunsaturated Fatty Acids (PUFA)

Green, red and brown algae have distinct fatty acid which is independent of the geographical position of algae. *Undaria pinnatifida*, stearidonic and eicosapentaenoic acids were able to inhibit UV-induced dermal fibroblasts (Kim et al. 2005a), leukocyte-endothelial interactions, and inflammatory mediator release in the blood and splenocytes of mice (Ishihara et al. 1998). PUFAs also contain antioxidant activities. *Rhodomela confervoides* and *Symphyocladia latiuscula* showed high antioxidant activity. *Nannochloropsis gaditana*, *Pavlova lutheri*, *Thraustochytrium* and *Isochrysis galbana* sp. (rich in EPA and DHA), *Parietochloris incisa* (arachidonic acid), *Tetraselmis suecica* and *Rhodomonas salina* (α-linolenic acid) (Liu et al. 2013; Mu et al. 2019).

2.2.11 Tocopherol

Tocopherols are lipophilic compounds having α-tocopherol (Fig. 2.9) as the chief source of vitamin E in the body and contain the highest antioxidant activity. α-tocopherol produced by photosynthetic organisms protects against

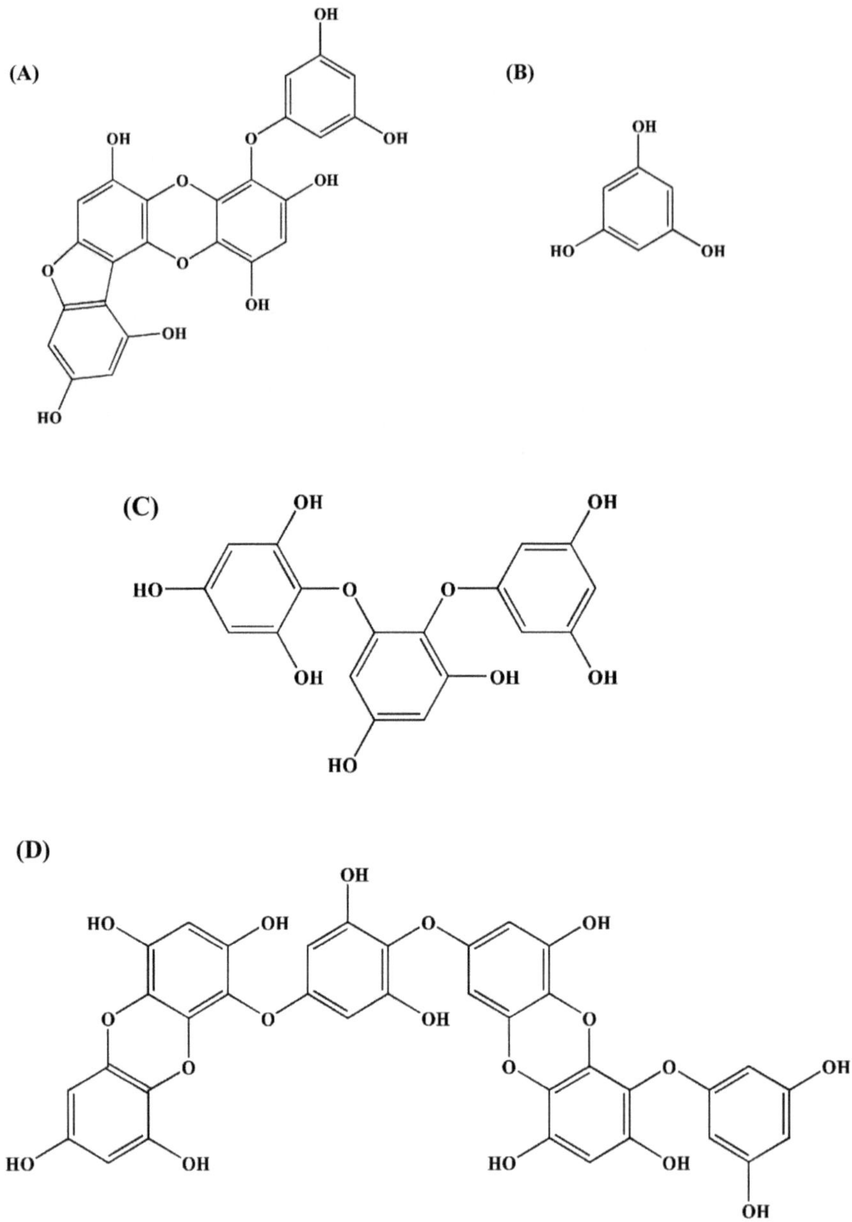

Fig. 2.8 Chemical structure of fucofuroeckol-A (**a**), phloroglucinol (**b**), triphlorethol-A (**c**) and dieckol (**d**)

UVR. α-tocopherol is used to prevent eye and skin diseases in humans that are induced by UV radiation due to the presence of its antioxidant properties (Hamid et al. 2010). *Tetraselmis suecica* and *Dunaliella tertiolecta* contain a high amount of

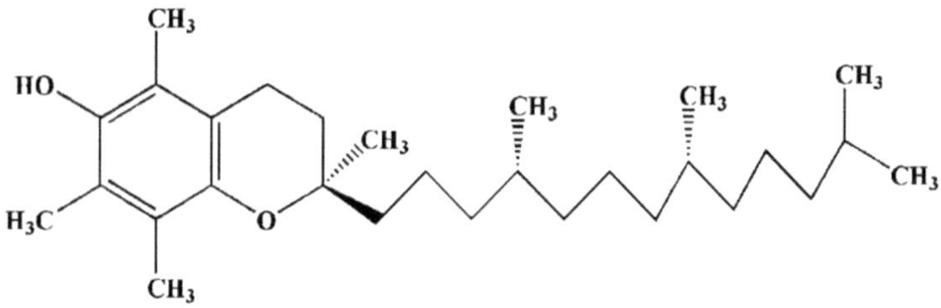

Fig. 2.9 Chemical structure of α-tocopherol

α-tocopherol and are broadly used as a feed for fish in the aquaculture industry (Kim et al. 2008). They are lipophilic and have free radical scavenging. So, it prevents lipoproteins and cellular membranes from oxidative damage.

It has antioxidant properties and donates hydrogen atoms to lipid and lipid peroxyl radicals to prevent oxidation. Glutathione and ascorbic acid increase the antioxidizing properties of vitamin E by donating hydrogen atoms to the tocopherol radical and converting it back to its unoxidized form. Interstitial collagenase (MMP-1) was decreased when fibroblasts were incubated with α-tocopherol. MMP-1 breaks collagenase, so α-tocopherol provides protection to the dermis from breakage (Zussman et al. 2010). In the presence of radiation, α-tocopherol gets converted into a UV-absorbing dimer and trimer which functions similar to UV-screening agents. Edema, erythema and skin sensitivity are reduced by d-α-tocopherol. Treatment of 5% tocopherol before UV-B exposure showed a 75% reduction in skin wrinkling and skin tumours.

2.2.12 *Terpenoids*

They are abundantly found in algae and also present in insects and microorganisms. Terpenoids and isoprenoids are knowns as terpenes. Terpenes are hydrocarbons whereas isoprenoids and terpenoids are oxygen-containing analogs of terpenes. They can be used in cosmeceuticals because they protect against UVR, they show very less toxicity and less irritation (Patil and Saraogi 2014; Chen et al. 2016). Fucosterol (Fig. 2.10) is a steroidal terpenoid found in brown algae *Pelvetia siliquosa, Sargassum carpophyllum, Ecklonia stolonifera, Laminaria ochroleuca, Himanthalia elongate* and *Undaria pinnatifid* (Lee et al. 2003; Jung et al. 2009). They have antioxidant properties and enhance the antioxidant enzymes such as catalase, glutathione peroxidase (GSH-px) and superoxide dismutase (SOD), which regulate the concentration of cellular H_2O_2.

Fucosterol also protects cell membrane oxidation by scavenging H_2O_2 and restoring SOD activity. Fucosterol also restores the internal antioxidants (glutathione) by enhancing the GSH-px activity (Lee et al. 2003; Pillai et al. 2005). Enzymatic and non-enzymatic antioxidants are present in human cells to provide protection against

Fig. 2.10 Chemical structure of fucosterol

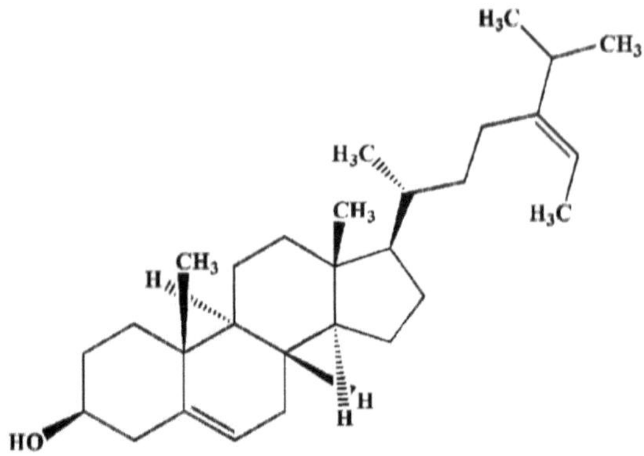

oxidative stress. Enzymatic antioxidants include glutathione reductase, SOD, GSH-px, catalase and non-enzymatic antioxidants are β-carotene, coenzyme Q10 (ubiquinone, ubidecarenone) and vitamins E and C (Pillai et al. 2005). Oxidized glutathione is reduced to glutathione (substrate for glutathione peroxidase) by glutathione reductase. Peroxides and superoxides are immediately converted to harmless species by SOD, GSH-px, and catalase. Enzymatic and non-enzymatic antioxidants concentrations are decreased in photoaged skins. They are less able to fight oxidative stress and Reactive Oxygen Species (ROS). Photo oxidative damage is caused by UVR due to the formation of ROS (Tapiero et al. 2004).

2.3 Applications of Photoprotective Phytochemicals Derived from Algae in Cosmetics

Algae show various biological properties which can be used in cosmetic products like anti-ageing, photo-protection, hair care, moisturizers (Table 2.1). Photoprotective phytochemicals such as polysaccharides, phlorotannins, carotenoids, fucoids have the ability to improve the products' organoleptic qualities, stabilise them.

2.3.1 *Anti-ageing Properties*

Asparagopsis armata containing MAAs is already used in anti-ageing lotions and cream. MAAs contains anti-inflammatory, immunomodulatory and antioxidant properties (Lawrence et al. 2018). *Chlamydomonas hedleyi* (green algae) contain shinorine and porphyra-334 and reduce ageing by increasing expression of procollagen C proteinase enhancer (PCOLCE) and elastin. PCOLCE regulates the deposition of collagen in the skin (Berthon et al. 2017). *Chlorella*, *Dunaliella* and *Spirulina* decreased damage caused by UV. MAAs are broadly used in sunscreen creams and lotions such as Helionori™ (Biosil Technologies, France) and Helioguard 365™

Table 2.1 Applications of photoprotective compounds derived from algae in the cosmeceutical industries

Algae	Photoprotective compounds	Mechanism	Properties	References
Corallina pilulifera	Phenolic compounds	Anti-photo-ageing activity and matrix metalloproteinase is inhibited	Sunscreen	Ryu et al. (2009)
Sargassum confusum	Fucoidan	Photo-oxidative stress is inhibited		Fernando et al. (2020)
Sargassum spp.	Fucoxanthin	Cell damage is prevented induced by UV-B		Heo et al. (2009a, b)
Ecklonia cava	Dioxinodehydroeckol	Hair growth is stimulated	Hair care	Bak et al. (2013)
Laminaria japonica	Fucoxanthin	Antityrosinase activity	Whitening	Wang et al. (2015)
Rhizoclonium hieroglyphicum	Polysaccharides	Moisturizing effects same as glycerin and hyaluronic acid	Moisturizer	Leelapornpisid et al. (2014)
Fucus vesiculosus	Fucoidan	Hyaluronidase enzyme is inhibited	Moisturizer	Pozharitskaya et al. (2020)
Ecklonia stolonifera	Phlorotannins	Metalloproteinase-1 expression is suppressed	Skin ageing	Joe et al. (2006)

(Mibelle AG Biochemistry, Switzerland). Fucoidans isolated from *Fucus vesiculosus* have anti-wrinkle and anti-ageing properties. It inhibits matrix enzymes against collagenase, hyaluronidase, phospholipase A2, tyrosine kinase, heparanase. Additionally, it increases the production of dermal fibroblasts and collagen synthesis, which promotes the facial elasticity and collagen tightness. It also contains antioxidant activities, antiviral, anti-inflammatory and anti-tumour anticoagulant properties (Kim et al. 2005a).

In vitro, fucoidan from *Costania costata* inhibits UV-B-induced expression of the MMP-1 promoter, mRNA, and protein in human keratinocyte cells. Human leukocyte elastase activity can be decreased by fucoidan, which also protects the skin's elastic fibre network against enzymatic proteolysis. The presence of sulfated polysaccharides in red algae hyaluronic acid can be replaced with *porphyridium* as a biolubricant since it has antioxidant qualities that prevent linoleic acid from oxidizing on its own (Borah et al. 2021). Fucoidan obtained from *Turbinaria conoides* (brown algae) contains antioxidative properties and protects skin ageing. Polysaccharides present in *Pyropia yezoensis* (red algae) show antifatigue, antitumour, anti-inflammatory and antioxidant properties. It also provides protection against UV-A-induced photo-ageing (Lee et al. 2016). PYP1-5 isolated from *P. yezoensis* helps in the synthesis of collagen and protects ageing (Kim et al. 2017). Fucoidan found in *Fucus vesiculosus* and *Undaria pinnatifida* contains antioxidant properties and reduces spots, protects the skin (Fitton et al. 2015).

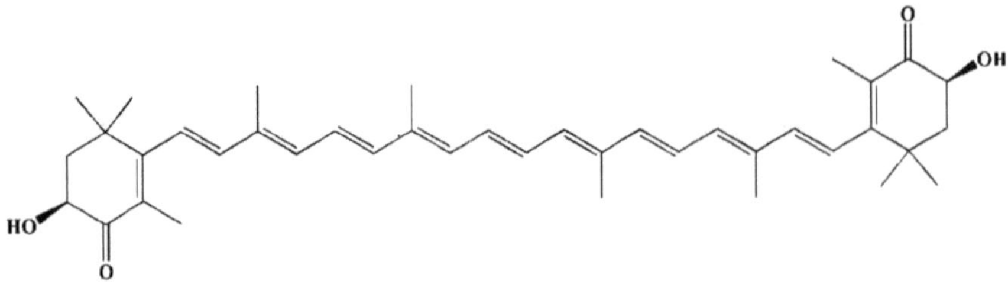

Fig. 2.11 Chemical structure of astaxanthin

Astaxanthin (Fig. 2.11) is a red-coloured pigment that has antioxidant properties stronger than β-carotene and vitamin E. It blocks the production of proinflammatory cytokines. The astaxanthin present in *Haematococcus pluvialis* plays important role in improving skin texture and skin wrinkle and protects against photooxidative damage. Alternatively, *Chlorella zofingiensis* can be used as a source of astaxanthin (Liu et al. 2014). Fucoxanthin isolated from *Sargassum saliquastrum* has antioxidant properties. Brown algae called *Hijika fusiformis* provided the fucoxanthin, which had strong antioxidant activity against the scavenging of 1-diphenyl-2-picrylhydrazyl (DPPH) radicals. Fucoxanthin obtained from *Laminaria japonica* has the potential to inhibit tyrosinase activity and melanogenesis in UV-B-irradiated pigs and mice, respectively. Phloroglucinol from brown algae inhibits tyrosinase activity because it is able to chelate copper. *Sargassum horneri* reduced lipid peroxidation, membrane protein oxidation, intracellular formation of ROS and collagenases such as MMP-1, MMP-2 and MMP-9 which are produced by UV-A exposure (Lee et al. 2022). In comparison to hyaluronidase inhibitors like epigallocatechin gallate, disodium cromoglycate and catechin, some other derivatives of Ecklonia kurome and E. bicyclis such as phlorotannins phloroglucinol, eckol, dieckol, phlorofucofuroeckol A 8, 80-bieckol show more hyaluronidase inhibiting activity. Zeaxanthin (Fig. 2.12) isolated from *Nannochloropsis oculata* showed anti-tyrosinase activity and may be used in skin whitening (Shen et al. 2011).

2.3.2 Hair Care

Sericin protein playing important role in skin formulations and hair conditioning is isolated from *Arthrospira platensis* and *Chlorella vulgaris* (Bari et al. 2017). In outer root sheath cells (ORS) and dermal papilla cells (DPCs), 7-phloroeckol stimulates the growth of hair (Zhou et al. 2018). Algae's Docosahexaenoic Acid (DHA) (Fig. 2.13a) and Eicosapentaenoic Acid (EPA) (Fig. 2.13b) give the hair follicles and scalp deep nourishment, resulting in strong, healthy hair. So, they are used in hair gels, hair serum, spray and hair oils.

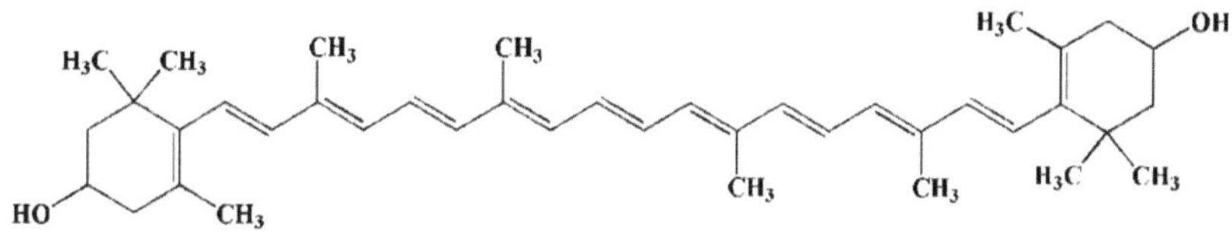

Fig. 2.12 Chemical structure of zeaxanthin

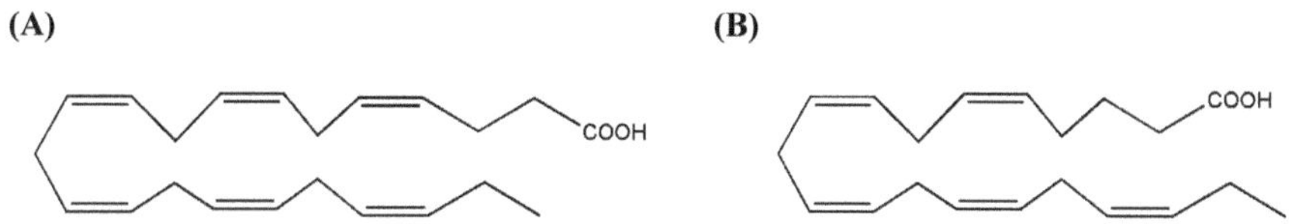

Fig. 2.13 Chemical structure of docosahexaenoic acid (**a**) and eicosapentaenoic acid (**b**)

2.4 Role of Photoprotective Phytochemicals Derived from Algae in Pharmaceutical Industries

Photoprotective compounds obtained from various algae show anticancer and anti-inflammatory, antioxidant activities (Table 2.2). Blood pressure is lowered by γ-linolenic acid isolated from *Arthrospira* sp. Lutein (Fig. 2.14a) and zeaxanthin accumulate in the macula of the retina and inhibit photooxidative damage (Neelam et al. 2005). Carotenoids containing antioxidant properties can decrease light-mediated diseases (Astley et al. 2004). Astaxanthin is effective against various diseases like diabetic nephropathy, cancer, inflammatory diseases and diabetes. It also contains neuroprotective properties and is used to treat neurodegenerative diseases such as Parkinson's or Alzheimer's disease (Oliyaei et al. 2023; Fumia et al. 2022). Violaxanthin (Fig. 2.14b) obtained from *Chlorella ellipsoidea* possesses anti-inflammatory qualities through preventing the generation of NO and PGE2 and nuclear factor (NF-jB) activation (Soontornchaiboon et al. 2012).

Grateloupia turuturu has antimicrobial properties against *Escherichia coli* and *Staphylococcus aureus* (Cardoso et al. 2019). Phlorotannins isolated from *S. muticum* showed antimicrobial, anticancer, anti-inflammatory and antioxidant activities (Casas et al. 2019; Pérez-Larrán et al. 2019). Phlorotannins isolated from *Halidrys siliquos* showed bactericidal activities against *Staphylococcus aureus*, *Pseudomonas aeruginosa* and *Escherichia coli* (Le Lann et al. 2016). Phlorotannin obtained from *Eisenia bicyclis* showed antiviral activity (Eom et al. 2015). *Ecklonia* showed anti-diabetic, anticancer and antioxidant activities because of the presence of various phlorotannin like phloroglucinol, eckol, fucodiphlorethol G, dieckol and phlorofucofuroeckol A (Men et al. 2022; Imchen and Singh 2022). Folk medicine practitioners have utilized *Eisenia arborea* because it has anti-allergic properties due to the presence of phlorofucofuroeckol B compound (Heo et al. 2009b). *Himanthalia elongate* showed high antioxidant and anti-microbial properties owing to the phloroglucinol (Yegdaneh et al. 2016). Fucosterol isolated from *Padina sanctae-Crucis*

Table 2.2 Photoprotective phytochemicals derived from different algae and their applications

Organism	Photoprotective compound	Applications	References
Brown algae			
Stoechospermum marginatum	Diterpenoids	Anticancer	Velatooru et al. (2016)
Turbinaria conoides	Fucoidan	Antioxidant	Wang et al. (2023)
Turbinaria ornata	Fucoxanthin	Antioxidant, anti-ageing	Urikura et al. (2011), Kelman et al. (2012)
Undaria pinnatifida	Polyunsaturated fatty acid, fucoxanthin	Anti-inflammatory, antioxidant	Kalasariya et al. (2022)
Saccharina latissima	Phenol	Antioxidant	Sappati et al. (2019)
Sargassum fulvellum	Fucoxanthin	Antioxidant	Yan et al. (1999)
Sargassum hemiphyllum	Sulfated polysaccharide	Anti-inflammation	Li et al. (2022)
Sargassum henslowianum	Sulfated polysaccharide	Anticancer	Ale et al. (2011)
Sargassum horneri	Sargachromanol E	Anti-ageing	Kim et al. (2013a)
Sargassum horneri	Alginic acid	Anti-inflammation	Fernando et al. (2018)
Sargassum muticum	Tetraprenyltoluquinol,chromane meroterpenoid	Anti-ageing	Balboa et al. (2015)
Sargassum serratifolium	Sargachromanol	Anti-melanogenic	Azam et al. (2017)
Sargassum siliquastrum	Fucoxanthin	Antioxidant	Heo et al. (2019)
Sargassum siliquastrum	Fucosterol	Anti-ageing	Hwang et al. (2014)
Hizikia fusiformis	Fucoxanthin	Antioxidant	Yan et al. (1999)
Ishige foliacea	Phlorotannin	Anti-melanogenic	Kim et al. (2013b)
Ishige okamurae	Diphlorethohydroxycarmalol	Anti-inflammation	Heo et al. (2010a)
Laminaria japonica	Fucoxanthin	Anti-melanogenic	Shimoda et al. (2010)
Laminaria ochroleuca	Polyphenol, phlorotannin	Antioxidant	Ariede et al. (2017), del Olmo et al. (2019)
Macrocystis pyrifera	Hyaluronic acid	Anti-ageing	Yoon et al. (2009)
Padina pavonica	Polyphenol	Antimicrobial	Saidani et al. (2012)

(continued)

Table 2.2 (continued)

Organism	Photoprotective compound	Applications	References
Padina tetrastromatic	Diterpenes	Antioxidant	Antony and Chakraborty (2019)
Padina tetrastromatic	Sulphated polysaccharide, dieckol	Anti-inflammation	Jung et al. (2009), Mohsin and Kurup (2011)
Ecklonia cava	Phlorotannin	Anti-melanogenic, antioxidant	Yoon et al. (2009), Kang et al. (2005)
Ecklonia kurome	Holorotannin	Anti-inflammation	Shibata et al. (2002)
Ecklonia kurome	Phlorotannin	Anti-ageing	Joe et al. (2006)
Ecklonia stolonifera	Phlorofucofuroeckol A and B	Anti-inflammation	Lee et al. (2012)
Eisenia arborea	Phlorotannin	Anti-inflammation	Sugiura et al. (2006)
Eisenia bicyclis	Phlorotannin	Anti-inflammation	Shibata et al. (2002)
Fucus evanescens	Fucoidan	Anticancer	Anastyuk et al. (2012)
Fucus vesiculosus	Fucoidan	Anti-melanogenic, anticancer	Wang et al. (2017a), Teas and Irhimeh (2017)
Fucus vesiculosus	Fucoxanthin	Antioxidant	Zaragozá et al. (2008)
Himanthalia elongota	Fatty acid and phenol	Antimicrobial	Plaza et al. (2010)
Bifurcaria bifurcata	Bifurcadiol	Cytotoxic activity	Capon et al. (1998)
Sargassum tortile	Meroterpenoids, sargol	Cytotoxic activity	Capon et al. (1998)
Cystoseira tamariscifolia	Meroditerpenoid	Anti-fungal, anti-bacterial	Kim et al. (2005b)
Dictyota dichotoma	Dictyol, diterpenes, dictyolactone	Algicidal activity	Capon et al. (1998)
Ecklonia stolonifera	Phloroglucinol	Hepatoprotective agents	Yan et al. (1999)
Sargasum siliquastrum	Fucoxanthin	Anti-inflammatory	Wang et al. (2012)
Chnoospora minima	Fucoidan	Anti-inflammatory	Fernando et al. (2017)
Myagropsis myagroides	Fucoxanthin	Anti-inflammatory	Heo et al. (2010b)

(continued)

Table 2.2 (continued)

Organism	Photoprotective compound	Applications	References
Cladosiphon okamuranus	Fucoxanthin	Antioxidant	Mise et al. (2011)
Cystoseira hakodatensis	Phenol and fucoxanthin	Antioxidant	Airanthi et al. (2011)
Red algae			
Schizymenia dubyi	Phenol	Anti-melanogenic	Azam et al. (2017)
Laurencia rigida	Sesquiterpenes	Antimicrobial	Thomas and Kim (2013)
Palmaria palmata	MAA	Anti-ageing	Hartmann et al. (2015)
Polysiphonia howei	Fucoxanthin	Antioxidant	Kelman et al. (2012)
Porphyra haitanensis	Sulfated polysaccharide	Antioxidant	Lajili et al. (2019)
Porphyra umbilicalis	MAA	Anti-ageing	Wang et al. (2017b)
Porphyra sp.	MAA	Anti-ageing	Hartmann et al. (2015)
Porphyra yezoensis	MAA	Antioxidant	Sakai et al. (2011)
Porphyra yezoensis	Polyphenol	Anticancer	Sakai et al. (2011)
Porphyra yezoensis	Phycoerythrin	Anti-inflammation	Sakai et al. (2011)
Pterocladia capillacea	Sulphated polysaccharide	Antimicrobial	Pimentel et al. (2018)
Pyropia columbia	Phenol	Antioxidant	Cian et al. (2019)
Pyropia yezoensis	Polysaccharide	Anti-ageing	Kim et al. (2017)
Rhodomela confervoides	Polyphenol	Antimicrobial	Saidani et al. (2012)
Rhodomela confervoides	Bromophenol	Antioxidant	Li et al. (2012)
Chondrus canaliculatus	Polysaccharide	Antioxidant	Jaballi et al. (2019)
Gelidium crinaale	Carotenoids	Antioxidant	Panayotova et al. (2017)
Gracilaria gracilis	Phenol	Antioxidant	Francavilla et al. (2013)
Gracilariopsis lemaneiformis	Sulphated polysaccharide	Antioxidant	Wang et al. (2019)
Gracilaria salicornia	2H-chromenyl	Antioxidant	Antony and Chakraborty (2019)

(continued)

Table 2.2 (continued)

Organism	Photoprotective compound	Applications	References
Laurencia caspica	Phenol	Antioxidant	Moshfeġh et al. (2019)
Laurencia luzonensis	Sesquiterpenes	Antimicrobial	Thomas and Kim (2013)
Laurenicia obtusa	Polysaccharide	Antioxidant	Lajili et al. (2019)
Porphyra sp.	MAA (Aminocyclohexene imine)	Anti-inflammatory	Suh et al. (2014)
Catenella repens	MAA (Aminocyclohexene imine)	Antioxidant	Gao and Garcia-Pichel (2011)
Green algae			
Ulva prolifera	Phenol and flavonoid	Antioxidant	Farasat et al. (2013)
Ulva rigida	Phenol	Antioxidant	Fernandes et al. (2019)
Ulva sp.	Sulphated polysaccharide	Anti-ageing	Adrien et al. (2017)
Bryopsis plumose	Polysaccharide	Antioxidant	Zhang et al. (2010)
Chaetomorpha antennia	Fucoxanthin	Antioxidant	Premalatha et al. (2011)
Chlamydomonas hedleyi	MAA	Antioxidant, anti-ageing, anti-inflammation	Suh et al. (2014)
Codium fragile	Sterol	Anti-inflammation	Lee et al. (2013)
Enteromorpha linza	Polysaccharide	Antioxidant	Zhang et al. (2010)
Gayralia oxysperma	Fucoxanthin	Antioxidant	Kelman et al. (2012)
Ulva fasciata	Fucoxanthin	Antioxidant	Premalatha et al. (2011)
Ulva pertusa	Polysaccharide	Antioxidant	Zhang et al. (2010)
Chlamydomonas hedleyi	MAA (Aminocyclohexene imine)	Anti-ageing	Gao and Garcia-Pichel (2011)
Tetraselmis sp.	Polyunsaturated ω-3 fatty acids	Anti-inflammatory	Francavilla et al. (2013)
Nannochloropsis sp.	Polyunsaturated ω-3 fatty acids	Anti-photo-ageing	Francavilla et al. (2013)
Haematococcus pluvialis	Asthaxanthin	Antioxidant	Wang et al. (2012)
Chlorella zofigiensis	Asthaxanthin	Anti-inflammatory	Wang et al. (2012)

(continued)

Table 2.2 (continued)

Organism	Photoprotective compound	Applications	References
Chlorococcum sp.	Asthaxanthin	Anti-photo-ageing, anti-melanogenic	Wang et al. (2012)
Cyanobacteria			
Nitzschia sp.	Fucoxanthin	Antioxidant	Wu et al. (2019)
Nostoc sp.	MAA	Antioxidant	Rastogi et al. (2016)
Odontella aurita	EPA	Antioxidant	Haimeur et al. (2012)
Planktochlorella nurekis	Fatty acid	Antimicrobial	Shannon and Abu-Ghannam (2016)
Porphyridium sp.	Sulphated polysaccharide	Anti-inflammation, antioxidant	de Jesus Raposo et al. (2015)
Rhodella reticulata	Sulphated polysaccharide	Antioxidant	de Jesus Raposo et al. (2015)
Skeletonema marinoi	Polyunsaturated fatty acid	Anticancer	Lauritano et al. (2016)
Skeletonema marinoi	β-Carotene	Antioxidant	Lauritano et al. (2016)
Spirulina platensis	Phycocyanin	Anti-inflammation	Wu et al. (2016)
Synechocystis sp.	Fatty acids and phenols	Antimicrobial	Stirk and van Staden (2022)
Anabaena vaginicola	Lycopene	Antioxidant, anti-ageing	Mourelle et al. (2017)
Arthrospira platensis	Exopolysaccharides	Antioxidant	Mourelle et al. (2017)
Chlorella fusca	Sporopollenin	Anti-ageing	de Andrade and de Andrade (2017)
Chlorella minutissima	MAA	Anti-ageing	de Andrade and de Andrade (2017)
Chlorella sorokiniana	MAA	Anti-ageing	de Andrade and de Andrade (2017)
Chlorella sorokiniana	Lutein	Anti-ageing	Berthon et al. (2017)
Chlorella vulgaris	β-Carotene	Antioxidant	Murthy et al. (2005)
Dunaliella salina	β-Cryptoxanthin	Anti-inflammation	Yang et al. (2013)
Haematococcus pluvialis	Astaxanthin	Anti-ageing, anticancer	Shin et al. (2017)), Rao et al. (2013)
Nannochloropsis granulata	Carotenoid	Antioxidant	Banskota et al. (2019)

(continued)

Table 2.2 (continued)

Organism	Photoprotective compound	Applications	References
Nannochloropsis oculata	Zeaxanthin	Anti-melanogenic	Shen et al. (2011)
Spirulina platensis	Polyunsaturated fatty acid	Anti-inflammatory	Francavilla et al. (2013)
Scytonema	Scytonemin	Anti-inflammatory, anti-proliferation	Stevenson et al. (2002)

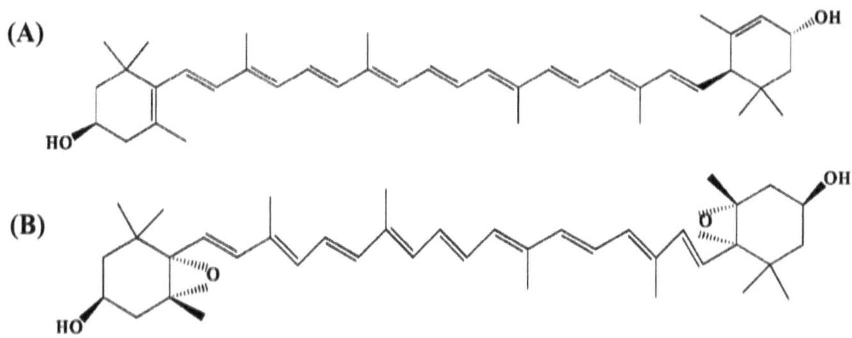

Fig. 2.14 Chemical structure of leutin (**a**) and violaxanthin (**b**)

and *Dictyota ciliolate* showed anti-proliferative and cytotoxic activities (Murugan and Iyer 2014).

Fucoidans obtained from brown algae *Padina sanctae-crucis, Sargassum fluitans* and *Dictyota ciliolata* contain high antioxidant activity protecting HepG2 cells from oxidative stress. The fucoidan obtained from *Sargassum polycystum* showed strong anticancer, antiproliferative and antioxidant activities against human breast cancer cell line MCF 7 at 150 g/mL and an IC50 of 50 g/mL (Palanisamy et al. 2017). Fucoidans extracted from *Fucus evanescens* (800 μg/mL) showed anticancer properties by inhibiting the proliferation of melanoma SK-MEL-28 cells (Menshova et al. 2016). Fucoidan has greater anti-inflammatory activity as compared to heparin. It can be used to inhibit eosinophilia due to its greater stability (Teixeira and Hellewell 1997).

Polysaccharides present in *Porphyridium cruentum* increases immunity and shows anti-inflammatory activity by down-regulating the NF-κB in the nucleus and inhibiting pro-inflammatory modulators such as TNF-a, COX-2, IL-6 and IL-1b (Çetin et al. 2021). Phlorotannins *extracted from Enteromorpha* contain good antioxidant properties and can be used as natural antioxidants in various pharmaceutical products (Ganesan et al. 2011). By scavenging free radicals (hydroxyl, superoxide, and 1,1-diphenyl-2-picrylhydrazyl (DPPH)-radicals), sulphated polysaccharides extracted from green seaweeds show antioxidant activity (Wang et al. 2014). A sulphated polysaccharide, porphyran obtained from *Porphyra yezoensis* and *Porphyra*

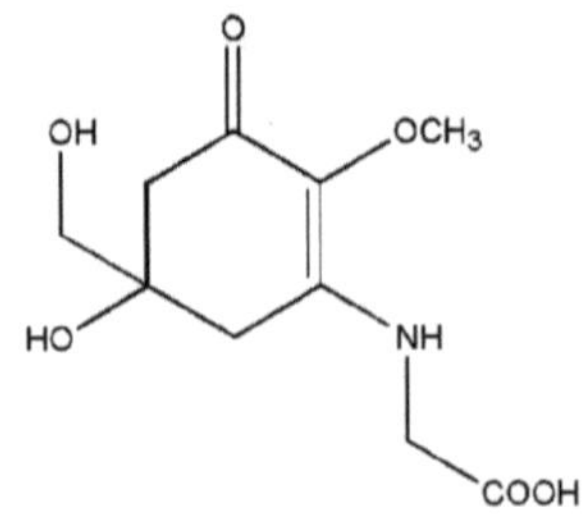

Fig. 2.15 Chemical structure of mycosporine-glycine

tenera possess antiallergic activities. Porphyran also contains anti-inflammatory activity by scavenging reactive oxygen species (Fleurence and Levine 2016).

Inflammation is decreased by heterofucans from *Decidua menstrualis* that bind directly to leukocytes, particularly polymorphonuclear cells (PMNs) (Berthon et al. 2017). *Laurencia* (Red algae) produce various sesquiterpenoids such as 9-hydroxy-3-epi-perforenone A, 4-hydroxy-1,8-epiisotenerone and 3-epi-perforenone (Cotas et al. 2020). They show cytotoxic and antimicrobial activity. *Chondrococcus* and *Plocamium* produce polyhalogenated monoterpenes which contain anti-tubercular, anti-tumour and antimicrobial properties (Cotas et al. 2020). Brominated cyclic diterpenes produced by *Sphaerococcus coronopifolius* showed bactericidal efficacy against methicillin- and multidrug-resistant strains of *Staphylococcus aureus* (Rodrigues et al. 2015). Nitric oxide generation and the expression of pro-inflammatory cytokines were both suppressed by *Sargassum muticum* in RAW 264.7 macrophages (Yang et al. 2014).

Mycosporine-Glycine (M-Gly) (Fig. 2.15) protects against cell death caused by UVR. It also reduces COX-2 mRNA caused by UVR and shows anti-inflammatory activity. Fucoidan also effectively inhibits the chemokine receptor type 4 (CXCR4).

Xylofucans obtained from *Punctaria plantaginea* inhibited clot formation (Ustyuzhanina et al. 2016). It shows similar anticoagulant and antithrombotic properties exhibited by heparinoid clexane and the native fucoidan from *Saccharina latissima*. Polyphenols obtained from *E. cava* showed anti-inflammatory, antihyperlipidaemic and antihyperglycaemic antioxidant properties (Murray et al. 2018).

2.5 Role of Photoprotective Phytochemicals Derived from Algae in Food and Food Colorant

β-carotene is used as a food additive to improve the health and fertility of grain-fed livestock as well as the coloring of fish and egg yolks (Borowitzka 1988). Additionally, it is utilised in cosmetics and food items like fruit juices, margarine, cheese, and dairy products as well as baked foods (Dufossé et al. 2005). β-carotene isolated from *Dunaliella salina* (green alga) has higher antioxidative activity than synthetic. Astaxanthin is used as a food additive for trout, shrimp and salmon for various aquaculture and poultry industries and food colouring agents (Ambati et al.

2014). As a nutraceutical, astaxanthin is taken as an encapsulated substance. As a nutritional supplement for people, *Haematococcus pluvialis*, which is high in astaxanthin, is offered for sale in the market (Bishop and Zubeck 2012). Carrageenan is used in meat processing, the dairy industry and other products like air freshener gels, pet food and toothpaste. *Grateloupia turuturu* is used to produce R-phycoerythrin which is a water-soluble pink-purple compound used in the industrial sector (Le Guillard et al. 2015).

2.6 Conclusion

Demands of algal bioactive compounds are increasing as they can be used as a key ingredient in pharmaceutical and cosmetic industries. Although natural compounds from algae are less damaging to the skin than synthetic compounds, they can be employed efficiently in skincare and cosmeceuticals. The UV-screening role of photoprotective compounds isolated from many algae has been reported. So, this can provide a gateway for the production of UV-protective supplements or pharmaceuticals isolated from algae. They can be considered a valuable source of bioactive substances that, because of their anti-proliferative, anti-inflammatory, antibacterial, antiviral, and antioxidant properties, can improve human health by avoiding or shortening the recovery time for a number of disorders. The use of novel biologically produced substances in cosmetics has been greatly boosted by breakthroughs in biotechnology, genetic enhancement of the organism and enormous microbial variety.

Acknowledgements N. Kumari (09/013(0819)/2018-EMR-I) is grateful to the CSIR, New Delhi, India, for the financial assistance in the form of a senior research fellowship. The University Grants Commission (UGC), New Delhi, India, provided junior research fellowship and senior research fellowship to S. Mishra (Joint CSIR-UGC JRF-2019/NTA Ref. No.: 191620046790) and J. Jaiswal (926/CSIR-UGC-JRF DEC, 2018), respectively. A. Gupta (09/013(0912)/2019-EMR-I) and V. K. Singh (09/0013(12862)/2021-EMR-I) are appreciative of the financial assistance in the form of a senior research fellowship and junior research fellowship respectively provided by the CSIR, New Delhi, India. Incentive grant received from IoE (Scheme No. 6301), Banaras Hindu University, Varanasi, India, to Rajeshwar P. Sinha is highly acknowledged.

References

Adrien A, Bonnet A, Dufour D, Baudouin S, Maugard T, Bridiau N (2017) Pilot production of ulvans from *Ulva* sp. and their effects on hyaluronan and collagen production in cultured dermal fibroblasts. Carbohydr Polym 157:1306–1314

Airanthi MWA, Hosokawa M, Miyashita K (2011) Comparative antioxidant activity of edible Japanese brown seaweeds. J Food Sci 76(1):C104–C111

Ale MT, Maruyama H, Tamauchi H, Mikkelsen JD, Meyer AS (2011) Fucose-containing sulfated polysaccharides from brown seaweeds inhibit proliferation of melanoma cells and induce apoptosis by activation of caspase-3 in vitro. Mar Drugs 9(12):2605–2621

Ambati RR, Phang SM, Ravi S, Aswathanarayana RG (2014) Astaxanthin: sources, extraction, stability, biological activities and its commercial applications—a review. Mar Drugs 12(1):128–152

Anastyuk SD, Shevchenko NM, Ermakova SP, Vishchuk OS, Nazarenko EL, Dmitrenok PS, Zvyagintseva TN (2012) Anticancer activity in vitro of a fucoidan from the brown alga *Fucus evanescens* and its low-molecular fragments, structurally characterized by tandem mass-spectrometry. Carbohydr Polym 87(1):186–194

Antony T, Chakraborty K (2019) Xenicanes attenuate pro-inflammatory 5-lipoxygenase: prospective natural anti-inflammatory leads from intertidal brown seaweed *Padina tetrastromatica*. Med Chem Res 28(4):591–607

Ariede MB, Candido TM, Jacome ALM, Velasco MVR, de Carvalho JCM, Baby AR (2017) Cosmetic attributes of algae—a review. Algal Res 25:483–487

Astley SB, Hughes DA, Wright AJ, Elliott RM, Southon S (2004) DNA damage and susceptibility to oxidative damage in lymphocytes: effects of carotenoids in vitro and in vivo. Br J Nutr 91(1):53–61

Azam MS, Choi J, Lee MS, Kim HR (2017) Hypopigmenting effects of brown algae-derived phytochemicals: a review on molecular mechanisms. Mar Drugs 15(10):297

Bak SS, Ahn BN, Kim JA, Shin SH, Kim JC, Kim MK, Sung YK, Kim SK (2013) *Ecklonia cava* promotes hair growth. Clin Exp Dermatol 38(8):904–910

Balboa EM, Li YX, Ahn BN, Eom SH, Domínguez H, Jiménez C, Rodríguez J (2015) Photodamage attenuation effect by a tetraprenyltoluquinol chromane meroterpenoid isolated from *Sargassum muticum*. J Photochem Photobiol B Biol 148:51–58

Banskota AH, Sperker S, Stefanova R, McGinn PJ, O'Leary SJ (2019) Antioxidant properties and lipid composition of selected microalgae. J Appl Phycol 31(1):309–318

Barceló-Villalobos M, Figueroa FL, Korbee N, Álvarez-Gómez F, Abreu MH (2017) Production of mycosporine-like amino acids from *Gracilaria vermiculophylla* (Rhodophyta) cultured through one year in an integrated multi-trophic aquaculture (IMTA) system. Mar Biotechnol 19(3):246–254

Bari E, Arciola CR, Vigani B, Crivelli B, Moro P, Marrubini G, Sorrenti M, Catenacci L, Bruni G, Chlapanidas T, Lucarelli E (2017) In vitro effectiveness of microspheres based on silk sericin and *Chlorella vulgaris* or *Arthrospira platensis* for wound healing applications. Materials 10(9):983

Bedoux G, Hardouin K, Burlot AS, Bourgougnon N (2014) Bioactive components from seaweeds: cosmetic applications and future development. Adv Bot Res 71:345–378

Berthon JY, Nachat-Kappes R, Bey M, Cadoret JP, Renimel I, Filaire E (2017) Marine algae as attractive source to skin care. Free Radic Res 51(6):555–567

Bhatia S, Sharma K, Namdeo AG, Chaugule BB, Kavale M, Nanda S (2010) Broad-spectrum sun-protective action of Porphyra-334 derived from *Porphyra vietnamensis*. Pharm Res 2(1):45

Bishop WM, Zubeck HM (2012) Evaluation of microalgae for use as nutraceuticals and nutritional supplements. J Nutr Food Sci 2(5):1–6

Borah D, Gopalakrishnan S, Nooruddin T (2021) Carbohydrate biolubricants from algae and cyanobacteria. J Polym Environ 29:3444–3458

Borowitzka MA (1988) Vitamins and fine chemicals from micro-algae. Cambridge University Press, pp 153–196

Capon RJ, Barrow RA, Rochfort S, Jobling M, Skene C, Lacey E, Gill JH, Friedel T, Wadsworth D (1998) Marine nematocides: tetrahydrofurans from a southern Australian brown alga, Notheia Anomala. Tetrahedron 54(10):2227–2242

Cardoso I, Cotas J, Rodrigues A, Ferreira D, Osório N, Pereira L (2019) Extraction and analysis of compounds with antibacterial potential from the red alga *Grateloupia turuturu*. J Mar Sci Eng 7(7):220

Cardozo KH, Guaratini T, Barros MP, Falcão VR, Tonon AP, Lopes NP, Campos S, Torres MA, Souza AO, Colepicolo P, Pinto E (2007) Metabolites from algae with economical impact. Comp Biochem Physiol Part C Toxicol Pharmacol 146(1–2):60–78

Carreto JI, Carignan MO (2011) Mycosporine-like amino acids: relevant secondary metabolites. Chemical and ecological aspects. Mar Drugs 9(3):387–446

Casas MP, Conde E, Domínguez H, Moure A (2019) Ecofriendly extraction of bioactive fractions from *Sargassum muticum*. Process Biochem 79:166–173

Çetin Z, Saygili Eİ, Benlier N, Ozkur M, Sayin S (2021) Omega-3 polyunsaturated fatty acids and cancer. In: Nutraceuticals and cancer signaling. Springer, Cham, pp 591–631

Chen J, Jiang QD, Chai YP, Zhang H, Peng P, Yang XX (2016) Natural terpenes as penetration enhancers for transdermal drug delivery. Molecules 21(12):1709

Cian RE, Bacchetta C, Rossi A, Cazenave J, Drago SR (2019) Red seaweed Pyropia columbina as antioxidant supplement in feed for cultured juvenile Pacú (*Piaractus mesopotamicus*). J Appl Phycol 31(2):1455–1465

Cotas J, Leandro A, Pacheco D, Gonçalves AM, Pereira L (2020) A comprehensive review of the nutraceutical and therapeutic applications of red seaweeds (Rhodophyta). Life 10(3):19

Daniel S, Cornelia S, Fred Z (2004) UV-A sunscreen from red algae for protection against premature skin aging. In: Cosmet toiletries manufacture worldwide, pp 139–143

de Andrade CJ, de Andrade LM (2017) An overview on the application of genus *Chlorella* in biotechnological processes. J Adv Res Biotechnol 2:1–9

del Olmo A, Picon A, Nuñez M (2019) High pressure processing for the extension of *Laminaria ochroleuca* (kombu) shelf-life: a comparative study with seaweed salting and freezing. Innov Food Sci Emerg Technol 52:420–428

Dufossé L, Galaup P, Yaron A, Arad SM, Blanc P, Murthy KNC, Ravishankar GA (2005) Microorganisms and microalgae as sources of pigments for food use: a scientific oddity or an industrial reality? Trends Food Sci Technol 16(9):389–406

Eom SH, Moon SY, Lee DS, Kim HJ, Park K, Lee EW, Kim TH, Chung YH, Lee MS, Kim YM (2015) In vitro antiviral activity of dieckol and phlorofucofuroeckol-A isolated from edible brown alga *Eisenia bicyclis* against murine norovirus. Algae 30(3):241–246

Farasat M, Khavari-Nejad RA, Nabavi SMB, Namjooyan F (2013) Antioxidant properties of two edible green seaweeds from northern coasts of the Persian Gulf. Jundishapur J Nat Pharm Prod 8(1):47

Fernandes H, Salgado JM, Martins N, Peres H, Oliva-Teles A, Belo I (2019) Sequential bioprocessing of *Ulva rigida* to produce lignocellulolytic enzymes and to improve its nutritional value as aquaculture feed. Bioresour Technol 281:277–285

Fernando IS, Sanjeewa KA, Samarakoon KW, Lee WW, Kim HS, Kang N, Ranasinghe P, Lee HS, Jeon YJ (2017) A fucoidan fraction purified from *Chnoospora minima*; a potential inhibitor of LPS-induced inflammatory responses. Int J Biol Macromol 104:1185–1193

Fernando IS, Jayawardena TU, Sanjeewa KA, Wang L, Jeon YJ, Lee WW (2018) Anti-inflammatory potential of alginic acid from *Sargassum horneri* against urban aerosol-induced inflammatory responses in keratinocytes and macrophages. Ecotoxicol Environ Saf 160:24–31

Fernando IPS, Dias MKHM, Madusanka DMD, Han EJ, Kim MJ, Jeon YJ, Ahn G (2020) Fucoidan refined by *Sargassum confusum* indicate protective effects suppressing photo-oxidative stress and skin barrier perturbation in UVB-induced human keratinocytes. Int J Biol Macromol 164:149–161

Fitton JH, Dell'Acqua G, Gardiner VA, Karpiniec SS, Stringer DN, Davis E (2015) Topical benefits of two fucoidan-rich extracts from marine macroalgae. Cosmetics 2(2):66–81

Fleurence J, Levine I (eds) (2016) Seaweed in health and disease prevention. Academic Press, London, pp 389–420

Francavilla M, Franchi M, Monteleone M, Caroppo C (2013) The red seaweed *Gracilaria gracilis* as a multi products source. Mar Drugs 11(10):3754–3776

Fu W, Nelson DR, Yi Z, Xu M, Khraiwesh B, Jijakli K, Chaiboonchoe A, Alzahmi A, Al-Khairy D, Brynjolfsson S, Salehi-Ashtiani K (2017) Bioactive compounds from microalgae: current development and prospects. Stud Nat Prod Chem 54:199–225

Fumia A, Cicero N, Gitto M, Nicosia N, Alesci A (2022) Role of nutraceuticals on neurodegenerative diseases: neuroprotective and immunomodulant activity. Nat Prod Res 36(22):5916–5933

Ganesan K, Kumar KS, Rao PS (2011) Comparative assessment of antioxidant activity in three edible species of green seaweed, *Enteromorpha* from Okha, Northwest coast of India. Innov Food Sci Emerg Technol 12(1):73–78

Gao Q, Garcia-Pichel F (2011) Microbial ultraviolet sunscreens. Nat Rev Microbiol 9(11):791–802

Garcia-Pichel F, Castenholz RW (1991) Characterization and biological implications of scytonemin, a cyanobacterial sheath pigment. J Phycol 27:395–409

Haimeur A, Ulmann L, Mimouni V, Guéno F, Pineau-Vincent F, Meskini N, Tremblin G (2012) The role of *Odontella aurita*, a marine diatom rich in EPA, as a dietary supplement in dyslipidemia, platelet function and oxidative stress in high-fat fed rats. Lipids Health Dis 11(1):1–13

Hakim MM, Patel IC (2020) A review on phytoconstituents of marine brown algae. Futur J Pharm Sci 6(1):1–11

Hamid AA, Aiyelaagbe OO, Usman LA, Ameen OM, Lawal A (2010) Antioxidants: its medicinal and pharmacological applications. Afr J Pure Appl Chem 4(8):142–151

Hartmann A, Gostner J, Fuchs JE, Chaita E, Aligiannis N, Skaltsounis L, Ganzera M (2015) Inhibition of collagenase by mycosporine-like amino acids from marine sources. Planta Med 81(10):813–820

Hashimoto H, Sugai Y, Uragami C, Gardiner AT, Cogdell RJ (2015) Natural and artificial light-harvesting systems utilizing the functions of carotenoids. J Photochem Photobiol C: Photochem Rev 25:46–70

Heo SJ, Ko SC, Cha SH, Kang DH, Park HS, Choi YU, Kim D, Jung WK, Jeon YJ (2009a) Effect of phlorotannins isolated from *Ecklonia cava* on melanogenesis and their protective effect against photo-oxidative stress induced by UV-B radiation. Toxicol In Vitro 23(6):1123–1130

Heo SJ, Hwang JY, Choi JI, Han JS, Kim HJ, Jeon YJ (2009b) Diphlorethohydroxycarmalol isolated from *Ishige okamurae*, a brown alga, a potent α-glucosidase and α-amylase inhibitor, alleviates postprandial hyperglycemia in diabetic mice. Eur J Pharmacol 615(1–3):252–256

Heo SJ, Hwang JY, Choi JI, Lee SH, Park PJ, Kang DH, Oh C, Kim DW, Han JS, Jeon YJ, Kim HJ (2010a) Protective effect of diphlorethohydroxycarmalol isolated from *Ishige okamurae* against high glucose-induced-oxidative stress in human umbilical vein endothelial cells. Food Chem Toxicol 48(6):1448–1454

Heo SJ, Yoon WJ, Kim KN, Ahn GN, Kang SM, Kang DH, Oh C, Jung WK, Jeon YJ (2010b) Evaluation of anti-inflammatory effect of fucoxanthin isolated from brown algae in lipopolysaccharide-stimulated RAW 264.7 macrophages. Food Chem Toxicol 48(8–9):2045–2051

Heo JC, Park JA, Kim DK, Lee JH (2019) Photobiomodulation (660 nm) therapy reduces oxidative stress and induces BDNF expression in the hippocampus. Sci Rep 9(1):10114

Hupel M, Lecointre C, Meudec A, Poupart N, Gall EA (2011) Comparison of photoprotective responses to UV radiation in the brown seaweed *Pelvetia canaliculata* and the marine angiosperm *Salicornia ramosissima*. J Exp Mar Biol Ecol 401(1–2):36–47

Hwang E, Park SY, Sun ZW, Shin HS, Lee DG, Yi TH (2014) The protective effects of fucosterol against skin damage in UVB-irradiated human dermal fibroblasts. Mar Biotechnol 16(3):361–370

Ignat I, Volf I, Popa VI (2011) A critical review of methods for characterisation of polyphenolic compounds in fruits and vegetables. Food Chem 126(4):1821–1835

Imchen T, Singh KS (2022) Marine algae colorants: antioxidant, anti-diabetic properties and applications in food industry. Algal Res 69:102898

Ishihara K, Murata M, Kaneniwa M, Saito H, Shinohara K, Maeda-Yamamoto M (1998) Inhibition of icosanoid production in MC/9 mouse mast cells by n-3 polyunsaturated fatty acids isolated from edible marine algae. Biosci Biotechnol Biochem 62(7):1412–1415

Jaballi I, Sallem I, Feki A, Cherif B, Kallel C, Boudawara O, Jamoussi K, Mellouli L, Nasri M, Amara IB (2019) Polysaccharide from a Tunisian red seaweed *Chondrus canaliculatus*: structural characteristics, antioxidant activity and in vivo hemato-nephroprotective properties on maneb induced toxicity. Int J Biol Macromol 123:1267–1277

de Jesus Raposo MF, De Morais AMB, De Morais RMSC (2015) Marine polysaccharides from algae with potential biomedical applications. Mar Drugs 13(5):2967–3028

Joe MJ, Kim SN, Choi HY, Shin WS, Park GM, Kang DW, Kim YK (2006) The inhibitory effects of eckol and dieckol from *Ecklonia stolonifera* on the expression of matrix metalloproteinase-1 in human dermal fibroblasts. Biol Pharm Bull 29(8):1735–1739

Jung WK, Heo SJ, Jeon YJ, Lee CM, Park YM, Byun HG, Choi YH, Park SG, Choi IW (2009) Inhibitory effects and molecular mechanism of dieckol isolated from marine brown alga on COX-2 and iNOS in microglial cells. J Agric Food Chem 57(10):4439–4446

Kalasariya HS, Pereira L, Patel NB (2022) Pioneering role of marine macroalgae in cosmeceuticals. Phycology 2:172–203

Kang KA, Lee KH, Chae S, Koh YS, Yoo BS, Kim JH, Ham YM, Baik JS, Lee NH, Hyun JW (2005) Triphlorethol-A from *Ecklonia cava* protects V79-4 lung fibroblast against hydrogen peroxide induced cell damage. Free Radic Res 39(8):883–892

Kang KA, Zhang R, Piao MJ, Ko DO, Wang ZH, Lee K, Kim BJ, Shin T, Park JW, Lee NH, Yoo BS (2008) Inhibitory effects of triphlorethol-A on MMP-1 induced by oxidative stress in human keratinocytes via ERK and AP-1 inhibition. J Toxic Environ Health A 71(15):992–999

Kelman D, Posner EK, McDermid KJ, Tabandera NK, Wright PR, Wright AD (2012) Antioxidant activity of Hawaiian marine algae. Mar Drugs 10(2):403–416

Kim HH, Shin CM, Park CH, Kim KH, Cho KH, Eun HC, Chung JH (2005a) Eicosapentaenoic acid inhibits UV-induced MMP-1 expression in human dermal fibroblasts. J Lipid Res 46(8):1712–1720

Kim YC, An RB, Yoon NY, Nam TJ, Choi JS (2005b) Hepatoprotective constituents of the edible brown alga *Ecklonia stolonifera* on tacrine-induced cytotoxicity in Hep G2 cells. Arch Pharm Res 28(12):1376–1380

Kim SK, Ravichandran YD, Khan SB, Kim YT (2008) Prospective of the cosmeceuticals derived from marine organisms. Biotechnol Bioprocess Eng 13(5):511–523

Kim KC, Piao MJ, Cho SJ, Lee NH, Hyun JW (2012) Phloroglucinol protects human keratinocytes from ultraviolet B radiation by attenuating oxidative stress. Photodermatol Photoimmunol Photomed 28(6):322–331

Kim JA, Ahn BN, Kong CS, Kim SK (2013a) The chromene sargachromanol E inhibits ultraviolet A-induced ageing of skin in human dermal fibroblasts. Br J Dermatol 168(5):968–976

Kim KN, Yang HM, Kang SM, Kim D, Ahn G, Jeon YJ (2013b) Octaphlorethol A isolated from *Ishige foliacea* inhibits α-MSH-stimulated induced melanogenesis via ERK pathway in B16F10 melanoma cells. Food Chem Toxicol 59:521–526

Kim CR, Kim YM, Lee MK, Kim IH, Choi YH, Nam TJ (2017) *Pyropia yezoensis* peptide promotes collagen synthesis by activating the TGF-β/Smad signaling pathway in the human dermal fibroblast cell line Hs27. Int J Mol Med 39(1):31–38

Kim YI, Oh WS, Song PH, Yun S, Kwon YS, Lee YJ, Ku SK, Song CH, Oh TH (2018) Anti-photoaging effects of low molecular-weight fucoidan on ultraviolet B-irradiated mice. Mar Drugs 16(8):286

Ko SC, Cha SH, Heo SJ, Lee SH, Kang SM, Jeon YJ (2011) Protective effect of *Ecklonia cava* on UVB-induced oxidative stress: in vitro and in vivo zebrafish model. J Appl Phycol 23(4):697–708

Komaristaya VP, Gorbulin OS (2006) Sporopollenin in the composition of cell walls of *Dunaliella salina* Teod. (Chlorophyta) zygotes. Int J Algae 8(1):43

Kumari N, Pathak J, Dwivedy AK, Sinha RP (2021) Bioprospection of UV-screening compounds from lichens inhabiting the Indian state of Sikkim. Plant Arch 21:1168–1177

Kumari N, Pandey A, Gupta A, Mishra S, Sinha RP (2023) Characterization of UV-screening pigment scytonemin from cyanobacteria inhabiting diverse habitats of Varanasi, India. Biologia 78:1–12

Lajili S, Ammar HH, Mzoughi Z, Amor HBH, Muller CD, Majdoub H, Bouraoui A (2019) Characterization of sulfated polysaccharide from *Laurencia obtusa* and its apoptotic, gastroprotective and antioxidant activities. Int J Biol Macromol 126:326–336

Lauritano C, Andersen JH, Hansen E, Albrigtsen M, Escalera L, Esposito F, Helland K, Hanssen KØ, Romano G, Ianora A (2016) Bioactivity screening of microalgae for antioxidant, anti-inflammatory, anticancer, anti-diabetes, and antibacterial activities. Front Mar Sci 3:68

Lawrence KP, Long PF, Young AR (2018) Mycosporine-like amino acids for skin photoprotection. Curr Med Chem 25(40):5512–5527

Le Guillard C, Dumay J, Donnay-Moreno C, Bruzac S, Ragon JY, Fleurence J, Bergé JP (2015) Ultrasound-assisted extraction of R-phycoerythrin from *Grateloupia turuturu* with and without enzyme addition. Algal Res 12:522–528

Le Lann K, Surget G, Couteau C, Coiffard L, Cérantola S, Gaillard F, Larnicol M, Zubia M, Guérard F, Poupart N, Stiger-Pouvreau V (2016) Sunscreen, antioxidant, and bactericide capacities of phlorotannins from the brown macroalga *Halidrys siliquosa*. J Appl Phycol 28(6):3547–3559

Lee S, Lee YS, Jung SH, Kang SS, Shin KH (2003) Anti-oxidant activities of fucosterol from the marine algae *Pelvetia siliquosa*. Arch Pharm Res 26(9):719–722

Lee MS, Kwon MS, Choi JW, Shin T, No HK, Choi JS, Byun DS, Kim JI, Kim HR (2012) Anti-inflammatory activities of an ethanol extract of *Ecklonia stolonifera* in lipopolysaccharide-stimulated RAW 264.7 murine macrophage cells. J Agric Food Chem 60(36):9120–9129

Lee C, Park GH, Ahn EM, Kim BA, Park CI, Jang JH (2013) Protective effect of *Codium fragile* against UVB-induced pro-inflammatory and oxidative damages in HaCaT cells and BALB/c mice. Fitoterapia 86:54–63

Lee JH, Kim HH, Ko JY, Jang JH, Kim GH, Lee JS, Nah JW, Jeon YJ (2016) Rapid preparation of functional polysaccharides from *Pyropia yezoensis* by microwave-assistant rapid enzyme digests system. Carbohydr Polym 153:512–517

Lee MK, Ryu H, Lee JY, Jeong HH, Baek J, Van JY, Kim MJ, Jung WK, Lee B (2022) Potential beneficial effects of *Sargassum* spp. in skin aging. Mar Drugs 20(8):540

Leelapornpisid P, Mungmai L, Sirithunyalug B, Jiranusornkul S, Peerapornpisal Y (2014) A novel moisturizer extracted from freshwater macroalga [*Rhizoclonium hieroglyphicum* (C. Agardh) Kützing] for skin care cosmetic. Chiang Mai J Sci 41(5.2):1195–1207

Li K, Li XM, Gloer JB, Wang BG (2012) New nitrogen-containing bromophenols from the marine red alga *Rhodomela confervoides* and their radical scavenging activity. Food Chem 135(3):868–872

Li R, Zhou QL, Chen ST, Tai MR, Cai HY, Ding R, Liu XF, Chen JP, Luo LX, Zhong SY (2022) Chemical characterization and immunomodulatory activity of fucoidan from *Sargassum hemiphyllum*. Mar Drugs 21(1):18

Liu J, Sommerfeld M, Hu Q (2013) Screening and characterization of *Isochrysis* strains and optimization of culture conditions for docosahexaenoic acid production. Appl Microbiol Biotechnol 97(11):4785–4798

Liu J, Sun Z, Gerken H, Huang J, Jiang Y, Chen F (2014) Genetic engineering of the green alga *Chlorella zofingiensis*: a modified norflurazon-resistant phytoene desaturase gene as a dominant selectable marker. Appl Microbiol Biotechnol 98(11):5069–5079

Martins A, Vieira H, Gaspar H, Santos S (2014) Marketed marine natural products in the pharmaceutical and cosmeceutical industries: tips for success. Mar Drugs 12(2):1066–1101

Matsui M, Tanaka K, Higashiguchi N, Okawa H, Yamada Y, Tanaka K, Taira S, Aoyama T, Takanishi M, Natsume C, Takakura Y (2016) Protective and therapeutic effects of fucoxanthin against sunburn caused by UV irradiation. J Pharmacol Sci 132(1):55–64

Men X, Han X, Lee SJ, Oh G, Jin H, Oh HJ, Kim E, Kim J, Lee BY, Choi SI, Lee OH (2022) In-depth understanding of *Ecklonia stolonifera Okamura*: a review of its bioactivities and bioactive compounds. Mar Drugs 20(10):607

Menshova RV, Shevchenko NM, Imbs TI, Zvyagintseva TN, Malyarenko OS, Zaporoshets TS, Besednova NN, Ermakova SP (2016) Fucoidans from brown alga *Fucus evanescens*: structure and biological activity. Front Mar Sci 3:129

Mise T, Ueda M, Yasumoto T (2011) Production of fucoxanthin-rich powder from *Cladosiphon okamuranus*. Adv J Food Sci Technol 3(1):73–76

Mohsin S, Kurup GM (2011) Mechanism underlying the anti-inflammatory effect of sulphated polysaccharide from *Padina tetrastromatica* against carrageenan induced paw edema in rats. Biomed Prev Nutr 1(4):294–301

Moshfegh A, Salehzadeh A, Shandiz SAS, Shafaghi M, Naeemi AS, Salehi S (2019) Phytochemical analysis, antioxidant, anticancer and antibacterial properties of the Caspian Sea red macroalgae, *Laurencia caspica*. Iran J Sci Technol Trans A Sci 43(1):49–56

Mourelle ML, Gómez CP, Legido JL (2017) The potential use of marine microalgae and cyanobacteria in cosmetics and thalassotherapy. Cosmetics 4(4):46

Mu N, Mehar JG, Mudliar SN, Shekh AY (2019) Recent advances in microalgal bioactives for food, feed, and healthcare products: commercial potential, market space, and sustainability. Compr Rev Food Sci Food Saf 18(6):1882–1897

Murray M, Dordevic AL, Ryan L, Bonham MP (2018) An emerging trend in functional foods for the prevention of cardiovascular disease and diabetes: marine algal polyphenols. Crit Rev Food Sci Nutr 58(8):1342–1358

Murthy KC, Vanitha A, Rajesha J, Swamy MM, Sowmya PR, Ravishankar GA (2005) In vivo antioxidant activity of carotenoids from *Dunaliella salina*—a green microalga. Life Sci 76(12):1381–1390

Murugan K, Iyer VV (2014) Antioxidant activity and gas chromatographic-mass spectrometric analysis of extracts of the marine algae, *Caulerpa peltata* and *Padina gymnospora*. Indian J Pharm Sci 76(6):548

Neelam K, O'Gorman N, Nolan J, O'Donovan O, Wong HB, Eong KGA, Beatty S (2005) Measurement of macular pigment: Raman spectroscopy versus heterochromatic flicker photometry. Investig Ophthalmol Vis Sci 46(3):1023–1032

Oliyaei N, Moosavi-Nasab M, Tanideh N, Iraji A (2023) Multiple roles of fucoxanthin and astaxanthin against Alzheimer's disease: their pharmacological potential and therapeutic insights. Brain Res Bull 193:11–21

Oren A, Gunde-Cimerman N (2007) Mycosporines and mycosporine-like amino acids: UV protectants or multipurpose secondary metabolites? FEMS Microbiol Lett 269(1):1–10

Osthoff KS, Wiermann R (1987) Phenols as integrated compounds of sporopollenin from *Pinus* pollen. J Plant Physiol 131(1–2):5–15

Palanisamy S, Vinosha M, Marudhupandi T, Rajasekar P, Prabhu NM (2017) Isolation of fucoidan from *Sargassum polycystum* brown algae: structural characterization, in vitro antioxidant and anticancer activity. Int J Biol Macromol 102:405–412

Panayotova V, Merzdhanova A, Dobreva DA, Zlatanov M, Makedonski L (2017) Lipids of Black Sea algae: unveiling their potential for pharmaceutical and cosmetic applications. J IMAB Annu Proc Sci Pap 23(4):1747–1751

Pandey A, Pathak J, Singh DK, Ahmed H, Singh V, Kumar D, Sinha RP (2020) Photoprotective role of UV-screening pigment scytonemin against UV-B-induced damages in the heterocyst-forming cyanobacterium *Nostoc* sp. strain HKAR-2. Braz J Bot 43:67–80

Pathak J, Pandey A, Maurya PK, Rajneesh R, Sinha RP, Singh SP (2020) Cyanobacterial secondary metabolite scytonemin: a potential photoprotective and pharmaceutical compound. Proc Natl Acad Sci India Sect B Biol Sci 90(3):467–481

Patil UK, Saraogi R (2014) Natural products as potential drug permeation enhancer in transdermal drug delivery system. Arch Dermatol Res 306(5):419–426

Peinado NK, Abdala Díaz RT, Figueroa FL, Helbling EW (2004) Ammonium and UV radiation stimulate the accumulation of mycosporine-like amino acids in *Porphyra columbina* (Rhodophyta) from Patagonia, Argentina. J Phycol 40(2):248–259

Pérez-Larrán P, Torres MD, Flórez-Fernández N, Balboa EM, Moure A, Domínguez H (2019) Green technologies for cascade extraction of *Sargassum muticum* bioactives. J Appl Phycol 31(4):2481–2495
Piao MJ, Zhang R, Lee NH, Hyun JW (2012) Phloroglucinol attenuates ultraviolet B radiation-induced matrix metalloproteinase-1 production in human keratinocytes via inhibitory actions against mitogen-activated protein kinases and activator protein-1. Photochem Photobiol 88(2):381–388
Pillai S, Oresajo C, Hayward J (2005) Ultraviolet radiation and skin aging: roles of reactive oxygen species, inflammation and protease activation, and strategies for prevention of inflammation-induced matrix degradation—a review. Int J Cosmet Sci 27(1):17–34
Pimentel FB, Alves RC, Rodrigues F, Oliveira PP, M.B. (2018) Macroalgae-derived ingredients for cosmetic industry—an update. Cosmetics 5(1):2
Plaza M, Santoyo S, Jaime L, Reina GGB, Herrero M, Señoráns FJ, Ibáñez E (2010) Screening for bioactive compounds from algae. J Pharm Biomed Anal 51(2):450–455
Portwich A, Garcia-Pichel F (2003) Biosynthetic pathway of mycosporines (mycosporine-like amino acids) in the cyanobacterium *Chlorogloeopsis* sp. strain PCC 6912. Phycologia 42(4):384–392
Pozharitskaya ON, Obluchinskaya ED, Shikov AN (2020) Mechanisms of bioactivities of fucoidan from the brown seaweed *Fucus vesiculosus* L. of the Barents Sea. Mar Drugs 18(5):275
Premalatha M, Dhasarathan P, Theriappan P (2011) Phytochemical characterization and antimicrobial efficiency of seaweed samples, *Ulva fasciata* and *Chaetomorpha antennina*. Int J Pharm Biol Sci 2:288–293
Rao AR, Sindhuja HN, Dharmesh SM, Sankar KU, Sarada R, Ravishankar GA (2013) Effective inhibition of skin cancer, tyrosinase, and antioxidative properties by astaxanthin and astaxanthin esters from the green alga *Haematococcus pluvialis*. J Agric Food Chem 61(16):3842–3851
Rastogi RP, Sonani RR, Madamwar D (2015) Cyanobacterial sunscreen scytonemin: role in photoprotection and biomedical research. Appl Biochem Biotechnol 176(6):1551–1563
Rastogi RP, Sonani RR, Madamwar D, Incharoensakdi A (2016) Characterization and antioxidant functions of mycosporine-like amino acids in the cyanobacterium *Nostoc* sp. R76DM. Algal Res 16:110–118
Reef R, Kaniewska P, Hoegh-Guldberg O (2009) Coral skeletons defend against ultraviolet radiation. PLoS One 4(11):e7995
Rodrigues D, Alves C, Horta A, Pinteus S, Silva J, Culioli G, Thomas OP, Pedrosa R (2015) Antitumor and antimicrobial potential of bromoditerpenes isolated from the red alga, *Sphaerococcus coronopifolius*. Mar Drugs 13(2):713–726
Ryu B, Qian ZJ, Kim MM, Nam KW, Kim SK (2009) Anti-photoaging activity and inhibition of matrix metalloproteinase (MMP) by marine red alga, *Corallina pilulifera* methanol extract. Radiat Phys Chem 78(2):98–105
Saidani K, Bedjou F, Benabdesselam F, Touati N (2012) Antifungal activity of methanolic extracts of four Algerian marine algae species. Afr J Biotechnol 11(39):9496–9500
Sakai S, Komura Y, Nishimura Y, Sugawara T, Hirata T (2011) Inhibition of mast cell degranulation by phycoerythrin and its pigment moiety phycoerythrobilin, prepared from *Porphyra yezoensis*. Food Sci Technol Res 17(2):171–177
Sappati PK, Nayak B, VanWalsum GP, Mulrey OT (2019) Combined effects of seasonal variation and drying methods on the physicochemical properties and antioxidant activity of sugar kelp (*Saccharina latissima*). J Appl Phycol 31(2):1311–1332
Shannon E, Abu-Ghannam N (2016) Antibacterial derivatives of marine algae: an overview of pharmacological mechanisms and applications. Mar Drugs 14(4):81
Shen CT, Chen PY, Wu JJ, Lee TM, Hsu SL, Chang CMJ, Young CC, Shieh CJ (2011) Purification of algal anti-tyrosinase zeaxanthin from *Nannochloropsis oculata* using supercritical anti-solvent precipitation. J Supercrit Fluids 55(3):955–962
Shibata T, Fujimoto K, Nagayama K, Yamaguchi K, Nakamura T (2002) Inhibitory activity of brown algal phlorotannins against hyaluronidase. Int J Food Sci Technol 37(6):703–709

Shimoda H, Tanaka J, Shan SJ, Maoka T (2010) Anti-pigmentary activity of fucoxanthin and its influence on skin mRNA expression of melanogenic molecules. J Pharm Pharmacol 62(9):1137–1145

Shin J, Kim JE, Pak KJ, Kang JI, Kim TS, Lee SY, Yeo IH, Park JHY, Kim JH, Kang NJ, Lee KW (2017) A combination of soybean and *haematococcus* extract alleviates ultraviolet B-induced photoaging. Int J Mol Sci 18(3):682

Sinha RP, Klisch M, Gröniger A, Häder DP (1998) Ultraviolet-absorbing/screening substances in cyanobacteria, phytoplankton and macroalgae. J Photochem Photobiol B Biol 47(2–3):83–94

Smit AJ (2004) Medicinal and pharmaceutical uses of seaweed natural products: a review. J Appl Phycol 16(4):245–262

Soontornchaiboon W, Joo SS, Kim SM (2012) Anti-inflammatory effects of violaxanthin isolated from microalga *Chlorella ellipsoidea* in RAW 264.7 macrophages. Biol Pharm Bull 35(7):1137–1144

Stevenson C, Capper E, Roshak A, Marquez B, Grace K, Gerwick W, Jacobs R, Marshall L (2002) Scytonemin-a marine natural product inhibitor of kinases key in hyperproliferative inflammatory diseases. Inflamm Res 51(2):112

Stirk WA, van Staden J (2022) Bioprospecting for bioactive compounds in microalgae: antimicrobial compounds. Biotechnol Adv 59:107977

Su W, Wang L, Fu X, Ni L, Duan D, Xu J, Gao X (2020) Protective effect of a fucose-rich fucoidan isolated from *Saccharina japonica* against ultraviolet B-induced photodamage in vitro in human keratinocytes and in vivo in zebrafish. Mar Drugs 18(6):316

Sugiura Y, Takeuchi Y, Kakinuma M, Amano H (2006) Inhibitory effects of seaweeds on histamine release from rat basophile leukemia cells (RBL-2H3). Fish Sci 72(6):1286–1291

Suh SS, Hwang J, Park M, Seo HH, Kim HS, Lee JH, Moh SH, Lee TK (2014) Anti-inflammation activities of mycosporine-like amino acids (MAAs) in response to UV radiation suggest potential anti-skin aging activity. Mar Drugs 12(10):5174–5187

Tapiero H, Townsend DM, Tew KD (2004) The role of carotenoids in the prevention of human pathologies. Biomed Pharmacother 58(2):100–110

Teas J, Irhimeh MR (2017) Melanoma and brown seaweed: an integrative hypothesis. J Appl Phycol 29(2):941

Teixeira MM, Hellewell PG (1997) The effect of the selectin binding polysaccharide fucoidin on eosinophil recruitment in vivo. Br J Pharmacol 120(6):1059–1066

Thiyagarasaiyar K, Goh BH, Jeon YJ, Yow YY (2020) Algae metabolites in cosmeceutical: an overview of current applications and challenges. Mar Drugs 18(6):323

Thomas NV, Kim SK (2013) Beneficial effects of marine algal compounds in cosmeceuticals. Mar Drugs 11(1):146–164

Urikura I, Sugawara T, Hirata T (2011) Protective effect of fucoxanthin against UVB-induced skin photoaging in hairless mice. Biosci Biotechnol Biochem 75(4):757–760

Ustyuzhanina NE, Bilan MI, Gerbst AG, Ushakova NA, Tsvetkova EA, Dmitrenok AS, Usov AI, Nifantiev NE (2016) Anticoagulant and antithrombotic activities of modified xylofucan sulfate from the brown alga *Punctaria plantaginea*. Carbohydr Polym 136:826–833

Velatooru LR, Baggu CB, Janapala VR (2016) Spatane diterpinoid from the brown algae, *Stoechospermum marginatum* induces apoptosis via ROS induced mitochondrial mediated caspase dependent pathway in murine B16F10 melanoma cells. Mol Carcinog 55(12):2222–2235

Vo TS, Kim SK, Ryu B, Ngo DH, Yoon NY, Bach LG, Hang NTN, Ngo DN (2018) The suppressive activity of fucofuroeckol-A derived from brown algal *Ecklonia stolonifera* Okamura on UVB-induced mast cell degranulation. Mar Drugs 16(1):1

Wada N, Sakamoto T, Matsugo S (2013) Multiple roles of photosynthetic and sunscreen pigments in cyanobacteria focusing on the oxidative stress. Meta 3(2):463–483

Wang T, Jónsdóttir R, Liu H, Gu L, Kristinsson HG, Raghavan S, Ólafsdóttir G (2012) Antioxidant capacities of phlorotannins extracted from the brown algae *Fucus vesiculosus*. J Agric Food Chem 60(23):5874–5883

Wang L, Wang X, Wu H, Liu R (2014) Overview on biological activities and molecular characteristics of sulfated polysaccharides from marine green algae in recent years. Mar Drugs 12(9):4984–5020

Wang HMD, Chen CC, Huynh P, Chang JS (2015) Exploring the potential of using algae in cosmetics. Bioresour Technol 184:355–362

Wang ZJ, Xu W, Liang JW, Wang CS, Kang Y (2017a) Effect of fucoidan on B16 murine melanoma cell melanin formation and apoptosis. Afr J Tradit Complement Altern Med 14(4):149–155

Wang HMD, Li XC, Lee DJ, Chang JS (2017b) Potential biomedical applications of marine algae. Bioresour Technol 244:1407–1415

Wang X, Zhang Z, Wu Y, Sun X, Xu N (2019) Synthesized sulfated and acetylated derivatives of polysaccharide extracted from *Gracilariopsis lemaneiformis* and their potential antioxidant and immunological activity. Int J Biol Macromol 124:568–572

Wang L, Jayawardena TU, Hyun J, Wang K, Fu X, Xu J, Gao X, Park Y, Jeon YJ (2023) Antioxidant and anti-photoaging effects of a fucoidan isolated from *Turbinaria ornata*. Int J Biol Macromol 225:1021–1027

Wu Q, Liu L, Miron A, Klímová B, Wan D, Kuča K (2016) The antioxidant, immunomodulatory, and anti-inflammatory activities of *Spirulina*: an overview. Arch Toxicol 90(8):1817–1840

Wu HL, Fu XY, Cao WQ, Xiang WZ, Hou YJ, Ma JK, Wang Y, Fan CD (2019) Induction of apoptosis in human glioma cells by fucoxanthin via triggering of ROS-mediated oxidative damage and regulation of MAPKs and PI3K–AKT pathways. J Agric Food Chem 67(8):2212–2219

Xiong F, Komenda J, Kopecký J, Nedbal L (1997) Strategies of ultraviolet-B protection in microscopic algae. Physiol Plant 100(2):378–388

Yan X, Chuda Y, Suzuki M, Nagata T (1999) Fucoxanthin as the major antioxidant in *Hijikia fusiformis*, a common edible seaweed. Biosci Biotechnol Biochem 63(3):605–607

Yang DJ, Lin JT, Chen YC, Liu SC, Lu FJ, Chang TJ, Wang M, Lin HW, Chang YY (2013) Suppressive effect of carotenoid extract of *Dunaliella salina* alga on production of LPS-stimulated pro-inflammatory mediators in RAW264. 7 cells via NF-κB and JNK inactivation. J Funct Foods 5(2):607–615

Yang EJ, Moon JY, Kim SS, Yang KW, Lee WJ, Lee NH, Hyun CG (2014) Jeju seaweeds suppress lipopolysaccharide-stimulated proinflammatory response in RAW 264.7 murine macrophages. Asian Pac J Trop Biomed 4(7):529–537

Yegdaneh A, Ghannadi A, Dayani L (2016) Chemical constituents and biological activities of two Iranian *Cystoseira* species. Res Pharm Sci 11(4):311

Yoon NY, Eom TK, Kim MM, Kim SK (2009) Inhibitory effect of phlorotannins isolated from *Ecklonia cava* on mushroom tyrosinase activity and melanin formation in mouse B16F10 melanoma cells. J Agric Food Chem 57(10):4124–4129

Yuan YV, Westcott ND, Hu C, Kitts DD (2009) Mycosporine-like amino acid composition of the edible red alga, *Palmaria palmata* (dulse) harvested from the west and east coasts of Grand Manan Island, New Brunswick. Food Chem 112(2):321–328

Zaragozá MC, López D, Sáiz P, Poquet M, Pérez J, Puig-Parellada P, Marmol F, Simonetti P, Gardana C, Lerat Y, Burtin P (2008) Toxicity and antioxidant activity in vitro and in vivo of two *Fucus vesiculosus* extracts. J Agric Food Chem 56(17):7773–7780

Zhang Z, Wang F, Wang X, Liu X, Hou Y, Zhang Q (2010) Extraction of the polysaccharides from five algae and their potential antioxidant activity in vitro. Carbohydr Polym 82(1):118–121

Zhou L, Wang H, Jing J, Yu L, Wu X, Lu Z (2018) Regulation of hair follicle development by exosomes derived from dermal papilla cells. Biochem Biophys Res Commun 500(2):325–332

Zussman J, Ahdout J, Kim J (2010) Vitamins and photoaging: do scientific data support their use? J Am Acad Dermatol 63(3):507–525

Chapter 3
Bioprospection of Photoprotective Compounds from Cyanobacteria

Prashant R. Singh, Ashish P. Singh, Rajneesh, Amit Gupta, Rajeshwar P. Sinha, and Jainendra Pathak

Abstract Cyanobacteria are one of the oldest photosynthetic nitrogen fixers of the terrestrial as well as aquatic ecosystems. Cyanobacteria and microalgae produce a large number of secondary metabolites having biomedical, industrial, and biotechnological importance. Due to their presence in vast habitats, cyanobacteria and microalgae were exposed to variety of harsh environmental factors such as salt, desiccation, temperature, heavy metals, and ultraviolet radiation (UVR). In response, cyanobacteria have developed different mechanisms to cope these harsh environmental conditions. Photosynthetic nature of cyanobacteria continuously exposes them to lethal doses of UVR coming with solar radiation which affects their physiology, photosynthetic efficiency, productivity by reactive oxygen species (ROS) generation inside the cell and ultimately leads to cell death. In response to these stresses cyanobacteria have developed different protective mechanisms such as avoidance, enzymatic and non-enzymatic defence system and synthesis of novel secondary metabolites such as mycosporine-like amino acids (MAAs) and scytonemin. MAAs are water-soluble molecules that absorb short wavelength of solar UVR which release the energy in the form of heat. Scytonemin is a small hydrophobic alkaloid pigment present in the extracellular sheath of some cyanobacteria that acts as UVR protectant. Scytonemin and MAAs are highly photostable therefore, they primarily function as UV-screening compounds. They also show antioxidative

P. R. Singh · A. P. Singh · Rajneesh · A. Gupta
Laboratory of Photobiology and Molecular Microbiology, Centre of Advanced Study in Botany, Institute of Science, Banaras Hindu University, Varanasi, India

R. P. Sinha
Laboratory of Photobiology and Molecular Microbiology, Centre of Advanced Study in Botany, Institute of Science, Banaras Hindu University, Varanasi, India

University Center for Research & Development (UCRD), Chandigarh University, Chandigarh, India

J. Pathak (✉)
Department of Botany, Pt. Jawaharlal Nehru College (Affiliated to Bundelkhand University, Jhansi), Banda, India

V. K. Kannaujiya et al. (eds.), *Photoprotective Green Pharmacology: Challenges, Sources and Future Applications*,
https://doi.org/10.1007/978-981-99-0749-6_3

properties. The capability of cyanobacteria to produce large number of secondary metabolites which serve as natural sunscreens, antibiotic, antifungal, anticancer and antiviral agents make them economically important organisms. These are readily biotechnologically exploitable in the cosmetics and other industrial sectors for the creation of novel medications and drugs. Hence, bioprospection of these photoprotective compounds and other secondary metabolites from cyanobacteria and microalgae becomes crucial.

Keywords Cyanobacteria · Mycosporine-like amino acids (MAAs) · Scytonemin · Ultraviolet radiation (UVR)

3.1 Introduction

Cyanobacteria are the oldest group of Gram-negative oxygen evolving photosynthetic prokaryotes. Their physiological flexibility makes cyanobacteria widely distributed on the Earth's surface ranging from glaciers, hot springs to the Arctic and the Antarctic. They are the major primary biomass producers in the terrestrial and aquatic ecosystems (Häder et al. 2007; Rastogi and Sinha 2009). Cyanobacteria can fix atmospheric nitrogen in rice field, hence they are economically important and act as biofertilizers (Pathak et al. 2018). Cyanobacteria are the possible renewable option for bioactive secondary metabolites. They may be used in a variety of ways for commercial and biotechnological purposes (Rajneesh et al. 2017). Cyanobacteria produce these secondary metabolites in response to different abiotic and biotic stress as a strategy to protect cyanobacteria from these stresses (Singh et al. 2010; Pathak et al. 2019). The secondary metabolites produced by cyanobacteria under diverse environments are alkaloids, terpenoids, peptides, polyketides, and ultraviolet radiation (UVR) absorbing compounds. These compounds can be utilized in industrial biotechnology as pharmaceuticals, nutraceuticals and cosmeceuticals (Rajneesh et al. 2019).

The ozone layer is depleted, which causes solar UVR (280–400 nm) to be stronger (Crutzen 1992; Stolarski et al. 1992; Williamson et al. 2014) at the Earth's surface. Solar UVR mainly affects the protein and DNA of all forms of living organisms including cyanobacteria. UVR has an impact on cyanobacteria's growth, cellular morphology, survival, pigmentation, photosynthesis, and numerous important metabolic processes including N_2 fixation, CO_2 absorption, and ribulose 1,5-bisphosphate carboxylase/oxygenase (RuBISCO) activity (Vaishampayan et al. 2001; Gao et al. 2008; Häder et al. 2015; Rajneesh et al. 2019). Cyanobacteria have evolved a range of defense mechanisms to protect themselves from the damaging UVR which include migration and mat formation, production of antioxidants, UV-absorbing compounds, such as scytonemin and mycosporine-like amino acids (MAAs), light-dependent DNA repair by photolyase, excision repair and apoptosis (Rastogi and Sinha 2009; Singh et al. 2010; Pathak et al. 2019; Rajneesh et al. 2019). UVR photo-protectants such as MAAs and scytonemin are most common in

Table 3.1 Different members of the ROS family and their attributes (Adapted from Pathak et al. 2019)

ROS	$t_{1/2}$	Mode of action	Location
Superoxide ($O_2^{\bullet-}$)	1–4 μs	Reacts to compounds that contain double bond such as (Fe-S) proteins	Chloroplasts, mitochondria, membranes
Singlet oxygen ($^1O_2$)	1–4 μs	DNA, PUFAs, and oxidizes proteins	Chloroplasts, mitochondria, membranes
Hydroxyl radical (OH^-)	1 μs	Extremely reactive with all biomolecules	Chloroplasts, mitochondria, membranes
Hydrogen peroxide (H_2O_2)	1 ms	Uses $O_2^{\bullet-}$ to oxidize proteins and produce $OH^{\bullet}$	Chloroplasts, mitochondria, membranes

cyanobacteria and micro/macroalgae (Sinha and Häder 2008; Rastogi et al. 2013). MAAs has been found in variety of species, including fungi (Moliné et al. 2011), heterotrophic bacteria (Arai et al. 1992), micro/macroalgae, and cyanobacteria (Karsten et al. 1998; Shick and Dunlap 2002; Sinha et al. 2007; Richa and Sinha 2015). Scytonemin, a lipid soluble compound has been discovered from the extracellular polysaccharide sheath of around 300 cyanobacteria (Sinha and Häder 2008; Pathak et al. 2015, 2017a, b, 2019, 2020).

Light energy is used by cyanobacteria to fuel oxygenic photosynthesis, but along with photosynthesis photodamage in cyanobacteria also takes place. In cyanobacteria, photosystem II (PSII) is the major site that is affected by light and it results in photoinhibition. Because the reduction of plastoquinone by electrons generated by water oxidation at PSII initiates photosynthetic electron flow, the accumulation of photodamaged PSII reduces photosynthetic activity and results in production of reactive oxygen species (ROS). The different types of ROS produced during photodamage of PSII are stated below (Table 3.1).

3.2 Reactive Oxygen Species Generated in Cyanobacteria Due to Photodamage

3.2.1 *Singlet Oxygen ($^1O_2$)*

The initial excited electronic state of O_2 is represented by single oxygen ($^1O_2$), that is an uncommon ROS because it is not connected with the electron transfer to O_2 (Hatz et al. 2007). Chlorophyll (Chl) triplet state formation happens as a result of inadequate energy loss during photosynthesis. Singlet oxygen ($^1O_2$) is formed during photosynthesis, and it has an impact on the entire photosynthetic apparatus, especially PSI and PSII. According to some observations, $^1O_2$ has a ~3 μs cellular lifespan (Hatz et al. 2007). It can act as an oxidizing agent for a variety of biological compounds, including lipids, pigments, nucleic acids and proteins, and it is accountable for disintegration of PSII function caused by light, which may lead to cell death (Wagner et al. 2004; Krieger-Liszkay et al. 2008).

3.2.2 *Superoxide Radicals ($O_2^{\bullet -}$)*

The release of H_2O occurs when O_2 interacts with the terminal oxidases, cytochrome c oxidase and alternative oxidase, and four electrons are exchanged. Only one electron is transported when O_2 reacts with other ETC components, that generates $O_2^{\bullet -}$, a highly destructive ROS such as OH^- and maybe $^1O_2$ which may induce cellular weakness owing to peroxidation of membrane lipids is generally the first ROS to be formed (Elstner 1987; Halliwell 2006).

3.2.3 *Hydrogen Peroxide (H_2O_2)*

Hydrogen peroxide (H_2O_2) is generated by the univalent reduction of oxygen. It has a half-life of ~1 ms, which is considerably higher as compared to other ROS like $O_2^{\bullet -}$, OH^-, and $^1O_2$, that have significantly shorter half-lives of ~2–4 μs (Bhattacharjee 2005). Excessive amount of H_2O_2 inside the cell resulted in oxidative stress.

3.2.4 *Hydroxyl Radicals (OH^-)*

One of the most potently destructive ROS found in cells is the hydroxyl radical (OH^-). OH^- reacts with Fe and can also generate H_2O_2. $O_2^{\bullet -}$ at ambient temperatures and neutral pH by the Fe-catalysed $O_2^{\bullet -}$-driven Fenton reaction. These OH^- play a significant role in mediating in vivo oxygen toxicity. It interacts with lipids, proteins, DNA, and practically every other component of cells. Because there is no enzymatic mechanism to remove OH^-, it is produced excessively, which ultimately results in cell death (Vranová et al. 2002).

3.3 Cyanobacterial Defence

Carotenoids, α-tocopherol (vitamin E), ascorbate (vitamin C), and orange carotenoid proteins are the most common defence strategies adapted by cyanobacteria in response to UVR.

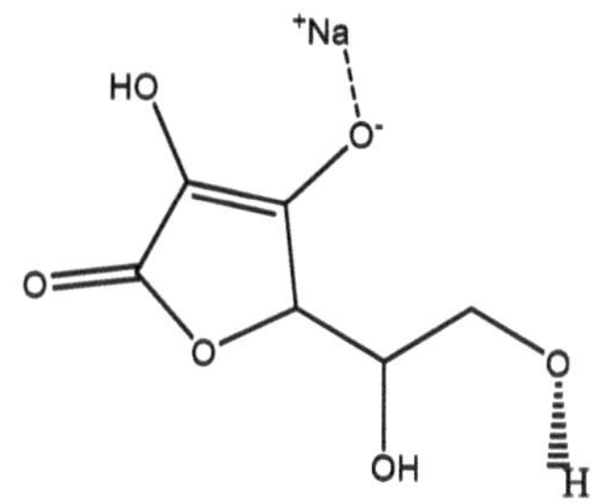

Fig. 3.1 Structure of ascorbate

3.3.1 *Ascorbate*

Ascorbate (Fig. 3.1) and ascorbate peroxidase (APX) function as a crucial line of defence towards H_2O_2 (Smirnoff and Wheeler 2000). Two molecules of ascorbate are utilized by APX to convert H_2O_2 into H_2O. Furthermore, ascorbate combines with other ROS such as peroxyl and hydroxyl radicals as well as singlet oxygen (Smirnoff and Wheeler 2000). The violaxanthin de-epoxidase and APX processes both utilized ascorbate as a substrate for the synthesis of α-tocopherol. Ascorbate reacts with ROS and produced malondialdehyde (MDA). Monodehydroascorbate reductase uses electrons from nicotinamide adenine dinucleotide phosphate hydrogen (NADPH) or ferredoxin to convert MDA back to ascorbate. If MDA is not back decreased quickly, it disproportionately forms dehydroascorbate (DHA) and ascorbate.

3.3.2 *α-Tocopherol*

α-Tocopherol is a lipid-soluble organic produced solely by oxygenic phototrophs, which includes certain cyanobacteria. ROS might oxidize tocopherols into tocopheryl radicals, which subsequently change singlet O_2 into hydroperoxide and these processes get reversed by ascorbate which recycles the tocopherol (Neely et al. 1988). Structure of tocopherol has been shown in Fig. 3.2. Both high-intensity PAR and $^1O_2$ buildup shows tolerance in *Synechocystis* PCC 6803 mutants that lack tocopherols exhibit that tocopherols are not the only antioxidants which protect PSs against the ROS (Maeda et al. 2005).

3.3.3 *Carotenoids*

Carotenoids are the largest group of naturally occurring pigments in several organisms. There are currently more than 640 carotenoids reported. They are responsible for majority red, yellow, orange pigments in microbes. Carotenoids are found in all

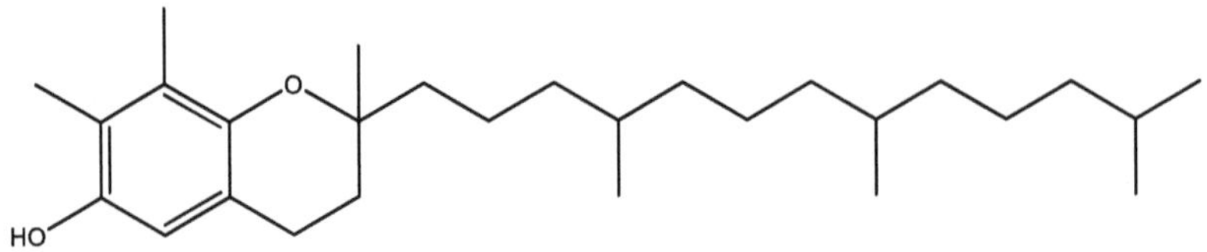

Fig. 3.2 α-Tocopherol

the photosynthetic organisms, as well as in certain non-photosynthetic bacteria (Pagels et al. 2021). The two main categories of carotenoids are xanthophylls, which include zeaxanthin and echinenone, and carotenes, which include α- and β-carotene (Hirschberg and Chamovitz 1994; Guedes et al. 2011). The majority of carotenoids have a C40 hydrocarbon backbone made up of eight C_5 isoprenoid units and a sequence of conjugated double bonds. Carotenes constitute of linear or cyclized molecules having one or two terminal rings that lacks oxygen atoms. Oxygenated carotenes are the by-products of xanthophylls. Numerous carotenoid esters and glycosylated carotenoids have been reported. The C40 backbone can be extended further to produce C45 or C50 carotenoids, or it can be truncated to produce apocarotenoids. C30 carotenoids are also produced by non-photosynthetic microorganisms. Background information about carotenoids may be found in (Goodwin 1980). Carotenoids accepts triplet state energy from chlorophyll, quenches the singlet state O_2 and protects the cells from damage caused by photooxidation. In contrast, α-tocopherol inhibits lipid peroxidation (LPO) by scavenging reactive oxygen species (Niyogi 1999). Cyanobacteria contain a wide range of carotenoids, including myxoxanthophyll, β-carotene, and its derivatives, zeaxanthin, and echinenone. The energy from photosensitized Chl or from $^1O_2$ is dissipated by these pigments, and multiple reports have found their antioxidant activity (Young and Frank 1996; Edge et al. 1997). In *Synechococcus* PCC 7942, zeaxanthin plays significant role in photoacclimation when exposed to UV-B radiation (Gotz et al. 1999), and photosynthetic activity of *Synechocystis* PCC 6803 mutants lacking zeaxanthin synthesis is found to be more susceptible to high-intensity PAR, and oxidative stress as compared to the wild-type strain (Schäfer et al. 2005). Four freshwater cyanobacteria were studied to see how they adapted to high-intensity PAR on the carotenoid composition, and the results showed that each cyanobacterium had a different type of overproduced carotenoid. When variations in the carotenoid's composition of *Nostoc commune* in response to UV-B irradiation were investigated and myxoxanthophyll and echinenone were discovered to function as UV-B photoprotectors that were present to the outer membrane of the cell (Ehling-Schulz et al. 2002).

3.3.4 Applications of Cyanobacterial Carotenoids

Carotenoids are bioactive chemicals with antioxidant, antibacterial, anti-inflammatory, anti-tumour property (Pagels et al. 2020). They may be employed as a colour enhancer in animal feed (Guedes et al. 2011) and as antioxidant and anti-ageing components in cosmetics (Morone et al. 2019). Table 3.2 shows different carotenoids extracted from cyanobacteria and their applications.

3.3.5 Orange Carotenoid Protein (OCP)

OCP has been identified in almost all the cyanobacteria that contain phycobilin proteins (PB). OCP is a 35 kD soluble protein made up of the ketocarotenoid, the 39-hydroxyechinenone (Kerfeld et al. 2003; Kerfeld 2004). It has two domains such as the N-terminal domain containing α-helical and an α/β C-terminal domain (Kerfeld et al. 2003). The N-terminal domain of OCP has a structure and sequence that is distinct to cyanobacteria, but the C-terminal domain fold belongs to the nuclear transport factor 2 superfamily. 39-hydroxyechinenon, a carotenoid, is mostly buried and covers both domains of the protein. In *Synechocystis* the *slr*1963 gene encodes OCP (Wu and Krogmann 1997), which is produced constitutively even in PBs lacking *Synechocystis* mutants (Wilson et al. 2007). OCP plays a function in photoprotection, according to research by Wilson et al. (2006) taking mutant of (ΔOCP). In the PB mutants of *Synechocystis*, photoprotective role was observed (Wilson et al. 2006), and eventually it was determined that OCP is photoactive in nature (Wilson et al. 2008). Absorption of blue-green light causes conformational

Table 3.2 Summarizes the application of carotenoids from cyanobacteria

Cyanobacteria	Main identified carotenoids	Application	References
Arthrospira platensis	β-Carotene and echinone	Antioxidant	Patias et al. (2017)
Cyanobium sp.	Zeaxanthin and β-carotene	Antioxidant	Pagels et al. (2020)
Pseudoanabaena sp. *Spirulina* sp. *Lyngbya* sp.	Myxoxanthophyll, zeaxanthin, canthaxanthin and carotenes	Antiurolithiasis	Paliwal et al. (2015)
Lyngbya sp.	Myxoxanthophyll, zeaxanthin, canthaxanthin and carotenes	Antioxidant	Paliwal et al. (2015)
Trichodesmium sp.	β-Carotene	Antioxidant	Kelman et al. (2009)
Nodosilinea (*Leptolyngbya antartica*)	β-Carotene and echinenone	Anti-inflammatory	Lopes et al. (2020)
Arthrospira platensis	Zeaxanthin, β-carotene and myxoxanthophyll	Colour enhancer (feed)	Mori et al. (1987), Belay et al. (1996)

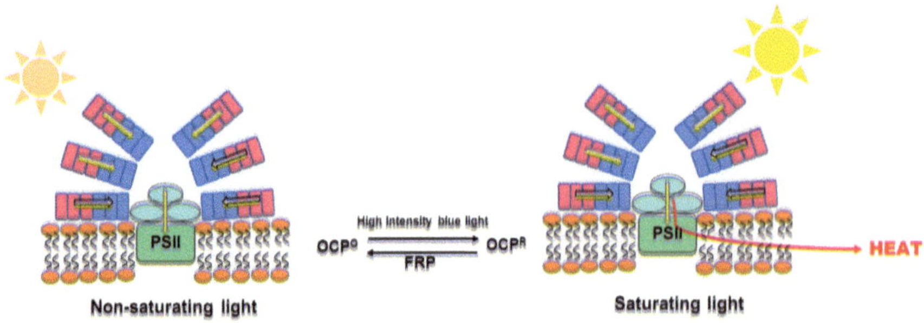

Fig. 3.3 Diagram of OCP and its role in decapitating high light radiation as heat through non-photochemical quenching

changes in the protein and the carotenoid, culminating in transformation of the stable orange dark OCP form (OCP^O) to the metastable red active form (OCP^R) (Wilson et al. 2008; Cogdell and Gardiner 2015). The presence of a ketocarotenoid was discovered to be necessary for its photoactivity. OCP interacts with zeaxanthin, that lacks the carbonyl group seen in ketocarotenoids, but zea-OCP complex is photo-inactive, and remains yellow when exposed to bright light. In cyanobacteria exposed to high irradiance, OCP^R promotes dissipation heat energy (Fig. 3.3), resulting in a decrease in PB fluorescence and the energy that received the reaction centres (Wilson et al. 2006; Cogdell and Gardiner 2015). In the dark, OCP^R transforms naturally into OCP^O (Wilson et al. 2006). Temperature plays important role in this reaction; recovery is directly proportional to the high temperature (Wilson et al. 2008). Temperature has no impact on how quickly OCP^O transforms into OCP^R in reaction to light. As a result, steady-state level of OCP^R during illumination decreases as the temperature increases (Wilson et al. 2008). It is possible that the OCP^R form is more stable in vivo than in vitro because the recovery kinetics are slower than the conversion from OCP^R to OCP^O in the dark (Wilson et al. 2006).

3.4 Ultraviolet Radiation Protective Compounds

Several cyanobacteria produce UVR-absorbing compounds to protect them from harmful UVR when they are exposed to it for long periods of time (Fig. 3.4). (Pathak et al. 2017a). UVR-screening compounds such as MAAs and scytonemin are well-known cyanobacterial photoprotective compounds that protect against harmful UV-A and UV-B radiations (Singh et al. 2010; Pathak et al. 2015; Richa and Sinha 2015). Photoprotective role of the novel UVR-screening compounds is briefly discussed below.

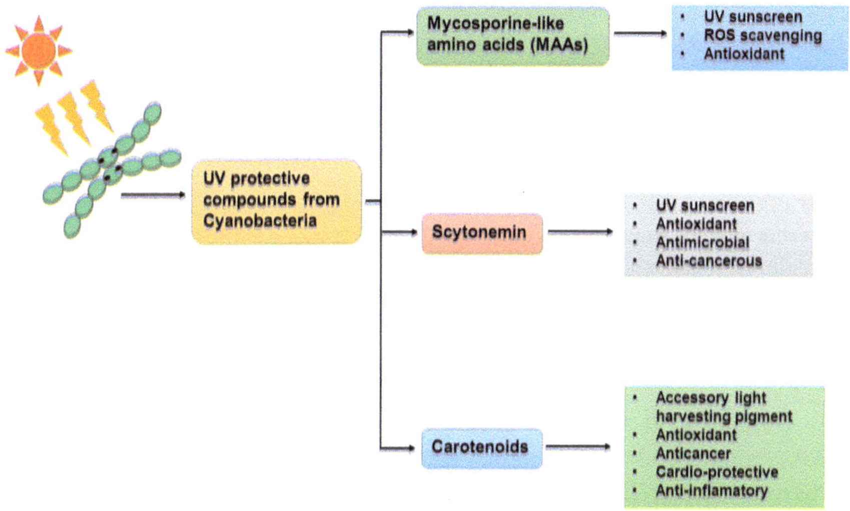

Fig. 3.4 Different cyanobacterial photoprotective compounds and their applications

3.4.1 *Mycosporine-Like Amino Acids (MAAs)*

MAAs are synthesized/accumulated by many taxonomic groups of organisms, including cyanobacteria, micro/macroalgae, symbiotic lichens and several aquatic animals (Sinha et al. 2007). MAAs are colourless, small, hydrophilic, compounds with a cyclohexenimine or cyclohexanone chromophore attached to an acid's nitrogen group or its imino alcohol. Their high molar extinction coefficients (ε = 28,100–50,000 M^{-1} cm^{-1}) and substantial UVR absorption (λ_{max}: 310–362 nm) indicate that they are UV-screening biomolecules. MAAs have been identified and characterized from cyanobacteria and other eukaryotic microalgae (Sinha et al. 2003; Rastogi et al. 2012; Pathak et al. 2017b). Some new glycosylated MAAs are also synthesized by cyanobacteria (Böhm et al. 1995; Nazifi et al. 2013). MAAs equip cyanobacteria with passive defensive mechanisms that allow them to collect photons and prevent them from interacting with proteins and DNA. MAAs can block three out of every ten photons from striking cytoplasmic targets in cyanobacteria (Garcia-Pichel and Castenholz 1993). MAAs have been shown to have photoprotective properties in various microalgae, including cyanobacteria (Garcia-Pichel and Castenholz 1993; Richa and Sinha 2015). Without generating ROS, MAAs effectively disperse absorbed radiation in the form of heat (Conde et al. 2000, 2007).

Novel glycosylated MAAs are also produced by certain cyanobacteria (Böhm et al. 1995; Nazifi et al. 2013). MAAs possess passive defense system that allow cyanobacteria to collect photons and inhibit them from interacting with DNA and proteins. Three out of every ten photons can be inhibited from reaching cytoplasmic targets by MAAs in cyanobacteria (Garcia-Pichel and Castenholz 1993). Different microalgae, particularly cyanobacteria, already have a well-established understanding of the photoprotective properties of MAAs (Garcia-Pichel and Castenholz 1993; Rastogi et al. 2015; Richa and Sinha 2015). MAAs effectively release the absorbed radiation as heat, without generation of ROS (Conde et al. 2000, 2007).

MAAs not only protects producer but also primary and secondary consumers along the food chain from UV radiation (Helbling et al. 2002). In microalgae and other organisms, MAA production may be exploited as a passive defense mechanism that enables solar light absorption and prevents UV rays from damaging essential cellular machinery like DNA and proteins. UV-screening compounds MAAs helps in reducing UV-induced thymine dimer formation and hence, maintain the integrity of the genome (Misonou et al. 2003; Rastogi et al. 2010), which are the most cytotoxic and genotoxic DNA lesions in the cell. MAAs might be employed as broad-spectrum UV absorbers/protectors in cosmetics and other cosmeceutical sectors. UV-induced skin cancer can be prevented using MAAs. Red algae-derived MAAs (porphyra-334 and shinorine) have been found to prevent the onset of premature skin ageing (Daniel et al. 2004). MAAs have been shown to protect human fibroblast cells against UV-induced cell death (Oyamada et al. 2008). Some artificial MAA analogues, such as tetrahydropyridine derivatives, have been produced for commercial use in sunscreen formulations (Bird et al. 1991; De la Coba 2007a, b) (Table 3.3).

3.4.2 Scytonemin

Scytonemin is exclusively produced by only some cyanobacteria such as *Scytonema* and *Lyngbya* etc. It is a dimeric phenolic pigment that is pale yellow to brown in colour and accumulates in the extracellular polysaccharide sheath of certain cyanobacteria and functions as a passive UV protector (Garcia-Pichel and Castenholz 1991; Pathak et al. 2017a). Scytonemin has an absorption maximum at 370 nm in vivo and at 386 nm in pure form, but it also absorbs considerably at 252, 278, and 300 nm and offers photoprotection to organisms living in UV-A/B exposed habitats (Pathak et al. 2017a; Rajneesh et al. 2019). Dimethoxyscytonemin, scytonemin-3a-imine, scytonin, and tetramethoxyscytonemin are a few of the scytonemin derivatives that have been identified from various cyanobacteria (Bultel-Poncé et al. 2004; Grant and Louda 2013; Rastogi and Incharoensakdi 2014a). Scytonemin exists in both a reduced (Mw 546 Da) and an oxidized (Mw 544 Da) form (Fig. 3.5). Several cyanobacterial genomes have been examined to find the genes necessary for scytonemin production (Soule et al. 2007, 2009; Sorrels et al. 2009). A gene cluster (NpR1276-NpR1259) with 18 unidirectionally transcribed ORFs was discovered in *Nostoc punctiforme* ATCC 29133 (Soule et al. 2007). The production of scytonemin is mostly facilitated by tyrosine and Trp derivatives. Balskus and Walsh (2009) proposed a potential pathway for scytonemin biosynthesis. They observed that the acycloin reaction was a crucial step in putting the scytonemin carbon structure together. Some additional UV-absorbing biomolecules, such as pteridines, prenostodione (λ_{max}: 318 nm), and biopteringlucoside (λ_{max}: 362 nm) besides scytonemin and

Table 3.3 MAAs present in cyanobacteria

S.no	Cyanobacteria	MAAs	References
1	*Euhalothece* sp.	Mycosporine-2-glycine	Kedar et al. (2002)
2	*Euhalothece* sp. strain LK-1	Euhalothece-362	Volkmann and Gorbushina (2006)
3	*Anabaena variabilis* PCC 7937	Shinorine	Singh et al. (2008a)
4	*Anabaena doliolum*	Mycosporine-glycine Porphyra-334	Singh et al. (2008b)
5	*Anabaena flos-aquae*	Shinorine	Singh (2009)
6	*Nostoc commune*	Porphyra-334 Palythine-threonine	Nazifi et al. (2013)
7	*Anabaena* sp.	MAAs Shinorine	Singh et al. (2014)
8	*Rivularia* sp. HKAR-4	Mycosporin-glycine	Rastogi et al. (2014)
9	*Gloeocapsa* sp. CU-2556	MAA M-307	Rastogi and Incharoensakdi (2014a)
10	*Lyngbya* sp. CU2555	Palythine Asterina	Rastogi and Incharoensakdi (2014b)
11	*Arthrospira* sp. CU2556	Mycosporine-glycine	Rastogi and Incharoensakdi (2014c)
12	*Microcystis aeruginosa* PCC 7806	Shinorine	Hu et al. (2015)
13	*Aphanothece halophytica*	Mycosporine-2-glycine	Waditee-Sirisattha et al. (2015)
14	*Nostoc* sp. strain HKAR-2	Porphyra-334 Shinorine	Richa and Sinha (2015)
15	*Nostoc* sp. strain HKAR-6	Shinorine Porphyra-334	Richa and Sinha (2015)
16	*Scytonema* cf. *Crispum*	Shinorine	D'Agostino et al. (2016)
17	*Nostoc sphaericum*	13-O-βgalactosyl-porphyra-334(β-gal-P334)	Ishihara et al. (2017)
18	*Nostoc verrucosum*	Prophyra-334	Inoue-Sakamoto et al. (2018)
19	*Anabaena* sp. HKAR-7	Phorphyra 334	Singh et al. (2021)
20	*Frischerella* sp. strain HKAR-14	Shinorine	Ahmed et al. (2021)
21	*Frischerella* sp. strain HKAR-13	Shinorine	Singh et al. (2022)

MAAs, have also been found in some cyanobacteria (Matsunaga et al. 1993; Ploutno and Carmeli 2001). Microalgae include additional chemicals that help with photoprotection, such as polyamines (PAs). Some frequent PAs found inside cells are the triamine spermidine, diamine putrescine, and tetramine spermine. Spermidine, however, is the main component of cyanobacteria (Jantaro et al. 2003; Incharoensakdi et al. 2010).

Microspectrophotometric analyses of the pigmented sheath's transmittance and the UVR excitation of phycocyanin fluorescence reveal that scytonemin is efficient

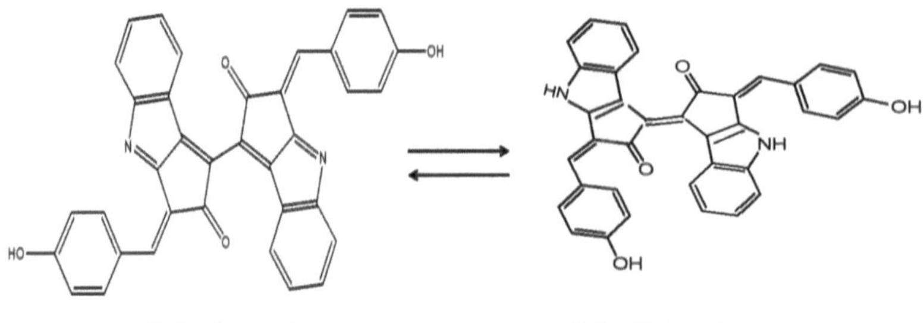

Fig. 3.5 Structures of oxidized and reduced forms of scytonemin

at protecting cells from UVR and may decrease the level of UV-A radiation that enters cells by approximately 90% (Garcia-Pichel et al. 1992). Scytonemin is adequate for offering real defence against UV-C deterioration (Dillon and Castenholz 1999). It has been determined that scytonemin makes up more than 5% of cyanobacteria's entire cellular weight. Scytonemin is resistant to a variety of stressors, including temperature and high UVR, and can survive for long periods of time in terrestrial cyanobacterial crusts or dried mats, (Brenowitz and Castenholz 1997) and perform its screening function without requiring any additional metabolic investment, even when physiologically inactive (during desiccation) (Sinha and Häder 2008). Scytonemin's photoprotective function has been demonstrated in a variety of cyanobacteria from varied hostile environments (Garcia-Pichel and Castenholz 1991; Garcia-Pichel et al. 1992; Sinha et al. 1999; Pathak et al. 2017a).

3.5 Conclusion

All organisms exposed to sunlight suffer from the detrimental effects of UVR that reaches the Earth's surface, which hinders their growth and survival. The synthesis UVR protective metabolites such as MAAs and scytonemin in cyanobacteria and algae act as defensive mechanisms that have developed to support their successful development and survival in high solar UVR. These compounds are not only effective photoprotectants, but also versatile biological substances with a wide range of uses. These compounds are extremely useful in the pharmaceutical industry since they offer great economic potential in the fields of biomedical, cosmetic, and agricultural research as well as in the production of specialized and new medications. Use of sunscreens has grown tremendously since more and more people are worried about the exposure of sun that may cause photo-ageing and skin cancer, Biosunscreens from cyanobacteria can be used as a substitute of chemical-based sunscreens. UV-screening compounds such as scytonemin and MAAs synthesized by cyanobacteria can protect them in harsh environments. Cyanobacteria have various benefits over existing macroalgae (another source of biosunscreen chemicals), including the potential to be produced sustainably and the elimination of the requirement to gather from the wild. Cyanobacteria are suitable micro-factories for the synthesis of UV-screening chemicals because they can be grown in a regulated in vitro environment with particular growth and induction conditions. Biotechnological applications of these natural chemicals are presently restricted, and more study is needed to improve the analysis, synthesis, and use of these unique compounds for the benefit of living beings in a variety of settings. As a result, additional research and study are needed to fully realize the potential of these versatile and new cyanobacterial UV-screening compounds. Bioprospection of these photoprotective compounds and other secondary metabolites from cyanobacteria and microalgae is becoming an important field of research, and these green photoprotective compounds may be employed with fewer side effects and at a lower cost.

Acknowledgements Prashant R. Singh (09/013(0795)/2018-EMR-I) and Amit Gupta (09/013(0912)/2019-EMR-I) are thankful to Council of Scientific and Industrial Research, New Delhi, India, for the financial support in the form of senior research fellowship. Ashish P. Singh is thankful to University Grant Commission (UGC), New Delhi, India (NTA Ref. No. 191620014505) for providing the senior research fellowship. Incentive grant received from IoE (Scheme No. 6031), Banaras Hindu University, Varanasi, India, to Rajeshwar P. Sinha is highly acknowledged.

References

Ahmed H, Pathak J, Sonkar PK, Ganesan V, Häder DP, Sinha RP (2021) Responses of a hot spring cyanobacterium under ultraviolet and photosynthetically active radiation: photosynthetic performance, antioxidative enzymes, mycosporine-like amino acid profiling and its antioxidative potentials. 3 Biotech 11(1):1–23

Arai T, Nishijima M, Adachi K, Sano H (1992) Isolation and structure of a UV absorbing substance from the marine bacterium *Micrococcus* sp. AK-334. Marine Biotechnology Institute, Tokyo, pp 88–94

Balskus EP, Walsh CT (2009) An enzymatic cyclopentyl [b] indole formation involved in scytonemin biosynthesis. J Am Chem Soc 131(41):14648–14649

Belay A, Kato T, Ota Y (1996) *Spirulina* (*Arthrospira*): potential application as an animal feed supplement. J Appl Phycol 8(4):303–311

Bhattacharjee S (2005) Reactive oxygen species and oxidative burst: roles in stress, senescence and signal transducation in plants. Curr Sci 89(7):1113–1121

Bird G, Fitzmaurice N, Dunlap WC, Chalker BE, Bandaranayake WM (1991) Google patents. In: Australia T (ed) Sunscreen compositions and compounds for use therein. Operations Pty Ltd and Australian Institute of Marine Science, Australia

Böhm GA, Pfleiderer W, Böger P, Scherer S (1995) Structure of a novel oligosaccharide-mycosporine-like amino acid ultraviolet A/B sunscreen pigment from the terrestrial cyanobacterium *Nostoc commune*. J Biol Chem 270(15):8536–8539

Brenowitz S, Castenholz RW (1997) Long-term effects of UV and visible irradiance on natural populations of a scytonemin-containing cyanobacterium (*Calothrix* sp.). FEMS Microbiol Ecol 24(4):343–352

Bultel-Poncé V, Félix-Théodose F, Sarthou C, Ponge JF, Bodo B (2004) New pigments from the terrestrial cyanobacterium *Scytonema* sp. collected on the MitarakaInselberg, French Guyana. J Nat Prod 67(4):678–681

Cogdell RJ, Gardiner AT (2015) Activated OCP unlocks nonphotochemical quenching in cyanobacteria. Proc Natl Acad Sci U S A 112(41):12547–12548

Conde FR, Churio MS, Previtali CM (2000) The photoprotector mechanism of mycosporine-like amino acids. Excited-state properties and photostability of porphyra-334 in aqueous solution. J Photochem Photobiol B Biol 56(2–3):139–144
Conde FR, Churio MS, Previtali CM (2007) Experimental study of the excited-state properties and photostability of the mycosporine-like amino acid palythine in aqueous solution. Photochem Photobiol Sci 6(6):669–674
Crutzen PJ (1992) Ultraviolet on the increase. Nature 356(6365):104–105
D'Agostino PM, Javalkote VS, Mazmouz R, Pickford R, Puranik PR, Neilan BA (2016) Comparative profiling and discovery of novel glycosylated mycosporine-like amino acids in two strains of the cyanobacterium *Scytonema cf. crispum*. Appl Environ Microbiol 82(19):5951–5959
Daniel S, Cornelia S, Fred Z (2004) UV-A sunscreen from red algae for protection against premature skin aging. In: Cosmet toiletries manufacture worldwide, pp 139–143
De la Coba F (2007a) Use of mycosporine-type amino acid Porphyra-334 as an antioxidant. Intl Patent WO2007/026035 A2
De la Coba F (2007b) Use of mycosporine-type amino acid shinorine as an antioxidant. Intl Patent WO2007/026038A2
Dillon JG, Castenholz RW (1999) Scytonemin, a cyanobacterial sheath pigment, protects against UVC radiation: implications for early photosynthetic life. J Phycol 35(4):673–681
Edge R, McGarvey DJ, Truscott TG (1997) The carotenoids as anti-oxidants—a review. J Photochem Photobiol B Biol 41(3):189–200
Ehling-Schulz M, Schulz S, Wait R, Görg A, Scherer S (2002) The UV-B stimulon of the terrestrial cyanobacterium *Nostoc commune* comprises early shock proteins and late acclimation proteins. Mol Microbiol 46(3):827–843
Elstner EF (1987) Metabolism of activated oxygen species. In: Biochemistry of metabolism. Academic Press, pp 253–315
Gao K, Li P, Watanabe T, Walter Helbling E (2008) Combined effects of ultraviolet radiation and temperature on morphology, photosynthesis, and DNA of *Arthrospira* (*Spirulina*) *platensis* (Cyanophyta). J Phycol 44(3):777–786
Garcia-Pichel F, Castenholz RW (1991) Characterization and biological implications of scytonemin, a cyanobacterial sheath pigment. J Phycol 27(3):395–409
Garcia-Pichel F, Castenholz RW (1993) Occurrence of UV-absorbing, mycosporine-like compounds among cyanobacterial isolates and an estimate of their screening capacity. Appl Environ Microbiol 59(1):163–169
Garcia-Pichel F, Sherry ND, Castenholz RW (1992) Evidence for an ultraviolet sunscreen role of the extracellular pigment scytonemin in the terrestrial cyanobacterium *Chiorogloeopsis* sp. Photochem Photobiol 56(1):17–23
Goodwin TW (1980) Nature and distribution of carotenoids. Food Chem 5(1):3–13
Gotz T, Windhovel U, Boger P, Sandmann G (1999) Protection of photosynthesis against ultraviolet-B radiation by carotenoids in transformants of the cyanobacterium *Synechococcus* PCC7942. Plant Physiol 120(2):599–604
Grant CS, Louda JW (2013) Scytonemin-imine, a mahogany-colored UV/Vis sunscreen of cyanobacteria exposed to intense solar radiation. Org Geochem 65:29–36
Guedes AC, Amaro HM, Malcata FX (2011) Microalgae as sources of carotenoids. Mar Drugs 9(4):625–644
Häder DP, Kumar HD, Smith RC, Worrest RC (2007) Effects of solar UV radiation on aquatic ecosystems and interactions with climate change. Photochem Photobiol Sci 6(3):267–285
Häder DP, Williamson CE, Wängberg SÅ, Rautio M, Rose KC, Gao K, Helbling EW, Sinha RP, Worrest R (2015) Effects of UV radiation on aquatic ecosystems and interactions with other environmental factors. Photochem Photobiol Sci 14(1):108–126
Halliwell B (2006) Reactive species and antioxidants. Redox biology is a fundamental theme of aerobic life. Plant Physiol 141(2):312–322
Hatz S, Lambert JD, Ogilby PR (2007) Measuring the lifetime of singlet oxygen in a single cell: addressing the issue of cell viability. Photochem Photobiol Sci 6(10):1106–1116

Helbling EW, Menchi CF, Villafañe VE (2002) Bioaccumulation and role of UV-absorbing compounds in two marine crustacean species from Patagonia, Argentina. Photochem Photobiol Sci 1(10):820–825

Hirschberg J, Chamovitz D (1994) Carotenoids in cyanobacteria. In: The molecular biology of cyanobacteria. Springer, Dordrecht, pp 559–579

Hu C, Völler G, Süßmuth R, Dittmann E, Kehr JC (2015) Functional assessment of mycosporine-like amino acids in *Microcystis aeruginosa* strain PCC 7806. Environ Microbiol 17(5):1548–1559

Incharoensakdi A, Jantaro S, Raksajit W, Mäenpää P (2010) Polyamines in cyanobacteria: biosynthesis, transport and abiotic stress response. In: Current research, technology and education topics in applied microbiology and microbial biotechnology, vol 1. Formatex Research Center, pp 23–32

Inoue-Sakamoto K, Nazifi E, Tsuji C, Asano T, Nishiuchi T, Matsugo S, Ishihara K, Kanesaki Y, Yoshikawa H, Sakamoto T (2018) Characterization of mycosporine-like amino acids in the cyanobacterium *Nostoc verrucosum*. J Gen Appl Microbiol 64(5):203–211

Ishihara K, Watanabe R, Uchida H, Suzuki T, Yamashita M, Takenaka H, Nazifi E, Matsugo S, Yamaba M, Sakamoto T (2017) Novel glycosylated mycosporine-like amino acid, 13-O-(β-galactosyl)-porphyra-334, from the edible cyanobacterium *Nostoc sphaericum*-protective activity on human keratinocytes from UV light. J Photochem Photobiol B Biol 172:102–108

Jantaro S, Mäenpää P, Mulo P, Incharoensakdi A (2003) Content and biosynthesis of polyamines in salt and osmotically stressed cells of *Synechocystis* sp. PCC 6803. FEMS Microbiol Lett 228(1):129–135

Karsten U, Sawall T, Wiencke C (1998) A survey of the distribution of UV-absorbing substances in tropical macroalgae. Phycol Res 46(4):271–279

Kedar L, Kashman Y, Oren A (2002) Mycosporine-2-glycine is the major mycosporine-like amino acid in a unicellular cyanobacterium (*Euhalothece* sp.) isolated from a gypsum crust in a hypersaline saltern pond. FEMS Microbiol Lett 208(2):233–237

Kelman D, Ben-Amotz A, Berman-Frank I (2009) Carotenoids provide the major antioxidant defence in the globally significant N_2-fixing marine cyanobacterium *Trichodesmium*. Environ Microbiol 11(7):1897–1908

Kerfeld CA (2004) Water-soluble carotenoid proteins of cyanobacteria. Arch Biochem Biophys 430(1):2–9

Kerfeld CA, Sawaya MR, Brahmandam V, Cascio D, Ho KK, Trevithick-Sutton CC, Krogmann DW, Yeates TO (2003) The crystal structure of a cyanobacterial water-soluble carotenoid binding protein. Structure 11(1):55–65

Krieger-Liszkay A, Fufezan C, Trebst A (2008) Singlet oxygen production in photosystem II and related protection mechanism. Photosynth Res 98(1):551–564

Lopes G, Clarinha D, Vasconcelos V (2020) Carotenoids from cyanobacteria: a biotechnological approach for the topical treatment of psoriasis. Microorganisms 8(2):302

Maeda H, Sakuragi Y, Bryant DA, DellaPenna D (2005) Tocopherols protect *Synechocystis* sp. strain PCC 6803 from lipid peroxidation. Plant Physiol 138(3):1422–1435

Matsunaga T, Burgess JG, Yamada N, Komatsu K, Yoshida S, Wachi Y (1993) An ultraviolet (UV-A) absorbing biopterin glucoside from the marine planktonic cyanobacterium *Oscillatoria* sp. Appl Microbiol Biotechnol 39(2):250–253

Misonou T, Saitoh J, Oshiba S, Tokitomo Y, Maegawa M, Inoue Y, Hori H, Sakurai T (2003) UV-absorbing substance in the red alga *Porphyra yezoensis* (Bangiales, Rhodophyta) block thymine photodimer production. Mar Biotechnol 5(2):194–200

Moliné M, Arbeloa EM, Flores MR, Libkind D, Farías ME, Bertolotti SG, Churio MS, van Broock MR (2011) UVB photoprotective role of mycosporines in yeast: photostability and antioxidant activity of mycosporine-glutaminol-glucoside. Radiat Res 175(1):44–50

Mori T, Muranaka T, Miki W, Yamaguchi K, Konosu S (1987) Pigmentation of cultured sweet smelt fed diets supplemented with a blue-green alga *Spirulina maxima*. Nippon Suisan Gakkaishi 53(3):433–438

Morone J, Alfeus A, Vasconcelos V, Martins R (2019) Revealing the potential of cyanobacteria in cosmetics and cosmeceuticals—a new bioactive approach. Algal Res 41:101541

Nazifi E, Wada N, Yamaba M, Asano T, Nishiuchi T, Matsugo S, Sakamoto T (2013) Glycosylated porphyra-334 and palythine-threonine from the terrestrial cyanobacterium *Nostoc commune*. Mar Drugs 11(9):3124–3154

Neely WC, Martin JM, Barker SA (1988) Products and relative reaction rates of the oxidation of tocopherols with singlet molecular oxygen. Photochem Photobiol 48(4):423–428

Niyogi KK (1999) Photoprotection revisited: genetic and molecular approaches. Annu Rev Plant Biol 50(1):333–359

Oyamada C, Kaneniwa M, Ebitani K, Murata M, Ishihara K (2008) Mycosporine-like amino acids extracted from scallop (*Patinopecten yessoensis*) ovaries: UV protection and growth stimulation activities on human cells. Mar Biotechnol 10(2):141–150

Pagels F, Salvaterra D, Amaro HM, Lopes G, Sousa-Pinto I, Vasconcelos V, Guedes AC (2020) Bioactive potential of *Cyanobium* sp. pigment-rich extracts. J Appl Phycol 32(5):3031–3040

Pagels F, Vasconcelos V, Guedes AC (2021) Carotenoids from cyanobacteria: biotechnological potential and optimization strategies. Biomol Ther 11(5):735

Paliwal C, Ghosh T, Bhayani K, Maurya R, Mishra S (2015) Antioxidant, anti-nephrolithe activities and *in vitro* digestibility studies of three different cyanobacterial pigment extracts. Mar Drugs 13(8):5384–5401

Pathak J, Richa R, Sonker AS, Kannaujiya VK, Sinha RP (2015) Isolation and partial purification of scytonemin and mycosporine-like amino acids from biological crusts. J Chem Pharm Res 7:362–371

Pathak J, Sonker AS, Kannaujiya VK, Singh V, Ahmed H, Sinha RP (2017a) Screening and partial purification of photoprotective pigment scytonemin from cyanobacterial crusts dwelling on the historical monuments in and around Varanasi, India. Microbiol Res 8(1):4–12

Pathak J, Ahmed H, Sinha P, Rajneesh (2017b) Metabolomic profiling of cyanobacterial UV-protective compounds. Curr Metabolomics 5(2):138–163

Pathak J, Maurya PK, Singh SP, Häder DP, Sinha RP (2018) Cyanobacterial farming for environment friendly sustainable agriculture practices: innovations and perspectives. Front Environ Sci 6:7

Pathak J, Ahmed H, Singh PR, Singh SP, Häder DP, Sinha RP (2019) Mechanisms of photoprotection in cyanobacteria. In: Cyanobacteria. Academic Press, pp 145–171

Pathak J, Pandey A, Maurya PK, Rajneesh R, Sinha RP, Singh SP (2020) Cyanobacterial secondary metabolite scytonemin: a potential photoprotective and pharmaceutical compound. Proc Natl Acad Sci India Sect B Biol Sci 90(3):467–481

Patias LD, Fernandes AS, Petry FC, Mercadante AZ, Jacob-Lopes E, Zepka LQ (2017) Carotenoid profile of three microalgae/cyanobacteria species with peroxyl radical scavenger capacity. Food Res Int 100:260–266

Ploutno A, Carmeli S (2001) Prenostodione, a novel UV-absorbing metabolite from a natural bloom of the cyanobacterium *Nostoc* species. J Nat Prod 64(4):544–545

Rajneesh, Singh SP, Pathak J, Sinha RP (2017) Cyanobacterial factories for the production of green energy and value-added products: an integrated approach for economic viability. Renew Sust Energ Rev 69:578–595

Rajneesh, Pathak J, Häder DP, Sinha RP (2019) Impacts of ultraviolet radiation on certain physiological and biochemical processes in cyanobacteria inhabiting diverse habitats. Environ Exp Bot 161:375–387

Rastogi RP, Incharoensakdi A (2014a) UV radiation-induced biosynthesis, stability and antioxidant activity of mycosporine-like amino acids (MAAs) in a unicellular cyanobacterium *Gloeocapsa* sp. CU2556. J Photochem Photobiol B Biol 130:287–292

Rastogi RP, Incharoensakdi A (2014b) Characterization of UV-screening compounds, mycosporine-like amino acids, and scytonemin in the cyanobacterium *Lyngbya* sp. CU2555. FEMS Microbiol Ecol 87(1):244–256

Rastogi RP, Incharoensakdi A (2014c) Analysis of UV-absorbing photoprotectant mycosporine-like amino acid (MAA) in the cyanobacterium *Arthrospira* sp. CU2556. Photochem Photobiol Sci 13(7):1016–1024

Rastogi RP, Sinha RP (2009) Biotechnological and industrial significance of cyanobacterial secondary metabolites. Biotechnol Adv 27(4):521–539

Rastogi RP, Kumar A, Tyagi MB, Sinha RP (2010) Molecular mechanisms of ultraviolet radiation-induced DNA damage and repair. J Nucleic Acids 2010:592980

Rastogi RP, Kumari S, Han T, Sinha RP (2012) Molecular characterization of hot spring cyanobacteria and evaluation of their photoprotective compounds. Can J Microbiol 58(6):719–727

Rastogi RP, Sinha RP, Incharoensakdi A (2013) Partial characterization, UV-induction and photoprotective function of sunscreen pigment, scytonemin from *Rivularia* sp. HKAR-4. Chemosphere 93(9):1874–1878

Rastogi RP, Singh SP, Incharoensakdi A, Häder DP, Sinha RP (2014) Ultraviolet radiation-induced generation of reactive oxygen species, DNA damage and induction of UV-absorbing compounds in the cyanobacterium *Rivularia* sp. HKAR-4. S Afr J Bot 90:163–169

Rastogi RP, Madamwar D, Incharoensakdi A (2015) Sun-screening bioactive compounds mycosporine-like amino acids in naturally occurring cyanobacterial biofilms: role in photoprotection. J Appl Microbiol 119(3):753–762

Richa, Sinha RP (2015) Biochemical characterization of sunscreening mycosporine-like amino acids from two *Nostoc* species inhabiting diverse habitats. Protoplasma 252(1):199–208

Schäfer L, Vioque A, Sandmann G (2005) Functional in situ evaluation of photosynthesis-protecting carotenoids in mutants of the cyanobacterium *Synechocystis* PCC6803. J Photochem Photobiol B Biol 78(3):195–201

Shick JM, Dunlap WC (2002) Mycosporine-like amino acids and related gadusols: biosynthesis, accumulation, and UV-protective functions in aquatic organisms. Annu Rev Physiol 64(1):223–262

Singh SP (2009) Study on mycosporine-like amino acids (MAAs) in cyanobacteria: a biochemical, bioinformatics and molecular biology approach (Doctoral dissertation, Friedrich-Alexander-Universität Erlangen-Nürnberg (FAU))

Singh SP, Klisch M, Sinha RP, Häder DP (2008a) Effects of abiotic stressors on synthesis of the mycosporine-like amino acid shinorine in the cyanobacterium *Anabaena variabilis* PCC 7937. Photochem Photobiol 84(6):1500–1505

Singh SP, Sinha RP, Klisch M, Häder DP (2008b) Mycosporine-like amino acids (MAAs) profile of a rice-field cyanobacterium *Anabaena doliolum* as influenced by PAR and UVR. Planta 229(1):225–233

Singh SP, Häder DP, Sinha RP (2010) Cyanobacteria and ultraviolet radiation (UVR) stress: mitigation strategies. Ageing Res Rev 9(2):79–90

Singh SP, Ha SY, Sinha RP, Häder DP (2014) Photoheterotrophic growth unprecedentedly increases the biosynthesis of mycosporine-like amino acid shinorine in the cyanobacterium *Anabaena* sp., isolated from hot springs of Rajgir (India). Acta Physiol Plant 36(2):389–397

Singh DK, Pathak J, Pandey A, Singh V, Ahmed H, Kumar D, Sinha RP (2021) Response of a rice-field cyanobacterium *Anabaena* sp. HKAR-7 upon exposure to ultraviolet-B radiation and ammonium chloride. Environ Sustain 4(1):95–105

Singh V, Pathak J, Pandey A, Ahmed H, Kumar D, Sinha RP (2022) UV-induced physiological changes and biochemical characterization of mycosporine-like amino acid in a rice-field cyanobacterium *Fischerella* sp. strain HKAR-13. S Afr J Bot 147:81–97

Sinha RP, Häder DP (2008) UV-protectants in cyanobacteria. Plant Sci 174(3):278–289

Sinha RP, Klisch M, Vaishampayan A, Häder DP (1999) Biochemical and spectroscopic characterization of the cyanobacterium *Lyngbya* sp. inhabiting Mango (*Mangifera indica*) trees: presence of an ultraviolet-absorbing pigment, scytonemin. Acta Protozool 38:291–229

Sinha RP, Ambasht NK, Sinha JP, Klisch M, Häder DP (2003) UV-B-induced synthesis of mycosporine-like amino acids in three strains of *Nodularia* (cyanobacteria). J Photochem Photobiol B Biol 71(1–3):51–58

Sinha RP, Singh SP, Häder DP (2007) Database on mycosporines and mycosporine-like amino acids (MAAs) in fungi, cyanobacteria, macroalgae, phytoplankton and animals. J Photochem Photobiol B Biol 89(1):29–35

Smirnoff N, Wheeler GL (2000) Ascorbic acid in plants: biosynthesis and function. Crit Rev Plant Sci 19(4):267–290

Sorrels CM, Proteau PJ, Gerwick WH (2009) Organization, evolution, and expression analysis of the biosynthetic gene cluster for scytonemin, a cyanobacterial UV-absorbing pigment. Appl Environ Microbiol 75(14):4861–4869

Soule T, Stout V, Swingley WD, Meeks JC, Garcia-Pichel F (2007) Molecular genetics and genomic analysis of scytonemin biosynthesis in *Nostoc punctiforme* ATCC 29133. J Bacteriol 189(12):4465–4472

Soule T, Palmer K, Gao Q, Potrafka RM, Stout V, Garcia-Pichel F (2009) A comparative genomics approach to understanding the biosynthesis of the sunscreen scytonemin in cyanobacteria. BMC Genomics 10(1):1–10

Stolarski R, Bojkov R, Bishop L, Zerefos C, Staehelin J, Zawodny J (1992) Measured trends in stratospheric ozone. Science 256(5055):342–349

Vaishampayan A, Sinha RP, Hader DP, Dey T, Gupta AK, Bhan U, Rao AL (2001) Cyanobacterial biofertilizers in rice agriculture. Bot Rev 67(4):453–516

Volkmann M, Gorbushina AA (2006) A broadly applicable method for extraction and characterization of mycosporines and mycosporine-like amino acids of terrestrial, marine and freshwater origin. FEMS Microbiol Lett 255(2):286–295

Vranová E, Atichartpongkul S, Villarroel R, Van Montagu M, Inzé D, Van Camp W (2002) Comprehensive analysis of gene expression in *Nicotiana tabacum* leaves acclimated to oxidative stress. Proc Natl Acad Sci 99(16):10870–10875

Waditee-Sirisattha R, Kageyama H, Fukaya M, Rai V, Takabe T (2015) Nitrate and amino acid availability affects glycine betaine and mycosporine-2-glycine in response to changes of salinity in a halotolerant cyanobacterium *Aphanothece halophytica*. FEMS Microbiol Lett 362(23)

Wagner D, Przybyla D, op den Camp R, Kim C, Landgraf F, Lee KP, Würsch M, Laloi C, Nater M, Hideg E, Apel K (2004) The genetic basis of singlet oxygen-induced stress responses of *Arabidopsis thaliana*. Science 306(5699):1183–1185

Williamson CE, Zepp RG, Lucas RM, Madronich S, Austin AT, Ballare CL, Norval M, Sulzberger B, Bais A, McKenzie RL, Robinson SA, Häder D-P, Paul ND, Bornman JF (2014) Solar ultraviolet radiation in a changing climate. Nat Clim Chang 4(6):434–441

Wilson A, Ajlani G, Verbavatz JM, Vass I, Kerfeld CA, Kirilovsky D (2006) A soluble carotenoid protein involved in phycobilisome-related energy dissipation in cyanobacteria. Plant Cell 18(4):992–1007

Wilson A, Boulay C, Wilde A, Kerfeld CA, Kirilovsky D (2007) Light-induced energy dissipation in iron-starved cyanobacteria: roles of OCP and IsiA proteins. Plant Cell 19(2):656–672

Wilson A, Punginelli C, Gall A, Bonetti C, Alexandre M, Routaboul JM, Kerfeld CA, van Grondelle R, Robert B, Kennis JT, Kirilovsky D (2008) A photoactive carotenoid protein acting as light intensity sensor. Proc Natl Acad Sci U S A 105(33):12075–12080

Wu YP, Krogmann DW (1997) The orange carotenoid protein of *Synechocystis* PCC 6803. Biochim Biophys Acta 1322(1):1–7

Young AJ, Frank HA (1996) Energy transfer reactions involving carotenoids: quenching of chlorophyll fluorescence. J Photochem Photobiol B Biol 36(1):3–15

Chapter 4
Photoprotective Compounds: Diversity, Biosynthetic Pathway and Genetic Regulation

Saumi Pandey and Vinod K. Kannaujiya

Abstract The rapid increment in atmospheric pollution causes loss of stratospheric ozone which enhances the exposure of harmful ultraviolet radiation (UV) on the earth's surface. The detrimental impact of UV-A and UV-B causes photodamage, alteration in normal physiological processes of aquatic and terrestrial organisms. However, certain inbuilt existence of photoprotective compounds makes organisms to counteract with harmful UV radiations. Most of the groups of organisms such as cyanobacteria, algae, lichens, fungi, bryophytes, higher plants, and certain animals have unique photoprotective strategies for survivability under harsh radiation. Various types of secondary metabolites such as mycosporine-like amino acids (MAAs), mycosporine, scytonemin, lycopodine, carotenoids, and melanin are inherently synthesized in most of the organisms for protection from deleterious ultraviolet radiation. In the last few decades, most of the genetic and metabolomics pathways for synthesis of photoprotective compounds have been well established. However, industrial-scale production is still undeveloped due to limited knowledge of gene expression and cost-effective rapid production in various group of organisms. This chapter briefly describes the presence of different photoprotective compounds in various taxa of organisms. This chapter also deals with the biosynthetic pathways and genetic regulation of certain compounds for production and commercial applications in various group of life sciences and cosmetic industries.

Keywords Cyanobacteria · Mycosporine-like amino acids · Pentose phosphate pathway · Photoprotection · Scytonemin · Shikimate pathway

S. Pandey · V. K. Kannaujiya (✉)
Department of Botany, MMV, Banaras Hindu University, Varanasi, India

V. K. Kannaujiya et al. (eds.), *Photoprotective Green Pharmacology: Challenges, Sources and Future Applications*,
https://doi.org/10.1007/978-981-99-0749-6_4

4.1 Introduction

The existence of different life forms on the earth's surface becomes possible due to obligate presence of solar energy. The solar spectrum mainly consists of photosynthetically active radiation (PAR) and ultraviolet radiation (UV). The UV (non-ionizing radiation) constitutes 8–9% of solar spectrum (Hollósy 2002). The visible range of solar spectrum drives the photosynthetic process whereas UV-A and UV-B cause several detrimental impacts on the survivability of organisms (Gao et al. 2007; Richa and Sinha 2013). The stratospheric ozone layer behaves as an umbrella to safeguard from biologically active harmful UV-B radiations coming from sun. Ozone harbors very high absorption coefficient but due to less amount (4 μm thick layer) and thinning of ozone layer, UV-B radiation passes to the biosphere (Björn 2007). Scientific literature revealed that every 1% reduction in ozone thickness leads to the increment of UV-B radiation about 1.3–1.8% on the earth's surface (Hollósy 2002). The exposure of ultraviolet radiation significantly induces biochemical, physiological, and morphological changes in an organism which altered natural community and structure and function of ecosystem (Paul and Gwynn-Jones 2003; Wong et al. 2019). Moreover, UV-B radiation has directly affected nucleic acids (DNA and RNA) and induced formation of harmful DNA lesions such as cyclobutane pyrimidine dimer and 6–4 photoproducts (Singh et al. 2010a; Richa and Sinha 2013). In comparison to UV-B, UV-A has indirect effect which includes photosensitization of biological compounds and production of reactive oxygen species (Hargreaves et al. 2007). In addition, UV radiations (UV-A, UV-B) also affect cell differentiation, morphology, pigmentation, motility, phycobiliprotein composition, protein profile, nitrogen fixation and membrane lipid oxidation pathways (Lesser 2006, 2008; Gao et al. 2007). In humans, it causes skin inflammation, skin cancer, cutaneous photoaging and development of pigmentation disorder (Passeron et al. 2021). However, most of the microorganisms including plants, animals and humans have the ability to cope up with fluctuations occurring in the ambient environment (Hemm et al. 2001; Edreva 2005; Wong et al. 2019). They reside and adapt through the production of photoprotective compounds in adverse environmental conditions (Rastogi et al. 2010; Le Lann et al. 2016; Lalegerie et al. 2019). The number of photoprotective compounds is already described in certain group of organisms such as scytonemin, mycosporine-like amino acids, melanins, flavonoids and phenyl propanoids (Pallela et al. 2010; Kannaujiya et al. 2014) (Fig. 4.1). However, there are plenty of photoprotective compounds still to be discovered. The study of other stress factors besides UV radiation has not been well studied in correlation with production of photoprotective compounds (Sen and Mallick 2022). This chapter briefly describes the various kinds of photoprotective compounds found in various group of organisms. We also describe the various biosynthetic mechanisms and genetic regulation for synthesis of photoprotective compounds. In addition, this chapter also deals the current advancement in production technology of photoprotective compounds and commercial applications in biomedical and cosmetic industries.

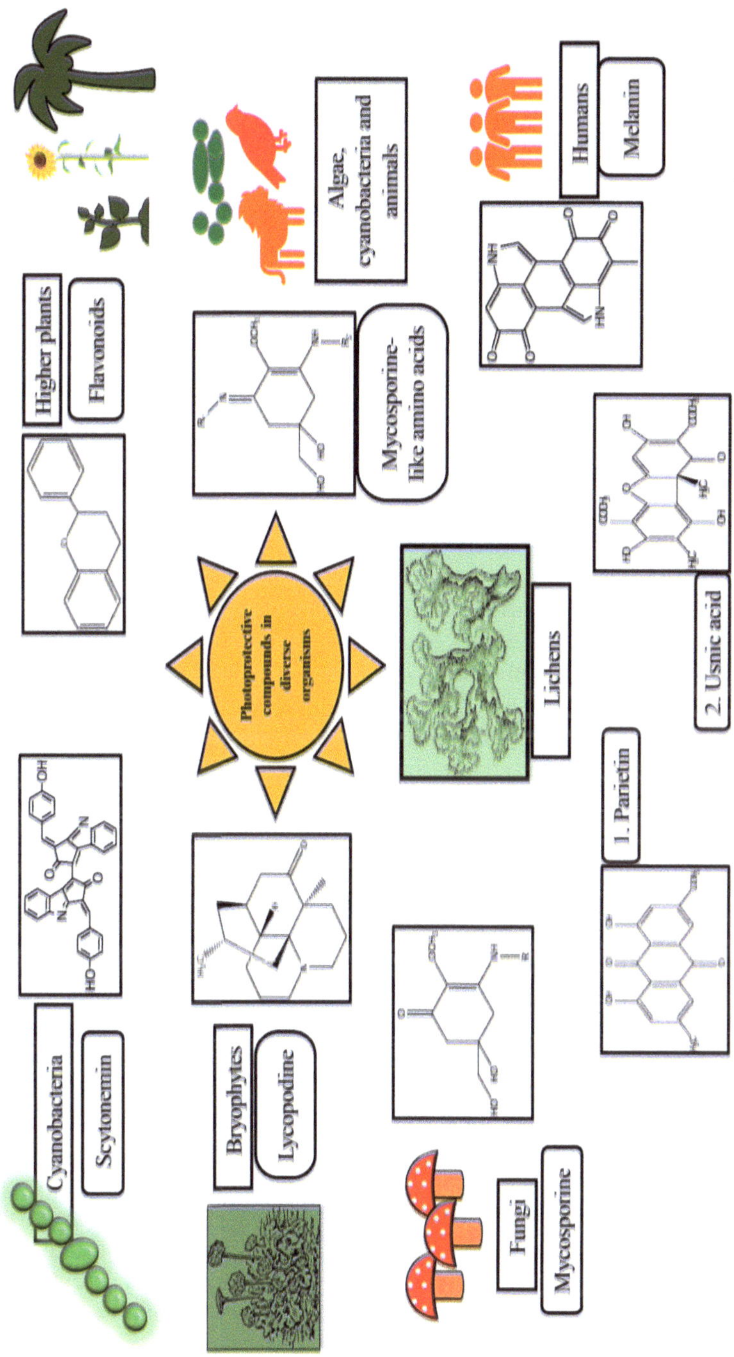

Fig. 4.1 Illustration of major photoprotective compounds found in different taxa with their molecular structures

4.2 Photoprotective Compounds and Their Role Against UV Radiation

4.2.1 Mycosporine-Like Amino Acids

Mycosporine-like amino acids (MAAs) belong to family of secondary metabolites, reported in many freshwater and marine organisms. They are water-soluble in nature with molecular weight >400 Da (Bhatia et al. 2011). They have absorption maxima from 310 to 360 nm, marked by the presence of cyclohexenimine or cyclohexenone chromophores (Häder et al. 2007; Bhatia et al. 2011; La Barre et al. 2014). They have been known for absorbing lethal UV radiations, act as cellular antioxidant and osmoprotectant in nature (Oren and Gunde-Cimerman 2007; La Barre et al. 2014). Corals predominantly exist in euphotic zone with endosymbiotic algal partners—*Symbiodinium* sp., zooxanthellae (Torregiani and Lesser 2007). Corals have sensitive MAAs which are inversely proportional to depth and also vary with season (Zeevi Ben-Yosef et al. 2006; Torregiani and Lesser 2007). Literature survey reveals that *Anabaena* sp. has produced a single kind of MAAs—shinorine (λ_{max} = 334 nm, Retention time (RT) = 2.3 min) (Sinha et al. 2007); however, other species of *Anabaena*, *Anabaena doliolum* possess different kinds of MAAs such as shinorine, porphyra-334 (λ_{max} = 334 nm, RT = 3.5 min) and mycosporine-glycine (λ_{max} = 310 nm, RT = 4.1 min) (Singh et al. 2008a). Sunscreen factor was found to be 0.3 for MAAs in single cells (Garcia-Pichel et al. 1993). In the cyanobacterium *A. variabilis* PCC 7937, the biosynthesis of shinorine was recently discovered and also found that biosynthesis of shinorine not only depends on UV radiation but also on ammonium concentration, temperature and salt (Singh et al. 2008b, 2010b).

4.2.2 Usnic Acid and Parietin

Lichens thrive in sun-exposed habitats especially Arctic, alpine, arid and warm deserts (Gauslaa and Solhaug 2004). The bright and non-melanic lichens residing in open habitats produce photoprotective compounds, namely, usnic acid, parietin or atranorin (Solhaug and Gauslaa 2004). Lichens screen out harmful UV radiations by the process of absorption through melanin and parietin and reflectance mechanism through atranorin (Solhaug and Gauslaa 2012). However, usnic acid is commonly found in fungal Lecanorales order whereas parietin in Teloschistales order (Solhaug et al. 2003; Gauslaa and Solhaug 2004). Usnic acid is a dibenzofuran derived biologically active lipophilic based, yellow coloured compound which exhibits pronounced absorption in UV radiation (Huneck and Yoshimura 1996; Luzina and Salakhutdinov 2018). It is also well demarcated that usnic acid synthesis depends upon environmental habitats. The level of solar radiation and usnic acid content can be positively correlated (Bjerke et al. 2002). Sun-exposed lichens has more amount of usnic acid than shade habitats (Bjerke and Dahl 2002). Interestingly, an enrichment of UV radiation in solar radiation is the primary cause of copious amount of usnic acid production in lichens (Bjerke and Dahl 2002). In addition, UV-B exposure of thalli of *Xanthoparmelia microspora* led to increase in usnic acid content from 1.9% to 4.16% (w/w) (Fernández et al. 2006). Recently, usnic acid has been compared with sunscreens available in the market along with some other chemical compounds such as Nivea sun spray LS 5 and 4-tert-butyl-4′-methoxy dibenzoylmethane (BM-DBM) (Rancan et al. 2002). The above results indicate that usnic acid can be used as ingredient for development of better UV filter as compared to Nivea sun spray LS-5 and BM-DBM (Rancan et al. 2002).

Parietin is an orange anthraquinone photoprotective compound (McEvoy et al. 2006). In literature, it is also mentioned that parietin functions as blue light screen filter rather than UV-B screen filter (Gauslaa and Ustvedt 2003). The parietin content is governed by habitat of lichens ranging from pale (closed forest) to orange colour (sea cliffs) (Solhaug and Gauslaa 2012). Interestingly, it was found that parietin content becomes double in mid-summer condition as compared in winter season (Gauslaa and McEvoy 2005). A study further confirms that parietin content in *Xanthoria parietina* was significantly higher in combined exposure of UV-A radiation and visible light than visible light alone (Solhaug et al. 2003).

4.2.3 Lycopodine

In *Lycopodium* (club moss), the various kinds of UV-absorbing compounds are present. The known compounds are lycopodine (alkaloid), flavonoids-luteonin and chrysoeriol. A lycopodine is a quinolizidine alkaloid and derivative of lycopodane hydrate. The confinement of flavonoids in reproductive parts like spores led to provide protection from UVR (Rozema et al. 2002). However, detailed study on lycopodine corelation with UV stress is not well explored.

4.2.4 Melanin

The skin colour of humans is governed by an amalgamation of oxyhemoglobin or deoxyhemoglobin, carotenoids and melanin. The types of melanin and its distribution in melansomes plays an important role in skin pigmentation (Brenner and Hearing 2008). However, epidemiological data supported that skin colour and skin cancer are inversely corelated (Gilchrest et al. 1999). White skin has 70% more tendency to develop skin cancer than black skin (Halder and Bang 1988). The melanin provides shielding effect from ultraviolet radiation mainly eumelanin by two ways: firstly by scattering the UVR and secondly by acting as absorbent filter (Kaidbey et al. 1979; Brenner and Hearing 2008). In the skin melanin composition, eumelanin is better UV protectant than pheomelanin. The melanosomes of darker skin colour are very resistant to lysosomal enzymes and persist in epidermal layers in the form of supranuclear caps in melanocytes and keratinocytes that help in protection from UV-induced damage whereas fair skin melanocytes degraded after exposure and existed in the form of melanin dust (Kobayashi et al. 1998). The degradation of melanocytes becomes causative factor for skin cancer (Bustamante et al. 1993).

4.2.5 *Phenylpropanoids*

Plants being sessile in nature respond to abiotic changes by modulating their cellular chemistry or physiology (Escobar-Bravo et al. 2017). Plants have specific UV-B photoreceptor-UVR8 (UV-resistant locus) (Rizzini et al. 2011) which regulates photomorphogenic responses as well as modulates expression of genes associated with DNA repair, inhibition of hypocotyl elongation, antioxidative defence and phenolic compounds production (Escobar-Bravo et al. 2017). In order to cut off harmful effect of UV radiations, plants accumulate phenylpropanoids and flavonoids in leaf epidermis as well as spongy and palisade mesophyll tissue (Agati et al. 2013). Phenylpropanoids belong to large class of secondary metabolites synthesized from aromatic amino acids such as tyrosine and phenylalanine in plants. These are produced through sequential enzymatic biosynthetic pathway (Deng and Lu 2017). The phenylpropanoid has been classified into five categories: (1) Monolignols, (2) Flavonoids, (3) Coumarins, (4) Stilbenes and (5) Phenolic acids. However, most common occurrence in plants are monolignols, flavonoids and phenolic acids (Noel et al. 2005; Liu et al. 2015). Coumarins are restricted to families Asteraceae, Apiaceae, Nyctaginaceae, Oleaceae, Fabaceae, Caprifoliaceae, Guttiferae, Rutaceae and Moraceae (Venugopala et al. 2013; Deng and Lu 2017). Stilbenes are confined to Myrtaceae, Polygonaceae, Poaceae, Vitaceae, and Gnetaceae (Kiselev et al. 2016). Phenylpropanoid biosynthetic enzymes are encoded by gene superfamilies such as NADPH-dependent reductase family, 2-oxoglutarate-dependent dioxygenase (2-ODD), type III polyketide synthase (PKS III), P450 gene family (Turnbull et al. 2004; Tohge et al. 2013; Deng and Lu 2017).

4.2.6 *Flavonoids*

Flavonoids have played an evolutionary role for establishment of plants from aquatic to land habitat (Frohnmeyer and Staiger 2003). Flavonoids are secondary metabolites, low molecular weight compounds and classified into different subclasses such as flavanones, flavones, flavonols, proanthocyanidins, anthocyanins, phlobaphenes and isoflavonoids (Liu et al. 2015). They are present in nucleus, cell vacuoles and chloroplast of mesophyll cells (Agati et al. 2013) and have an important biological function such as providing aroma and colour to flowers to attract pollinators for seed dispersion and germination, development and growth of seedlings.

Ultraviolet radiations have shown detrimental effect on plant growth, DNA content, protein and lipids (Deng and Lu 2017). Plants have well-defined photoproducts to counterfeit harmful effects of UV radiations namely flavonoids including flavones, anthocyanin, flavonol glycosides and phenylpropanoids such as coumarins, stilbenes, sinapate esters. These compounds act as ROS attenuator to inhibit deleterious effects of radiation (Heijde and Ulm 2012; Vidović et al. 2015). Thus, they act as strong UV-filter (Samanta et al. 2011). In addition, these compounds also protect plants from nutrient deficiency (low iron/nitrogen/phosphate), pathogen infection, herbivore attack, drought and low temperature (Simmonds 2003; Treutter 2005).

4.2.7 Scytonemin

Scytonemin is a natural compound present in cyanobacteria and has played an indispensable role in photoprotection from UV radiations since the origin of life (Rastogi and Sinha 2009). It is synthesized in extremophilic cyanobacteria of different groups (Garcia-Pichel and Castenholz 1991). It is small, hydrophobic, lipid-soluble yellow-brown to mahogany or dark red pigment molecules produced in extracellular sheath of approximately 300 species of cyanobacteria (Sinha and Häder 2008; Rastogi and Sinha 2009). The maximum UV absorption spectra of purified scytonemin is found at 386 ± 2 nm (Rastogi et al. 2013). Scytonemin is a dimer of phenolic and indolic subunits with molecular weight of 544 Da. (Rastogi et al. 2010). Scytonemin acts as natural sunscreen found in cyanobacteria (Proteau et al. 1993; Rastogi et al. 2010). Scytonemin also acts as a natural ROS scavenger (Takamatsu et al. 2003; Matsui et al. 2012). Moreover, scytonemin has also reduced thymine dimer formation and ROS generation in *Rivularia* sp. HKAR-4 under UV stress condition (Rastogi et al. 2015).

4.3 Biosynthetic Pathway

4.3.1 Shikimate Pathway

In the total UV radiation, mainly UV-B plays a keen role in the induction of biosynthetic pathways in photosynthetic species (Bhatia et al. 2011). MAAs biosynthesis occurs in most of prokaryotes including bacteria, blue-green algae (cyanobacteria), macroalgae and phytoplankton but not synthesized in animals due to absence of shikimate pathway (Fig. 4.2). In animal system, MAAs are transferred through diet, bacterial or symbiotic associations (Sinha et al. 2007; Bhatia et al. 2011). It was believed that MAAs biosynthesis occurred via first part of shikimate pathway but still strong proofs needed for validation (Sinha et al. 2007; Sinha and Häder 2008). The precursor molecule 3-dehydroquinate synthesizes fungal mycosporines and MAAs synthesized through gadusols (Bandaranayake 1998; Shick and Dunlap 2002).

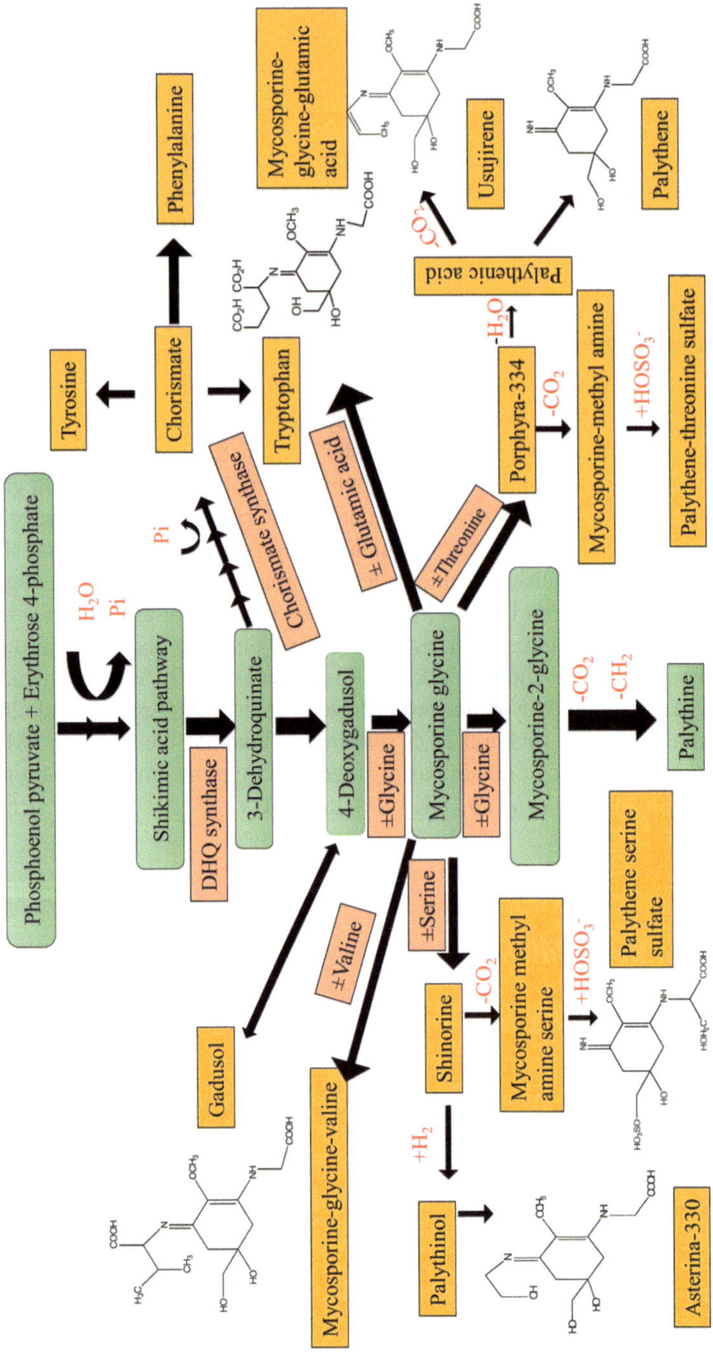

Fig. 4.2 Biosynthesis of different MAAs via shikimate pathway

4.3.2 Pentose Phosphate Pathway

Biosynthesis of MAAs is also carried out through pentose phosphate pathway. The core structure of MAAs is 4-deoxygadusol which is formed through intermediate sedoheptulose-7-phosphate by the enzymes O-methyltransferase (OMT) and 2-epi-5-epivaliolone synthase (EVS) in cyanobacteria (Pope et al. 2015; Jain et al. 2017). In literature survey, it was reported that deletion of EVS gene in *A. variabilis* ATCC 29413 has very little effect on MAAs biosynthesis which indicates pentose phosphate pathway is not sole pathway for biosynthesis (Jain et al. 2017). The genome mining also reveals EVS enzyme to be crucial for the synthesis of MAAs (Spence et al. 2012). A contradictory observation in relation to MAAs biosynthesis was elucidated in *Synechocystis* sp. PCC6803 in the absence of EVS enzyme; it synthesizes three novel mycosporine-like amino acids- M-tau, M-343 and dehydroxylusujirene (Spence et al. 2012; Zhang et al. 2007).

4.3.3 Phenylpropanoid Pathway

The pathway starts from phenylalanine precursor molecule to synthesize products through shikimate pathway. However, certain organisms of fungi, bacteria and monocots are used as tyrosine precursor to start the biosynthetic pathway (Deng and Lu 2017). The pathway starts from deamination of phenylalanine by PAL for conversion into cinnamic acid. The second enzymatic step involves hydroxylation of cinnamic acid to p-coumaric acid via C4H enzyme. In certain bacterial, monocots and fungal species, TAL—tyrosine ammonia lyase or PTAL—bifunctional ammonia lyase bypass the second enzymatic step; hence, tyrosine directly converts into p-coumaric acid (Watts et al. 2004). It is well documented that some enzymes of this pathway are UV light-inducible particularly phenylalanine ammonia lyase (PAL), chalcone reductase (CHR), cinnamic acid 4-hydroxylase (C4H), and flavanone 3-hydroxylase (F3H) (Meyer et al. 2021). The pathway is diagrammatically represented in Fig. 4.3.

4.4 Genetic Regulation

4.4.1 Scytonemin

Biosynthesis of scytonemin at genetic level has been well studied in cyanobacterium *N. punctiforme* ATCC 29133 through random transposon insertion mutagenesis (Soule et al. 2007). In the genome of *N. punctiforme* ATCC 29133, an 18-gene cluster (Npun_R1276 to Npun_R1259) was found linked with the scytonemin biosynthesis (Pathak et al. 2019). Moreover, it was also found that tyrosine and

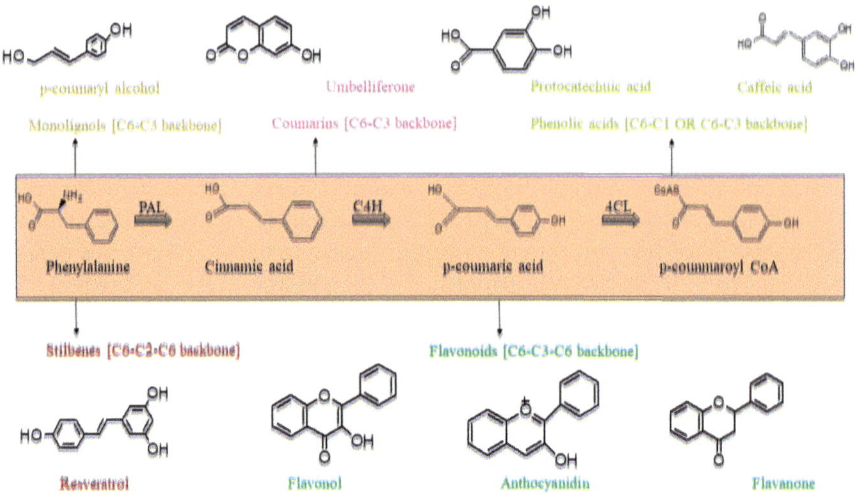

Fig. 4.3 General phenylpropanoid pathway found in plants. *PAL* phenylalanine ammonia lyase, *C4H* cinnamic acid 4-hydroxylase, *4-CL* 4-coumaroyl CoA ligase

tryptophan play as precursor molecules for synthesis of molecular structure of scytonemin (Proteau et al. 1993; Pathak et al. 2019). Certain identified gene clusters also play important role in encoding proteins for shikimic acid pathway such as tryptophan (TrpA, TrpB, TrpC, TrpD and TrpE) and tyrosine (TyrP) biosynthesis (Soule et al. 2007, 2009). Genomic analysis further validates the presence of additional copy of genes encoding for AroG, AroB, TrpA, TrpB, TrpC, TrpD and TyrP proteins present at another place in the genome of *Nostoc punctiforme* (Pathak et al. 2019). Scytonemin biosynthetic genes are also present in other cyanobacteria *Nodularia* sp. CCY 9414, *Chlorogloeopsis* sp. CCMEE 5094, *Anabaena* sp. PCC 7120 and *Lyngbya* sp. PCC 8106 (Pathak et al. 2019, 2020). Interestingly, heterologous expression in *E. coli* of three *N. punctiforme* genes, particularly *scyABC*, culminated into scytonemin monomer moiety upon supplementation of 1 mM of tyrosine and tryptophan (Malla and Sommer 2014). The above experiment elucidates ScyA, ScyB, and ScyC genes are solely responsible for scytonemin biosynthesis (Malla and Sommer 2014; Pathak et al. 2019). The comparative genomic analysis reveals that cyanobacterium *Nostoc punctiforme* ATCC 29133 has two-component regulatory system of proteins Npun_F1278 and Npun_F1277 upregulated in response to UV-A radiation (Naurin et al. 2016; Janssen and Soule 2016).

Scytonemin operon consists of five-gene cluster called as '*ebo*' cluster, discovered through comparative genomic analysis widely conserved in algal and bacterial phyla (Pathak et al. 2019). The function of ebo cluster was validated through in-frame deletion method and their analysis proves '*ebo*' cluster is responsible for scytonemin biosynthesis under UV-A exposure conditions (Klicki et al. 2018). Figures 4.4 and 4.5 depict scytonemin biosynthesis genes with annotations and scytonemin biosynthetic pathway, respectively.

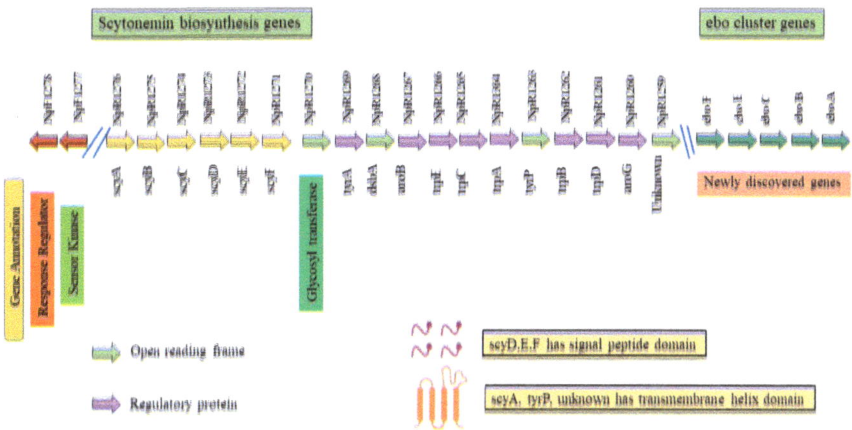

Fig. 4.4 Structure of gene cluster associated with scytonemin biosynthesis in *Nostoc punctiforme* ATCC 29133

4.4.2 Mycosporine-Like Amino Acids

Cyanobacterial ancestors are served as starting point for MAAs biosynthetic enzymatic machinery. These are widespread among all taxa due to evolutionary endosymbiotic events and lateral gene transfer mechanism (Singh et al. 2012; Richa and Sinha 2013). Multiple studies have been performed which proves that MAAs biosynthesis occurs through the shikimic acid pathway (Pathak et al. 2019). It was found that *A. variablis* PCC7937 has two gene locus YP_324357 and YP_324358 to carry biosynthesis of deoxygadusol for development of central core of mycosporine-like amino acids (Richa and Sinha 2013). The MAAs biosynthesis gets downregulated by exogenous supply of aromatic amino acids tyrosine at a concentration of 5 mM (Portwich and Garcia-Pichel 2003; Spence et al. 2012; Pathak et al. 2019). The cluster of four genes from *ava_3855* to *ava_3858* is responsible for the formation of shinorine in *Anabaena variabilis* ATCC 29413 (Singh et al. 2010b; Balskus and Walsh 2010). The shikimate and pentose phosphate pathways are completely linked for MAAs biosynthesis (Jain et al. 2017). The proteomic data clarifies shikimate pathway is predominant in UV-induced MAAs biosynthesis (Pope et al. 2015; Jain et al. 2017). The *ava_3856* gene expression in *Anabaena variabilis* confirmed the conversion of 6-deoxygadusol and glycine into mycosporine-glycine in the presence of Mg^{2+} cofactors and ATP (Balskus and Walsh 2010). MAAs biosynthesis involves NpR5598-5600 homologous genes in both cyanobacterium *Nostoc punctiforme* ATCC 29133 and *A. variabilis* (Katoch et al. 2016). The interlinked convergent pathway for the synthesis of MAAs is shown in Fig. 4.6.

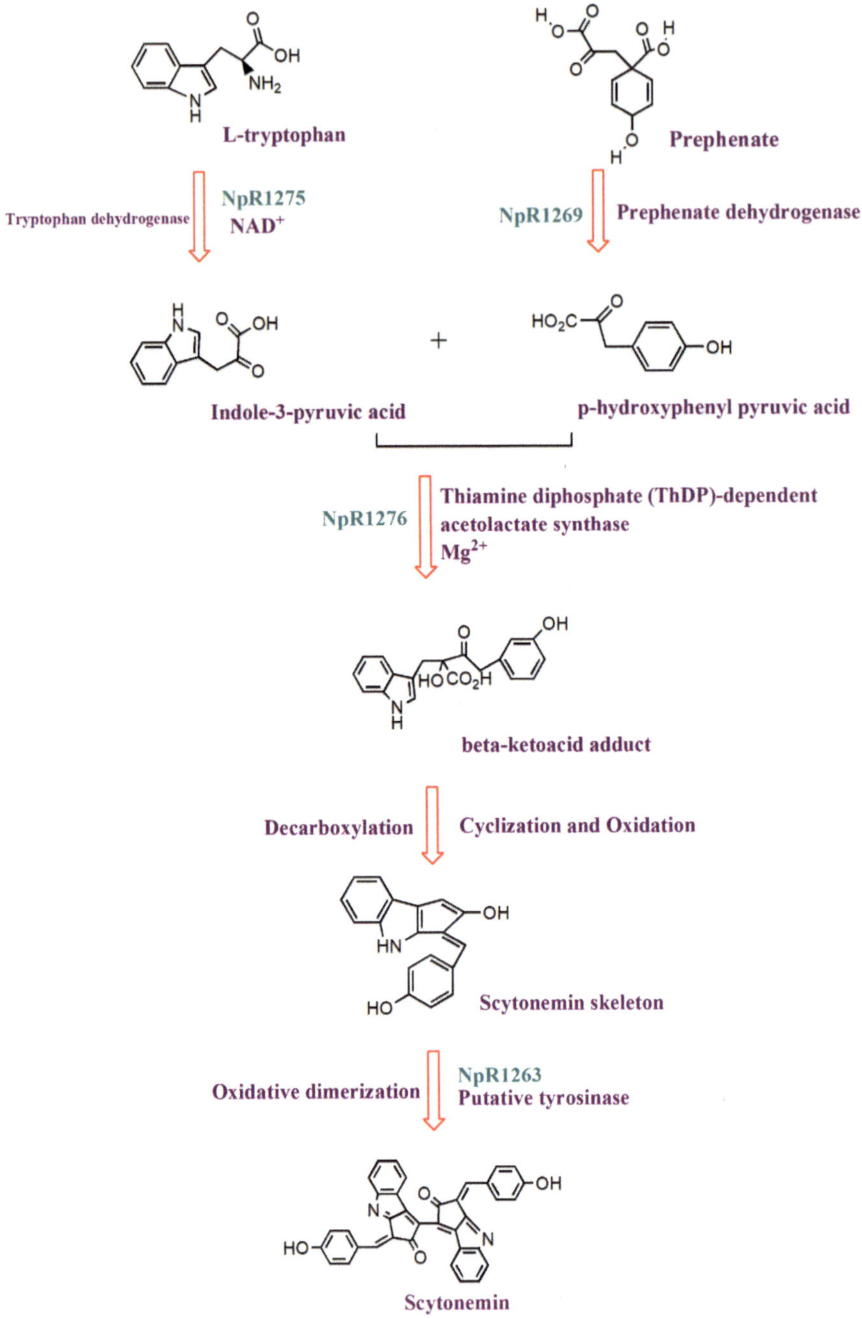

Fig. 4.5 Biosynthetic pathway of scytonemin biosynthesis in *Nostoc punctiforme* ATCC 29133

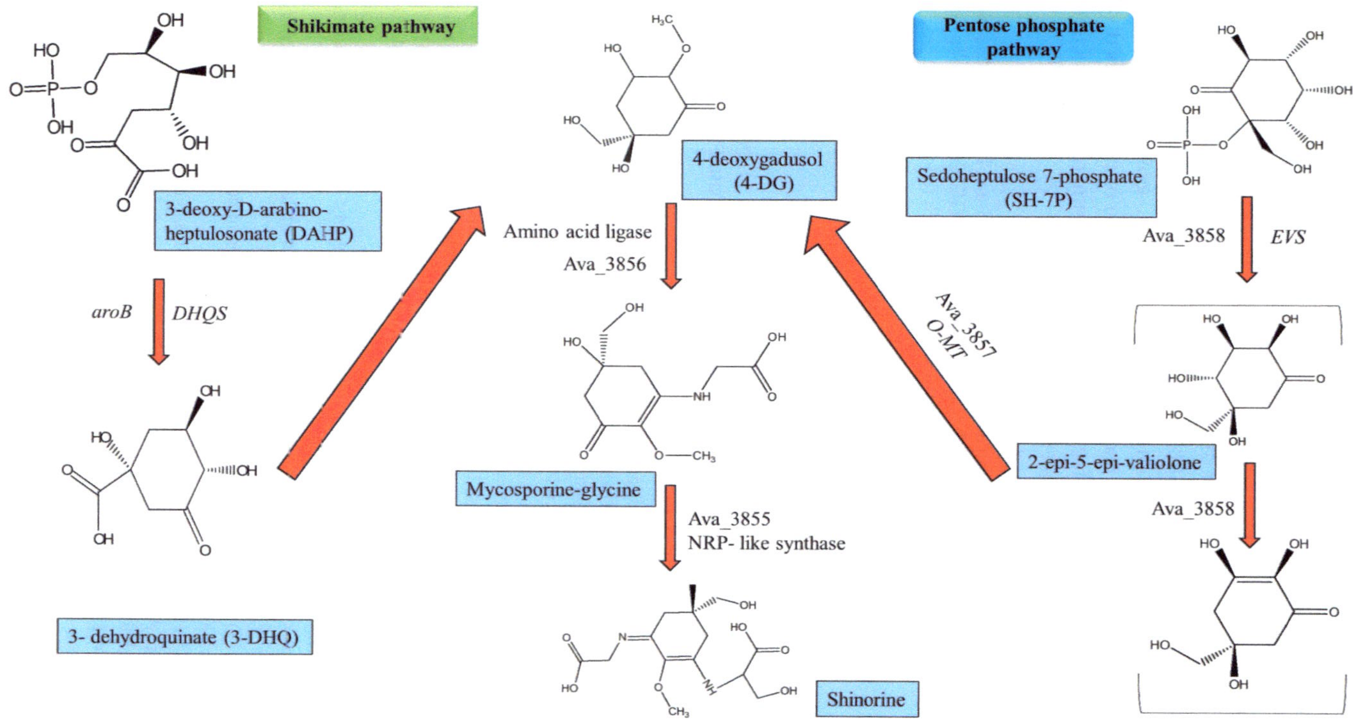

Fig. 4.6 Biosynthetic pathways involved in the synthesis of MAAs in cyanobacteria

4.5 Concluding Remarks

The organisms belonging to different taxa have unique kinds of photoprotective compounds. These photoprotective compounds have played an indispensable role in evolutionary context as well as for survival. Nowadays, extensive research work has been carried out in all species for the screening of valuable and stable photoprotective compounds to safeguard humans from the lethal effects of UV radiation. To complete studies on UV screening compounds, that is, genetic regulation and metabolomics, there is a current need for wide exploration of compounds diversity and the development of cost-effective strategies for commercial production and therapeutic applications for human welfare.

Acknowledgements Vinod K. Kannaujiya acknowleges IoE-Seed Grant, BHU, Varanasi (Scheme No. 6031) and Core Research Grant, DST-SERB, New Delhi (CRG/2020/001323) for financial support. Saumi is thankful to UGC and BHU for providing a research fellowship.

References

Agati G, Brunetti C, Di Ferdinando M, Ferrini F, Pollastri S, Tattini M (2013) Functional roles of flavonoids in photoprotection: new evidence, lessons from the past. Plant Physiol Biochem 72:35–45

Balskus EP, Walsh CT (2010) The genetic and molecular basis for sunscreen biosynthesis in cyanobacteria. Science 329(5999):1653–1656

Bandaranayake W (1998) Mycosporines: are they nature's sunscreens? Nat Prod Rep 15(2):159–172

Bhatia S, Garg A, Sharma K, Kumar S, Sharma A, Purohit AP (2011) Mycosporine and mycosporine-like amino acids: a paramount tool against ultra violet irradiation. Pharmacogn Rev 5(10):138

Bjerke JW, Dahl T (2002) Distribution patterns of usnic acid-producing lichens along local radiation gradients in West Greenland. Nova Hedwigia 75:487–506

Bjerke JW, Lerfall K, Elvebakk A (2002) Effects of ultraviolet radiation and PAR on the content of usnic and divaricatic acids in two arctic-alpine lichens. Photochem Photobiol Sci 1(9):678–685

Björn LO (2007) Stratospheric ozone, ultraviolet radiation, and cryptogams. Biol Conserv 135(3):326–333

Brenner M, Hearing VJ (2008) The protective role of melanin against UV damage in human skin. Photochem Photobiol 84(3):539–549

Bustamante J, Bredeston L, Malanga G, Mordoh J (1993) Role of melanin as a scavenger of active oxygen species. Pigment Cell Res 6(5):348–353

Deng Y, Lu S (2017) Biosynthesis and regulation of phenylpropanoids in plants. Crit Rev Plant Sci 36(4):257–290

Edreva A (2005) The importance of non-photosynthetic pigments and cinnamic acid derivatives in photoprotection. Agric Ecosyst Environ 106(2–3):135–146

Escobar-Bravo R, Klinkhamer PG, Leiss KA (2017) Interactive effects of UV-B light with abiotic factors on plant growth and chemistry, and their consequences for defense against arthropod herbivores. Front Plant Sci 8:278

Fernández E, Quilhot W, Rubio C, Hidalgo ME, Diaz R, Ojeda J (2006) Effects of UV radiation on usnic acid in *Xanthoparmelia microspora* (Müll. Arg. Hale). Photochem Photobiol 82(4):1065–1068

Frohnmeyer H, Staiger D (2003) Ultraviolet-B radiation-mediated responses in plants. Balancing damage and protection. Plant Physiol 133(4):1420–1428

Gao K, Yu H, Brown MT (2007) Solar PAR and UV radiation affects the physiology and morphology of the cyanobacterium *Anabaena* sp. PCC 7120. J Photochem Photobiol B Biol 89(2–3):117–124

Garcia-Pichel F, Castenholz RW (1991) Characterization and biological implications of scytonemin, a cyanobacterial sheath pigment 1. J Phycol 27(3):395–409

Garcia-Pichel F, Wingard CE, Castenholz RW (1993) Evidence regarding the UV sunscreen role of a mycosporine-like compound in the cyanobacterium *Gloeocapsa* sp. Appl Environ Microbiol 59(1):170–176

Gauslaa Y, McEvoy M (2005) Seasonal changes in solar radiation drive acclimation of the sun-screening compound parietin in the lichen *Xanthoria parietina*. Basic Appl Ecol 6(1):75–82

Gauslaa Y, Solhaug KA (2004) Photoinhibition in lichens depends on cortical characteristics and hydration. Lichenologist 36(2):133–143

Gauslaa Y, Ustvedt EM (2003) Is parietin a UV-B or a blue-light screening pigment in the lichen *Xanthoria parietina*? Photochem Photobiol Sci 2(4):424–432

Geraldes V, Pinto E (2021) Mycosporine-like amino acids (MAAs): biology, chemistry and identification features. Pharmaceuticals 14(1):63

Gilchrest BA, Eller MS, Geller AC, Yaar M (1999) The pathogenesis of melanoma induced by ultraviolet radiation. N Engl J Med 340(17):1341–1348

Häder DP, Kumar HD, Smith RC, Worrest RC (2007) Effects of solar UV radiation on aquatic ecosystems and interactions with climate change. Photochem Photobiol Sci 6(3):267–285

Halder RM, Bang KM (1988) Skin cancer in blacks in the United States. Dermatol Clin 6(3):397–405

Hargreaves A, Taiwo FA, Duggan O, Kirk SH, Ahmad SI (2007) Near-ultraviolet photolysis of β-phenylpyruvic acid generates free radicals and results in DNA damage. J Photochem Photobiol B Biol 89(2–3):110–116

Heijde M, Ulm R (2012) UV-B photoreceptor-mediated signalling in plants. Trends Plant Sci 17(4):230–237

Hemm MR, Herrmann KM, Chapple C (2001) AtMYB4: a transcription factor general in the battle against UV. Trends Plant Sci 6(4):135–136

Hollósy F (2002) Effects of ultraviolet radiation on plant cells. Micron 33(2):179–197

Huneck S, Yoshimura I (1996) Identification of lichen substances. In: Identification of lichen substances. Springer, Berlin, pp 11–123

Jain S, Prajapat G, Abrar M, Ledwani L, Singh A, Agrawal A (2017) Cyanobacteria as efficient producers of mycosporine-like amino acids. J Basic Microbiol 57(9):715–727

Janssen J, Soule T (2016) Gene expression of a two-component regulatory system associated with sunscreen biosynthesis in the cyanobacterium *Nostoc punctiforme* ATCC 29133. FEMS Microbiol Lett 363(2):fnv235

Kaidbey KH, Agin PP, Sayre RM, Kligman AM (1979) Photoprotection by melanin—a comparison of black and Caucasian skin. J Am Acad Dermatol 1(3):249–260

Katoch M, Mazmouz R, Chau R, Pearson LA, Pickford R, Neilan BA (2016) Heterologous production of cyanobacterial mycosporine-like amino acids mycosporine-ornithine and mycosporine-lysine in *Escherichia coli*. Appl Environ Microbiol 82(20):6167–6173

Kannaujiya VK, Richa and Sinha RP (2014) Peroxide scavenging potential of ultraviolet-B-absorbing mycosporine-like amino acids isolated from a marine red alga *Bryocladia* sp. Front Environ Sci 2: 26

Kiselev KV, Grigorchuk VP, Ogneva ZV, Suprun AR, Dubrovina AS (2016) Stilbene biosynthesis in the needles of spruce *Picea jezoensis*. Phytochemistry 131:57–67

Klicki K, Ferreira D, Hamill D, Dirks B, Mitchell N, Garcia-Pichel F (2018) The widely conserved ebo cluster is involved in precursor transport to the periplasm during scytonemin synthesis in *Nostoc punctiforme*. MBio 9(6):e02266–e02218

Kobayashi N, Nakagawa A, Muramatsu T, Yamashina Y, Shirai T, Hashimoto MW, Ishigaki Y, Ohnishi T, Mori T (1998) Supranuclear melanin caps reduce ultraviolet induced DNA photoproducts in human epidermis. J Investig Dermatol 110(5):806–810

La Barre S, Roullier C, Boustie J (2014) Mycosporine-like amino acids (MAAs) in biological photosystems. In: Outstanding marine molecules, pp 333–360

Lalegerie F, Lajili S, Bedoux G, Taupin L, Stiger-Pouvreau V, Connan S (2019) Photo-protective compounds in red macroalgae from Brittany: considerable diversity in mycosporine-like amino acids (MAAs). Mar Environ Res 147:37–48

Le Lann K, Surget G, Couteau C, Coiffard L, Cérantola S, Gaillard F, Larnicol M, Zubia M, Guérard F, Poupart N, Stiger-Pouvreau V (2016) Sunscreen, antioxidant, and bactericide capacities of phlorotannins from the brown macroalga *Halidrys siliquosa*. J Appl Phycol 28(6):3547–3559

Lesser MP (2006) Oxidative stress in marine environments: biochemistry and physiological ecology. Annu Rev Physiol 68:253–278

Lesser MP (2008) Effects of ultraviolet radiation on productivity and nitrogen fixation in the cyanobacterium, *Anabaena* sp. (Newton's strain). Hydrobiologia 598(1):1–9

Liu J, Osbourn A, Ma P (2015) MYB transcription factors as regulators of phenylpropanoid metabolism in plants. Mol Plant 8(5):689–708

Luzina OA, Salakhutdinov NF (2018) Usnic acid and its derivatives for pharmaceutical use: A patent review (2000–2017). Expert Opinion on Therapeutic Patents, 28(6):477–491

Malla S, Sommer MO (2014) A sustainable route to produce the scytonemin precursor using *Escherichia coli*. Green Chem 16(6):3255–3265

Matsui K, Nazifi E, Hirai Y, Wada N, Matsugo S, Sakamoto T (2012) The cyanobacterial UV-absorbing pigment scytonemin displays radical-scavenging activity. J Gen Appl Microbiol 58(2):137–144

McEvoy M, Nybakken L, Solhaug KA, Gauslaa Y (2006) UV triggers the synthesis of the widely distributed secondary lichen compound usnic acid. Mycol Prog 5(4):221–229

Meyer P, Van de Poel B, De Coninck B (2021) UV-B light and its application potential to reduce disease and pest incidence in crops. Hortic Res 8(1):194

Naurin S, Bennett J, Videau P, Philmus B, Soule T (2016) The response regulator Npun_F1278 is essential for scytonemin biosynthesis in the cyanobacterium *Nostoc punctiforme* ATCC 29133. J Phycol 52(4):564–571

Noel JP, Austin MB, Bomati EK (2005) Structure–function relationships in plant phenylpropanoid biosynthesis. Curr Opin Plant Biol 8(3):249–253

Oren A, Gunde-Cimerman N (2007) Mycosporines and mycosporine-like amino acids: UV protectants or multipurpose secondary metabolites? FEMS Microbiol Lett 269(1):1–10

Pallela R, Na-Young Y, Kim SK (2010) Anti-photoaging and photoprotective compounds derived from marine organisms. Mar Drugs 8(4):1189–1202

Passeron T, Lim HW, Goh CL, Kang HY, Ly F, Morita A et al (2021) Photoprotection according to skin phototype and dermatoses: practical recommendations from an expert panel. J Eur Acad Dermatol Venereol 35(7):1460–1469

Pathak J, Ahmed H, Singh SP, Häder DP, Sinha RP (2019) Genetic regulation of scytonemin and mycosporine-like amino acids (MAAs) biosynthesis in cyanobacteria. Plant Gene 17:100172

Pathak J, Pandey A, Maurya PK, Rajneesh R, Sinha RP, Singh SP (2020) Cyanobacterial secondary metabolite scytonemin: a potential photoprotective and pharmaceutical compound. Proc Natl Acad Sci India Sect B Biol Sci 90(3):467–481

Paul ND, Gwynn-Jones D (2003) Ecological roles of solar UV radiation: towards an integrated approach. Trends Ecol Evol 18(1):48–55

Pope MA, Spence E, Seralvo V, Gacesa R, Heidelberger S, Weston AJ, Dunlap WC, Shick JM, Long PF (2015) O-Methyltransferase is shared between the pentose phosphate and shikimate pathways and is essential for mycosporine-like amino acid biosynthesis in *Anabaena variabilis* ATCC 29413. Chembiochem 16(2):320–327

Portwich A, Garcia-Pichel F (2003) Biosynthetic pathway of mycosporines (mycosporine-like amino acids) in the cyanobacterium *Chlorogloeopsis* sp. strain PCC 6912. Phycologia 42(4):384–392

Proteau PJ, Gerwick WH, Garcia-Pichel F, Castenholz R (1993) The structure of scytonemin, an ultraviolet sunscreen pigment from the sheaths of cyanobacteria. Experientia 49(9):825–829
Rancan F, Rosan S, Boehm K, Fernández E, Hidalgo ME, Quihot W, Rubio C, Boehm F, Piazena H, Oltmanns U (2002) Protection against UVB irradiation by natural filters extracted from lichens. J Photochem Photobiol B Biol 68(2–3):133–139
Rastogi RP, Sinha RP (2009) Biotechnological and industrial significance of cyanobacterial secondary metabolites. Biotechnol Adv 27(4):521–539
Rastogi RP, Sinha RP, Singh SP, Häder DP (2010) Photoprotective compounds from marine organisms. J Ind Microbiol Biotechnol 37(6):537–558
Rastogi RP, Sinha RP, Incharoensakdi A (2013) Partial characterization, UV-induction and photoprotective function of sunscreen pigment, scytonemin from *Rivularia* sp. HKAR-4. Chemosphere 93(9):1874–1878
Rastogi RP, Sonani RR, Madamwar D (2015) Cyanobacterial sunscreen scytonemin: role in photoprotection and biomedical research. Appl Biochem Biotechnol 176(6):1551–1563
Richa, Sinha RP (2013) Biomedical applications of mycosporine-like amino acids. In: Kim S-K (ed) Marine microbiology in bioactive compounds and biotechnological applications, pp 509–534
Rizzini L, Favory JJ, Cloix C, Faggionato D, O'Hara A, Kaiserli E, Baumeister R, Schäfer E, Nagy F, Jenkins GI, Ulm R (2011) Perception of UV-B by the *Arabidopsis* UVR8 protein. Science 332(6025):103–106
Rozema J, Björn LO, Bornman JF, Gaberščik A, Häder DP, Trošt T, Germ M, Klisch M, Gröniger A, Sinha RP, Lebert M (2002) The role of UV-B radiation in aquatic and terrestrial ecosystems—an experimental and functional analysis of the evolution of UV-absorbing compounds. J Photochem Photobiol B Biol 66(1):2–12
Samanta A, Das G, Das SK (2011) Roles of flavonoids in plants. Carbon 100(6):12–35
Sen S, Mallick N (2022) Scytonemin: unravelling major progress and prospects. Algal Res 64:102678
Shick JM, Dunlap WC (2002) Mycosporine-like amino acids and related gadusols: biosynthesis, accumulation, and UV-protective functions in aquatic organisms. Annu Rev Physiol 64(1):223–262
Simmonds MS (2003) Flavonoid–insect interactions: recent advances in our knowledge. Phytochemistry 64(1):21–30
Singh SP, Sinha RP, Klisch M, Häder DP (2008a) Mycosporine-like amino acids (MAAs) profile of a rice-field cyanobacterium *Anabaena doliolum* as influenced by PAR and UVR. Planta 229(1):225–233
Singh SP, Klisch M, Sinha RP, Häder DP (2008b) Effects of abiotic stressors on synthesis of the mycosporine-like amino acid shinorine in the cyanobacterium *Anabaena variabilis* PCC 7937. Photochem Photobiol 84(6):1500–1505
Singh SP, Häder DP, Sinha RP (2010a) Cyanobacteria and ultraviolet radiation (UVR) stress: mitigation strategies. Ageing Res Rev 9(2):79–90
Singh SP, Klisch M, Sinha RP, Häder DP (2010b) Genome mining of mycosporine-like amino acid (MAA) synthesizing and non-synthesizing cyanobacteria: a bioinformatics study. Genomics 95(2):120–128
Singh SP, Häder DP, Sinha RP (2012) Bioinformatics evidence for the transfer of mycosporine-like amino acid core (4-deoxygadusol) synthesizing gene from cyanobacteria to dinoflagellates and an attempt to mutate the same gene (YP_324358) in *Anabaena variabilis* PCC 7937. Gene 500(2):155–163
Sinha RP, Häder DP (2008) UV-protectants in cyanobacteria. Plant Sci 174(3):278–289
Sinha RP, Singh SP, Häder DP (2007) Database on mycosporines and mycosporine-like amino acids (MAAs) in fungi, cyanobacteria, macroalgae, phytoplankton and animals. J Photochem Photobiol B Biol 89(1):29–35
Solhaug KA, Gauslaa Y (2004) Photosynthates stimulate the UV-B induced fungal anthraquinone synthesis in the foliose lichen *Xanthoria parietina*. Plant Cell Environ 27(2):167–176

Solhaug KA, Gauslaa Y (2012) Secondary lichen compounds as protection against excess solar radiation and herbivores. In: Progress in botany, vol 73. Springer, Berlin, pp 283–304
Solhaug KA, Gauslaa Y, Nybakken L, Bilger W (2003) UV-induction of sun-screening pigments in lichens. New Phytol 158(1):91–100
Soule T, Stout V, Swingley WD, Meeks JC, Garcia-Pichel F (2007) Molecular genetics and genomic analysis of scytonemin biosynthesis in *Nostoc punctiforme* ATCC 29133. J Bacteriol 189(12):4465–4472
Soule T, Garcia-Pichel F, Stout V (2009) Gene expression patterns associated with the biosynthesis of the sunscreen scytonemin in *Nostoc punctiforme* ATCC 29133 in response to UVA radiation. J Bacteriol 191(14):4639–4646
Spence E, Dunlap WC, Shick JM, Long PF (2012) Redundant pathways of sunscreen biosynthesis in a cyanobacterium. Chembiochem 13(4):531–533
Takamatsu S, Hodges TW, Rajbhandari I, Gerwick WH, Hamann MT, Nagle DG (2003) Marine natural products as novel antioxidant prototypes. J Nat Prod 66(5):605–608
Tohge T, Watanabe M, Hoefgen R, Fernie AR (2013) Shikimate and phenylalanine biosynthesis in the green lineage. Front Plant Sci 4:62
Torregiani JH, Lesser MP (2007) The effects of short-term exposures to ultraviolet radiation in the Hawaiian coral *Montipora verrucosa*. J Exp Mar Biol Ecol 340(2):194–203
Treutter D (2005) Significance of flavonoids in plant resistance and enhancement of their biosynthesis. Plant Biol 7(06):581–591
Turnbull JJ, Nakajima JI, Welford RW, Yamazaki M, Saito K, Schofield CJ (2004) Mechanistic studies on three 2-oxoglutarate-dependent oxygenases of flavonoid biosynthesis: anthocyanidin synthase, flavonol synthase, and flavanone 3β-hydroxylase. J Biol Chem 279(2):1206–1216
Venugopala KN, Rashmi V, Odhav B (2013) Review on natural coumarin lead compounds for their pharmacological activity. Biomed Res Int 2013:963248
Vidović M, Morina F, Milić S, Albert A, Zechmann B, Tosti T, Winkler JB, Jovanović SV (2015) Carbon allocation from source to sink leaf tissue in relation to flavonoid biosynthesis in variegated Pelargonium zonale under UV-B radiation and high PAR intensity. Plant Physiol Biochem 93:44–55
Watts KT, Lee PC, Schmidt-Dannert C (2004) Exploring recombinant flavonoid biosynthesis in metabolically engineered *Escherichia coli*. Chembiochem 5(4):500–507
Wong HJ, Mohamad-Fauzi N, Rizman-Idid M, Convey P, Alias SA (2019) Protective mechanisms and responses of micro-fungi towards ultraviolet-induced cellular damage. Pol Sci 20:19–34
Zeevi Ben-Yosef D, Kashman Y, Benayahu Y (2006) Response of the soft coral *Heteroxenia fuscescens* to ultraviolet radiation regimes as reflected by mycosporine-like amino acid biosynthesis. Mar Ecol 27(3):219–228
Zhang L, Li L, Wu Q (2007) Protective effects of mycosporine-like amino acids of *Synechocystis* sp. PCC 6803 and their partial characterization. J Photochem Photobiol B Biol 86(3):240–245

Chapter 5
Bioprospecting and Evolutionary Significance of Photoprotectors in Non-flowering Lower Plants

Amit Gupta, Ashish P. Singh, Niharika Sahu, Jyoti Jaiswal, Neha Kumari, Prashant R. Singh, and Rajeshwar P. Sinha

Abstract Environmental change as well as the continued increase in UV radiation (UVR; 280–400 nm) has a significant impact on terrestrial and aquatic ecosystems. The majority of sun-exposed species are negatively affected by solar UVR. This may be the strong reason behind the evolution of photoprotectants like phenylpropanoids, flavonoids, mycosporines, scytonemin, mycosporine-like amino acids (MAAs), parietin, xanthophyll, phycobiliproteins etc. It is mainly found in cyanobacteria, algae, fungi, lichens and other non-flowering lower plants. These photoprotectors have their own evolutionary significance. To reduce photochemical damage, carotenoids are being considered as UV-protective additives. Carotenoids are major light-harvesting and photoprotective components of the photosynthetic apparatus, and work as quencher of singlet oxygen species. Many other UV-absorbing compounds are also known for their multifunctional capabilities, such as flavonoids, which have antioxidative and antibacterial potentials. The study of these photoprotectants has resulted in the identification of new sunscreen classes and their distribution throughout various microbes and non-flowering lower plants. These natural photoprotectants play an important role in different forms such as scytonemin and MAAs operate as the third line of defence in cyanobacteria, mycosporines are critical for UV-induced photodamage in fungi. Lichens possess a wide range of primary and secondary metabolites, in which MAAs and parietin have major photoprotective effects. Moreover, these photoprotectants aid in diffusing heat from absorbed

A. Gupta · A. P. Singh · N. Sahu · J. Jaiswal · N. Kumari · P. R. Singh
Laboratory of Photobiology and Molecular Microbiology, Centre of Advanced Study in Botany, Institute of Science, Banaras Hindu University, Varanasi, India

R. P. Sinha (✉)
Laboratory of Photobiology and Molecular Microbiology, Centre of Advanced Study in Botany, Institute of Science, Banaras Hindu University, Varanasi, India

University Center for Research & Development (UCRD), Chandigarh University, Chandigarh, India

V. K. Kannaujiya et al. (eds.), *Photoprotective Green Pharmacology: Challenges, Sources and Future Applications*,
https://doi.org/10.1007/978-981-99-0749-6_5

radiation without generating reactive oxygen species (ROS). This chapter deals with the evolutionary significance of photoprotectors and their possible modes of action.

Keywords Algae · Antioxidants · Carotenoids · Cyanobacteria · Flavonoids · Fungi · Lichens · Mycosporine-like amino acids · Scytonemin

5.1 Introduction

The ancient organisms may adapt to the quality and quantity of light in their habitat by modulating antenna pigment complexes, photochemical reaction centers, and enzymes for CO_2 fixation to maximize light energy absorption and use at varied irradiance levels. However, rapid increase in urbanization and industrialization has resulted in an enhancement of solar ultraviolet (UV) radiation (280–400 nm) on the surface of Earth, due to release of various anthropogenic atmospheric pollutants such as chlorocarbons (CCs), chlorofluorocarbons (CFCs), organ bromides (OBs), and reactive nitrogen species (RNS) such as nitrous oxide (N_2O), nitric oxide (·NO), and peroxynitrite (ONOO) (Crutzen 1992; Ballaré et al. 2011). The incident has sparked widespread concern about the harmful effects of high-energy solar UV-B (280–315 nm) radiation on land and aquatic life forms such as micro/macroalgae, cyanobacteria and phytoplankton.

In response to the selection pressures produced by solar radiation, fungi have evolved a range of UV-protective systems. Pigment synthesis is the first line of defense for reducing or preventing UVR-induced intercellular damage. Melanin, carotenoids and mycosporines are three primary pigments found in fungi, acting as sunscreens or antioxidants to combat the damage caused by UV radiation (Almeida-Paes et al. 2012). In order to decrease the effects of UV-induced damages, any failure of these systems results in the activation of DNA repair pathways. Nucleotide excision repair (NER) and photoreactivation (Phr) are particularly important in fungi for repairing UV-induced damage of DNA like pyrimidine (6–4), pyrimidone photoproducts (6-4PPs) and cyclobutene pyrimidine dimers (CPDs) as well as maintaining DNA integrity and boosting cell survival (Rangel et al. 2011). In comparison to other carotenoids in fungi, phytoene and phytofluene, which are precursors to more complex carotenoid molecules, have been largely overlooked in terms of their possible functions in UV protection. The defensive apparatus used by most of the micro- and macrofungi in response to DNA damage as well as other biological and ecological variables that determine susceptibility to solar UVR, has been discussed in this chapter.

Lichens synthesize UV filters such as depsides, depsidones, diphenyl ether, bisxanthones, mycosporines and MAAs, scytonemin, as well as melanin and carotenoids. The UV-visible absorption spectra of some anthraquinones, notably perylene quinone pigments, are different. When exposed to UV and blue light, their conjugated heterocyclic aromatic quinone components allowed them to generate intense red fluorescence. The preponderance of UV spectra was collected in polar solvents such as methanol or ethanol, which might have a significant influence on UV profiles. While UV-B metabolites such as depsides, depsidones, dibenzofurans, chromones, and diphenyl ethers exhibit UV-B features, UV-A metabolites such as xanthones and pulvinic acid derivatives have extinction values surpassing 10,000 L mol^{-1} cm^{-1} for the lower energy absorption bands. They are cortical lichen compounds that operate as solar radiation shields, together with parietin, usnic acid, and atranorin (Boustie et al. 2011). The phenolic and indolic rings in these pigments serve as key building elements for the melanin polymeric matrix. The existence of these pigments shows that stress-tolerant lichens use defensive mechanisms similar to human skin to protect themselves from the harmful effects of UV radiation. Furthermore, the light-harvesting complex must be adaptable enough to avoid conditions that restrict the response center's effectiveness such as bright light and other stressful conditions. Non-photochemical quenching (NPQ) is the most effective way for safely dissipating surplus energy in the form of heat (Demmig-Adams et al. 2014). NPQ have a distinct evolutionary history in different species.

A unique, orange carotenoid protein (OCP)-independent PB quenching mechanism was recently proposed using single-molecule spectroscopy. According to studies, several intrinsic channels are found in the various subunits of PBs, and any of them can participate in the quenching process, although the protein core is the most common target. It functions as a fast-quenching process that occurs prior to the activation of the OCP-dependent mechanism. To prevent photooxidative damage in the organisms, xanthophyll cycle-related photoprotection through heat dissipation is as important as other photoprotective strategies such as the production of UV-absorbing compounds (Sinha et al. 1998) or the adaptation of chloroplast orientation. Xanthophyll cycle has its own evolutionary significance as these pigments are easily transformed under various environmental circumstances, and they also exhibit a high degree of plasticity, which has an influence on their evolutionary studies. Oxidative stress is nearly completely responsible for the increase in flavonoid synthesis. They have the ability to absorb damaging UV radiations such as UV-A and UV-B as well as to limit or quench the production of ROS (Agati et al. 2012).

5.2 Photoprotective Compounds in Cyanobacteria and Algae

Green algae are found in both freshwater and marine environments, and they are important biomass producers in both terrestrial and aquatic ecosystems. They are a huge source of a variety of useful natural compounds. The utilization of some green algal species as a nontraditional source of food and protein appears to be promising (Häder et al. 2011; Rastogi et al. 2010a). Green algae, like other photosynthetic organisms, rely on sunlight for their regular life processes and hence are exposed to damaging UV radiation (particularly UV-B). Growth and survival, buoyancy, photosynthesis, CO_2 absorption, ribulose 1,5-bisphosphate carboxylase/oxygenase (RuBisCO) activity and changes in the native structure of proteins and DNA molecules are all affected by increased solar UV radiation (Rastogi et al. 2010a).

Furthermore, chlorophytes are a prominent flora, and any negative effects on their existence might cause an environmental imbalance. Despite this, organisms have devised a number of defense mechanisms to counteract the direct or indirect adverse effects of UV radiation, such as mat formation, migration and antioxidant production. Biosynthesis of UV-absorbing compounds such as scytonemin, mycosporine-like amino acids (MAAs) production and intensive DNA repair mechanisms are proved to be the major processes to counteract the direct or indirect damaging effects of UV radiation (Rastogi et al. 2010b; Lee and Shiu 2009) (Table 5.1).

5.2.1 *Mycosporine-Like Amino Acids (MAAs)*

MAAs are notable UV-absorbing/screening chemicals that give photoprotection against UV-B and UV-A radiations (Garcia-Pichel et al. 1993). MAAs are small (400 Da), colorless, hydrophilic molecules having cyclohexenimine or cyclohexenone chromophores attached to an amino acid nitrogen group or its imino alcohol (Nakamura et al. 1982). Multi-beneficial mycosporine-like amino acids have a molecular weight of less than 1.2 kDa (Rastogi et al. 2010a). These compounds are recognized for dissipating absorbed energy as heat into their surroundings while avoiding the production of reactive oxygen species (ROS), and they play a key role in photoprotection (Conde et al. 2000). MAAs have a high molar extinction coefficient (28,100–50,000 M^{-1} cm^{-1}) and strong UV absorption maxima (310–362 nm). In recent years, MAAs have been discovered in a wide range of species, including cyanobacteria, heterotrophic bacteria, fungus, micro/macroalgae, lichens and numerous animal forms (Sinha et al. 2007; Llewellyn and Airs 2010). However, the existence of these ecologically significant chemicals in freshwater unicellular green algae has been reported infrequently. MAAs are categorized into two classes based on their core chromatophore structure: cyclohexenone ring exhibiting oxo-MAAs and cyclohexene imine ring showing imino-MAAs. Mycosporine-taurine and mycosporine-glycine are examples of oxo-MAAs, while shinorine, palythine, asterina-330, palythene and other imino-MAAs are examples of imino-MAAs (Shick and Dunlap 2002; Sinha et al. 1998) (Table 5.2).

In certain species, MAAs perform a supporting function by scavenging harmful oxygen radicals. They are chemically compatible solutes that defend against heat, desiccation and salt stress. In a few species, they also act as an intracellular nitrogen reserve (Oren and Gunde-Cimerman 2007). Certain environmental conditions including nutrient availability, stress conditions and UV radiation have an impact on MAAs levels inside cells. The formation of MAAs inside the cells is influenced equally by PAR (Photosynthetically Active Light), UV-A and UV-B radiation (Sinha and Häder 2008).

Table 5.1 Occurrence of MAAs in micro/macroalgae, cyanobacteria and lichens (Adapted from Rastogi et al. 2010a)

Organisms	MG	SH	AS	PL	PR	PT	PE	MT	PS	US	M328/360	SME	MGV	PNA	M2G	M 333	M 320	MMS	M335/360	References
Lichen	+																			De la Coba et al. (2009)
Cyanobacteria		+			+															Karsten and Garcia-Pichel (1996)
Macroalgae																				
Red algae	+	+	+	+	+	+	+			+						+	+		+	De la Coba et al. (2009)
Brown algae	+	+	+	+	+	+														Nakamura et al. (1982), Karentz et al. (1991)
Green algae	+	+	+	+	+	+														Karentz et al. (1991), Nakamura et al. (1982)
Microalgae/ phytoplankton	+	+	+	+	+	+	+	+	+	+	+	+	+	+	+	+	+	+	+	Carreto et al. (2005)

MG mycosporine-glycine, *SH* shinorine, *AS* asterina-330, *PL* palythinol, *PR* porphyra-334, *PT* palythine, *PE* palythene, *MT* mycosporine-taurine, *PS* palythine-serine, *US* usurijene, *M* mycosporine-like amino acids, *SME* shinorine methyl ester, *MGV* mycosporine glycine-valine, *PNA* palythenic acid, *M2G* mycosporine-2-glycine, *MMS* mycosporine-methylamine-serine

Table 5.2 Type of MAAs, microalgal/cyanobacterial source and applications (Raj et al. 2021)

Sl. No.	Types of MAAs	Microalgal/ cyanobacterial sources	Absorbance λ_{max} (nm)	Applications	References
1.	Mycosporine-glycine	*Chlamydomonas hedleyi*, *Alexandrium* sp. (*A. catenella*, *A. minutum*, *A. tamarense*, *A.excavatum*), *Gloeodinium viscum*, *Gymnodinium catenatum*, *Calothrix* sp., *Aphanothece halophytica*	310	Wound healing agent, sunscreens and cosmetics, anti-inflammatory agent, pharmaceutical products—cure for psoriasis-like conditions	Geraldes et al. (2020), Ngoennet et al. (2018)
2.	Palythine	*Alexandrium*sp. (*A. catenella*, *A. minutum*, *A. tamarense*, *A. excavatum*), *Acetabularia mediterranea*, *Lyngbya* sp., *Pseudococcomyxa* sp., *Gyrodinium dorsum*	320	Sunscreens and cosmetics, antioxidant	Geraldes et al. (2020), Fuentes-Tristan et al. (2019)
3.	Shinorine	*Chlamydomonas hedleyi*, *Alexandrium* sp. (*A. catenella*, *A. minutum*) *Gymnodinium catenatum*, *Gyrodinium dorsum*, *Microcystis aeruginosa*, *Scytonema*sp., *Calothrix* sp., *Acetabularia mediterranea*, *Lyngbya* sp.	334	Wound healing agent, sunscreens and cosmetics, anti-inflammatory agent, antioxidant, pharmaceutical products—cure for psoriasis-like conditions	Geraldes et al. (2020), Fuentes-Tristan et al. (2019)
4.	Asterina-330	*Lyngbya* sp., *Alexandrium* sp. (*A. catenella*, *A. minutum*, *A. tamarense*, *A. excavatum*)	330	Antioxidant	Fuentes-Tristan et al. (2019)

(continued)

Table 5.2 (continued)

Sl. No.	Types of MAAs	Microalgal/ cyanobacterial sources	Absorbance λ_{max} (nm)	Applications	References
5.	Mycosporine-γ aminobutyric acid	*Nostoc commune*	310	Pharmaceutical drugs help to reduce blood sugar levels, hypertension, and tumors	Nazifi et al. (2015), Li and Guo (2018)
6.	Mycosporine-taurine	*Synechocystis* sp.	309	Antioxidant, sunscreens	Zhang et al. (2007)
7.	Dehydroxylusujirene	*Synechocystis* sp.	357	UV-A filter, antioxidant	Zhang et al. (2007)
8.	7-O-(β-arabinopyranosyl)-porphyra334	*Nostoc commune*	335	Antioxidant, sunscreens	Nazifi et al. (2015), Li and Guo (2018)

5.2.2 *Applications of MAAs*

Photoprotective chemical MAAs have a good photostability and resistance to a variety of abiotic stresses (Rastogi et al. 2014). It is vital for photoprotection of organisms exposed to high radiation because of their photoprotective nature. MAAs production or accumulation in response to UV-A/B radiation in numerous species strongly implies that they have a role in decreasing the harmful effects of UV radiation. UVR protection is provided by MAAs not to their producers only, but also to primary and secondary levels of consumers across the food chain. In cyanobacteria, MAAs can block three out of every ten photons from striking their cytoplasmic targets (Garcia-Pichel et al. 1993). MAAs play a role in maintaining genome integrity by reducing the development of thymine dimers caused by UV exposure (Misonou et al. 2003).

MAAs might be employed as broad-spectrum UV protectors in cosmetics and other cosmeceutical sectors. UV-induced skin cancer can also be prevented using MAAs. Red algae MAAs (shinorine and porphyra-334) were found to protect premature skin ageing (Daniel et al. 2004) and also developed to protect human fibroblast cells against UV-induced cell death (Oyamada et al. 2008). Some MAA analogues, including tetrahydropyridine derivatives, have been developed for use in sun care products. The ability of MAAs to quench different oxygen derivatives makes them efficient antioxidants. Strong free radical scavengers were discovered in mycosporine-glycine obtained from *Arthrospira* strains and shinorine with an unknown MAA identified as M-307 recovered from *Gloeocapsa* sp. (Rastogi et al. 2014). Singlet oxygen produced by some endogenous photosensitizers was able to be scavenged by these compounds (Nakayama et al. 1999; Suh et al. 2017). Certain

MAAs such as shinorine, porphyra-334, palythine and asterina-330 were shown to have dose-dependent antioxidative activity in terms of scavenging hydrosoluble radicals, which increased with the basicity of the medium (pH 6–8.5) (De la Coba et al. 2009).

5.2.3 Scytonemin

In 1877, Na¨geli and Schwendener developed the name "scytonemin" to describe the cyanobacterial sheath's yellowish-brown pigment (Nägeli and Schwendener 1877). Scytonemin is a dimeric, lipid-soluble, aromatic indole alkaloid that is developed by the condensation of tyrosyl- and tryptophanyl-derived subunits that are linked through a carbon-carbon bond (Proteau et al. 1993). It is the second most important family of UV-absorbing chemicals in cyanobacteria and protects against UVR by absorption at 252, 278, 300 and 384 nm (Table 5.3). Based on redox circumstances, it occurs in two interconvertible forms, that is, reduced (Mw 546 Da) and oxidized (Mw 544 Da). It is a one-of-a-kind natural substance made up of phenolic and indolic subunits connected by an olefinic carbon (Wada et al. 2013). This connection gives scytonemin a novel ring structure, which Proteau et al. term "the scytoneman skeleton." It is partially soluble in organic solvents, and reduced form is brilliant red in color, which is more soluble in organic solvents (Garcia-Pichel and Castenholz 1991). Later, cyanobacteria produced a number of scytonemin derivatives, including scytonine, tetramethoxy-scytonemin, scytonemin-imine, and dimethoxy-scytonemin (Grant and Louda 2013).

Scytonemin is synthesized solely by cyanobacteria that can create extracellular polysaccharides. In vivo, scytonemin's complex ring structure allows high absorption over the UV-A, UV-B, UV-C and violet-blue spectral areas with a maximum of approximately 370 nm. Scytonemin has a high extinction coefficient

Table 5.3 Comparison between light-absorbing abilities and physical appearance of different classes of cyanobacterial scytonemin

Scytonemin	Absorbance at the wavelength (nm)	Color of pigment	References
Oxidized scytonemin	265, 278, 374, 393, 682, and 670	Yellowish brown	Varnali and Edwards (2014)
Reduced scytonemin	572, 474, 378, 314, 276, and 246	Bright red	Varnali and Edwards (2014)
Scytonine	270, 225, and 207	Brown	Varnali and Edwards (2014), Bultel-Poncé et al. (2004)
Dimethoxyscytonemin	422, 316, and 215	Dark red	Varnali and Edwards (2014)
Tetramethoxyscytonemin	562 and 212	Purple	Varnali and Edwards (2014), Bultel-Poncé et al. (2004)
Scytonemin-3a-imine	564, 437, 366, and 237	Reddish brown	Grant and Louda (2013)

(14,250 Lg^{-1} cm^{-1} at 384 nm), making it an effective photoprotective chemical (Wada et al. 2013). Pure scytonemin has a maximum absorption at 386 nm; other than that, it also absorbs substantially at 252, 278 and 300 nm, indicating that this could aid cyanobacteria in surviving UV exposure (Kumari et al. 2021, 2022). Scytonemin was found to decrease UV-A light penetration into the cell by 90%, and it alone has also been shown to provide significant protection against UV-C radiation (Rastogi et al. 2014). Furthermore, scytonemin long-term stability in cyanobacterial biocrusts or dried mats subjected to strong sun radiation implies that it has outstanding photostability and can be used in cosmetics (Fleming and Castenholz 2007).

5.2.4 *Applications of Scytonemin*

Scytonemin, like MAAs, plays an important function in photoprotection. Scytonemin UV-screening role in cyanobacteria has been extensively documented (Rastogi et al. 2015). Scytonemin can reduce cellular ROS and thymine dimer formation, and it has a lot of potential pharmacological properties, including anti-inflammatory and antiproliferative properties (Stevenson et al. 2002). Spectrophotometric analyses of the pigmented sheath's transmittance and the suppression of UV stimulation of phycocyanin fluorescence reveal that scytonemin is efficient in protecting the cells from incoming UVR, reducing UV-A radiation entrance to roughly 90% (Garcia-Pichel et al. 1993). Scytonemin has been found to account for more than 5% of the total cellular weight in cyanobacteria (Castenholz 1997). It is resistant to a variety of stressors, including strong UVR and temperature, and scytonemin also performs its screening function without requiring any additional metabolic investment, even when subjected to prolonged physiological activity (Sinha and Häder 2008).

Scytonemin as a sunscreen offers an advantage over manufactured sunscreen chemicals as it has developed via a selection process, making it suitable for human usage (Gao and Garcia-Pichel 2011). The use of scytonemin as a sunscreen chemical for skin protection, as demonstrated by Karlsson, has piqued dermatologists' curiosity. Scytonemin could be utilized as an antiproliferative and anti-inflammatory medication. It can decrease the proliferation of human endothelial, fibroblasts, and tumor cells in addition to UV protection.

5.3 Photoprotective Compounds in Fungi

In symbiotic interactions with plants and other microbes, fungi are microscopic creatures that are important for the cycling of nutrients in natural ecosystems. They thrive despite being exposed to a variety of abiotic and biotic stressors in the natural environment. Exposure to solar UV radiation has a significant impact on their growth, survival, conidia formation, pathogenicity, germination, virulence, and bioactive chemical synthesis. Different adaptation processes are utilized to preserve cells and retain DNA integrity, allowing them to survive in natural habitats when exposed to solar UV irradiation. The time of exposure, wavelength and irradiance of incoming photons that affects the cells determine the effects of solar radiation on fungus (Fuller et al. 2015). Damage can be caused at the cellular level by direct UV absorption or indirectly. Most of the time when intermediate molecules absorb UVR, many different types of reactive oxygen species (ROS) are produced, which destroy other biological components (Cadet et al. 2015). It leads to the formation of pyrimidine dimers, which can disrupt transcriptional and cellular activities (Yagura et al. 2017). Other components of the cell, such as aromatic amino acids, absorb UV light to a lower extent, resulting in photo-fragmentation, enzymatic specificity loss and other biological activities (Grégoire et al. 2009).

Prokaryotes and eukaryotes share many mechanisms to protect themselves from environmental stress, including (1) carotenoid and melanin (protective pigment) production (Cordero and Casadevall 2017), (2) modulation of saturated fatty acid in cell membranes (Keyhani 2018), (3) the presence of inducible DNA repair mechanisms (Sinha and Häder 2008), (4) making enzymes that protect against oxidising agents (Lushchak 2011), (5) the regulation of endogenous trehalose and mannitol and (6) different stress-related signalling proteins' expression (Hinnebusch and Natarajan 2002).

5.3.1 Melanins

Melanins are high-molecular-weight hydrophobic pigments made up of several types of phenolic or indolic monomers that are frequently found with proteins or carbohydrates. Eumelanin (black), allomelanin (yellow or red) and pheomelanin (yellow or red) are the three types of melanin found in mushrooms (Belozerskaya et al. 2017). The L-3,4-dihydroxyphenylalanine (L-DOPA) synthesis/shikimic acid pathway and 1,8-dihydroxynaphthalene (DHN) synthesis/acetate-malonate pathway are the two processes used by fungi to produce melanin. The tyrosinase or laccase firstly oxidize phenylalanine or tyrosin into DOPA quinone, then transform into dihydroxyindole (Belozerskaya et al. 2017; Singh et al. 2013). DOPA-quinones attach to cysteines and polymerize to create pheomelanin in the presence of cysteines. In *Yarrowia lipolytica, Aspergillus fumigatus* and *A. kawachii*, another DOPA route involving tyrosine transaminase is present. Recently, homogentisic acid has been discovered, which is responsible for the synthesis of soluble piomelanin. The acetate-malonate pathway is used by several filamentous melanogenic fungi to produce melanin (Braga et al. 2015). The existence of numerous kinds of melanins throughout a wide range of fungal taxa as well as other creatures strongly suggests that melanogenesis is important in evolution (Singh et al. 2013). Melanins may absorb a wide variety of wavelengths of radiation, including gamma rays, X-rays, and UV light. As a result, many fungi now use them as superb photoprotectants (Huijser et al. 2011).

Exophiala (Wangiella) dermatitidis, Cryptococcus neoformans, Aspergillus versicolor and *Alternaria alternata* are among the melanotic fungi that have achieved radiation adaptation by proceeding toward the radiation sources and using that energy for metabolic changes and growth (Shuryak et al. 2014). Two eumelanin-building components, 5,6-dihydroxyindole-2-carboxylic acid (DHICA) and 5,6-dihydroxyindole (DHI), have been shown to absorb and convert photon energy into other types of energy, minimizing UV radiation damage (Huijser et al. 2011). In the UVR-exposed melanotic fungus, *Cryptococcus neoformans* resulted in lower ATP levels, which might be due to NADH oxidized by irradiated melanin or the available ATP was used in cellular metabolism UV-induced DNA repair and growth (Bryan et al. 2011). Melanization occurs outside the plasma membrane in fungi and melanin granules are frequently connected with or found in the cell wall; the mechanism of depositing these extracellular granules, however, is yet not known (Jacobson and Ikeda 2005). According to new research, melanin production in *Aspergillus* begins in intracellular endosomes which are released to the cell wall utilizing an unusual mechanism, where the melanin biosynthesis enzymes, laccase and copper syntheses can be clustered (Upadhyay et al. 2016). Melanin has also been linked to increased UV resistance in *Sporothrix schenckii* and has been reported to have a role in preserving cell wall integrity and function in *A. fumigatus* and *Fonsecaea pedrosoi* (Bayry et al. 2014).

5.3.2 Carotenoids

Carotenoids are lipophilic tetraterpenoids with an eight-isoprenoid unit conjugated double-bond structure on a 40-carbon aliphatic polyene chain (Avalos and Carmen Limón 2015). The arrangement and chromophore length of the conjugated double bond control the colour and spectrum absorption of the carotenoid. The chemical structure and features of carotenoids, such as the oxygenation and hydrogenation of the carbon polyene chain as well as the length of the chromophore, are used to classify them (Meléndez-Martínez et al. 2015). The most studied carotenoid molecules are astaxanthin, carotene and neurosporaxanthin, which have been found in a variety of fungal phyla (Avalos and Carmen Limón 2015). The isoprenoid route, which involves many condensation stages from isopentenyl pyrophosphate (IPP) to geranylgeranyl pyrophosphate (GGPP), is used to make fungal carotenoids. In the presence of phytoene synthase (PSY), a 40-carbon phytoene is generated by isomerization and condensation of two GGPP units which is then cyclized with -ionone rings to yield the common structure of carotenoids (Saelices et al. 2007).

Astaxanthin is made from carotene by adding keto and hydroxyl groups to the molecule's rings. In the desaturation step, neurosporaxanthin production follows a different route than carotene and astaxanthin (Avalos and Carmen Limón 2015). Other carotenoid compounds, such as torularhodin, have been found in a variety of fungal species to have promise photoprotective characteristics, increasing the lifespan of yeast cells mostly due to their antioxidant capabilities (Moliné et al. 2010).

Most carotenoids absorb sun radiation largely in the visible band and appear to protect fungal cells mostly through quenching reactive oxygen species (ROS) (Yan et al. 2011). Although there are obvious variations in carotenoid accumulation between fungal species and treatments, higher carotenoid content enhances the resistance and survival rate of UVR-exposed cells. Phytoene and phytofluene are colourless compounds with linear hydrocarbons having short conjugated double bonds that absorb best at 347 and 365 nm (phytofluene) and 286 and 331 nm (phytoene) (Meléndez-Martínez et al. 2015). In comparison to other carotenoid compounds, the ability of these compounds to absorb UV wavelengths is exceptional. The administration of significant dosages of phytoene to guinea pigs over a 2-week period has been demonstrated to improve their tolerance to UV-induced erythema (Mathews-Roth and Pathak 1975).

5.3.3 Mycosporines

Mycosporines (MYCs) and mycosporine-like amino acids (MAAs) are colourless and have a molecular weight of 400 Da. Their absorption maxima range from 310 to 360 nm (maximum typically close to 310 nm). They were initially discovered in fungus during morphogenesis and sporulation when they were exposed to UV light (Bhatia et al. 2011). Due to their capacity to absorb UV wavelengths and antioxidant characteristics, they have lately been hypothesised to play a role in photoprotection. MYCs are distinguished from MAAs by the presence of an amino cyclohexenone unit rather than a cyclohexanone unit attached to an alcohol group or an amino acid. The shikimic acid route, which uses 3-dehydroquinate as a precursor, is one proposed method for mycosporine production in fungi (Bhatia et al. 2011).

Many microfungi appear to manufacture MYCs exclusively, with mycosporine-glutamicol-glucoside and mycosporine-glutaminol-glucoside found in the majority of the fungi investigated. At 310 nm, these two substances have comparable absorption maxima. MYCs have an absorption maximum at 310 nm, which makes them UV-protectant compounds, notably against damage caused by UV-B radiation. The negative correlation between CPD accumulation and MYC accumulation as well as the large coefficient of determination between MYC concentration and survival shows that other defensive mechanisms, presumably to protect against ROS, may be implicated (Libkind et al. 2011).

MYCs safeguard cells by dissipating photon energy from radiation without forming reactive oxygen species (ROS) (Conde et al. 2004). MYCs have also been shown to exhibit antioxidant activity by quenching singlet oxygen, and their performance in this regard is equivalent to that of the MAA, mycosporine-glycine (Moliné et al. 2010). MYCs are found in the extracellular polysaccharides (EPS), also known as mucilages, that wrap conidia of certain fungi, supporting survival by shielding conidia from sun radiation during aerial dispersal and avoiding premature germination (Leite and Nicholson 1992).

5.4 Photoprotective Compounds in Lichens

Lichens are symbiotic organisms with a complex structure that are exposed to UV radiation, extreme temperatures, desiccation etc. These poikilothermic species have developed photoprotection methods such as dispersion, radiation screening, heat dissipation, antioxidant defence membrane repair. With over 1000 secondary metabolites identified, these unique organisms synthesize a diverse range of compounds. UV-screening metabolites such as depsidones, depsides and diphenyl ethers as phenolic compounds, xanthones, anthraquinones, or shikimic acid derivatives are synthesized as key defensive strategies for lichens (parietin, mycosporines, scytonemin). The research for novel sunscreens is still essential due to the detrimental effects of UV-A wavelengths of sunlight. We provide here a review of UV protectants found in lichens and associated symbiotic partners such as lichenized fungi, cyanobacteria and a green alga. Lichens synthesise UV filters such as depsides, depsidones, diphenyl ether, bisxanthones, mycosporines and MAAs, scytonemin as well as melanin, carotenoids as traditional colors. We suggest classifying these compounds based on their chemical structures and reviewing the physicochemical qualities that make them UV filters. While UV-B screens are the most prevalent complex and multidimensional aromatic compounds produced by lichens, they also create UV-A filters such as parietin (anthraquinone derivative), bisxanthones (secalonic acids), together with that other common photoprotective compounds like scytonemin or mycosporines and MAAs were found.

5.4.1 *Orsellinic Acid Derivatives*

The polyketide pathway provides the primary aromatic compounds synthesized by lichens, which correspond to orsellinic acid derivatives, which are distinguished by the ortho-hydroxy carbonyl chromophore unit. The bonding of two or three orcinol or b-orcinol phenolic groups through ester (depsides, tridepsides), ether (diphenyl ethers, depsidones, dibenzofurans) or C–C links (depsones) produces these compounds (Elix and Stocker-Wörgötter 2008). Some studies have found that phenolic compounds accumulate more in lichens when exposed to sunlight or UV-B radiation especially in the higher regions, implying that they act as UV absorbers (Swanson and Fahselt 1997; Hall et al. 2002). Huneck and Yoshimura (1996) have detailed the UV characteristics of numerous types of compounds (Table 5.4) in an important review concentrating on the structural identification of lichen compounds. The majority of UV spectra were obtained in polar solvents such as methanol or ethanol, which may have a considerable impact on UV profiles. While the metabolites like depsides, depsidones, dibenzofurans, chromones, and diphenyl ethers have UV-B properties, xanthones and pulvinic acid derivatives have UV-A properties, with extinction values exceeding 10,000 L mol^{-1} cm^{-1} for the lower energy absorption bands.

Table 5.4 UV properties of various lichen compounds (solvents used: MeOH and EtOH) (Huneck and Yoshimura 1996)

Classes of compounds	λ_{max} (nm) (range)	Extinction coefficient (ε) (L mol^{-1} cm^{-1}) (range)	Chemical structures
Depsides	247–276 281–318	5012–32,360 3467–29,512	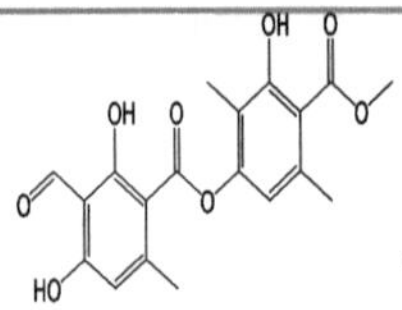 **Atranorin** 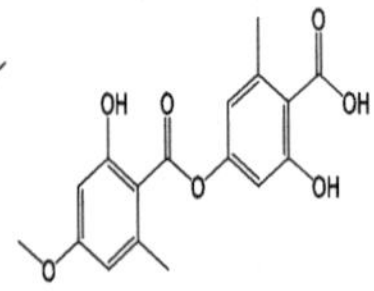 **Evernic acid** 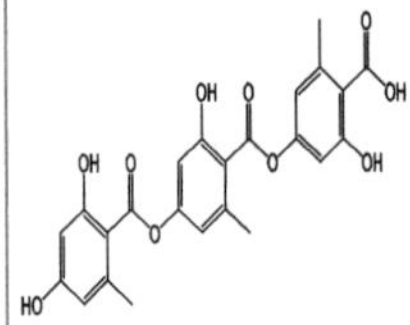 **Gyrophoric acid** 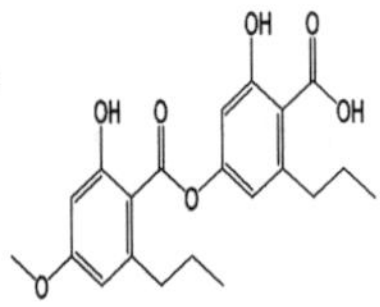 **Divaricatic acid**
Depsidones	250–270 295–333	8710–39,810 1995–18,620	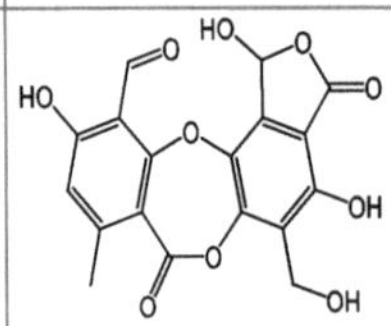 **Salazinic acid** 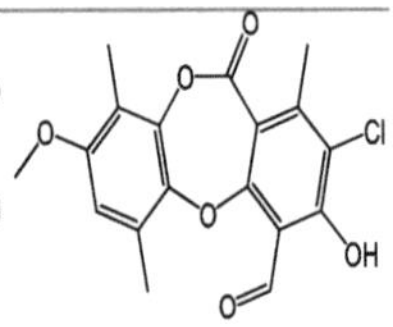 **Pannarin** 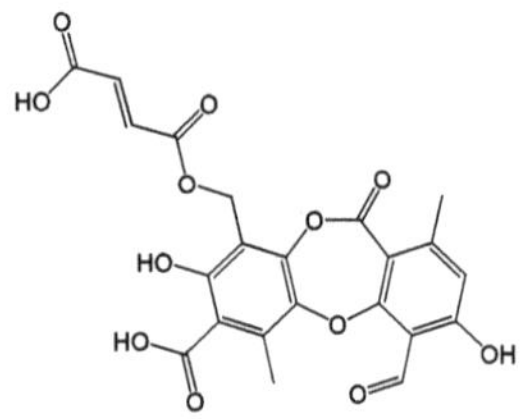 **Fumarprotocetraric acid**
Dibenzofurans	235–249 256–275 295–338 282–290	13,490–60,256 16,982–35,481 3715–19,055 21,380–32,359	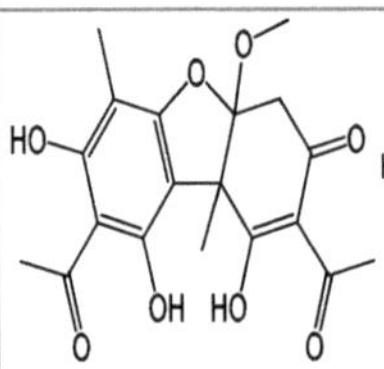 **Placodiolic acid** 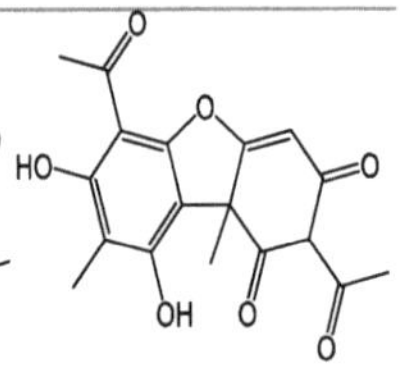 **Usnic acid**
Xanthones	240–256 260–284 308–323 340–372	21,878–70,795 5888–23,988 10,000–34,674 3981–32,359	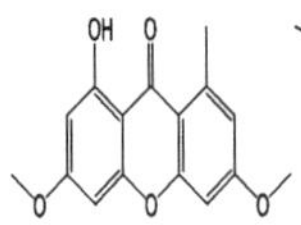 **Lichexanthone** 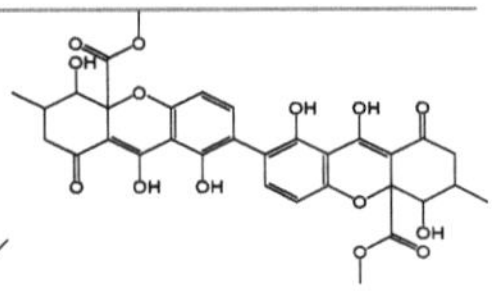 **Secalonic acid A**

(continued)

Table 5.4 (continued)

Classes of compounds	λ_{max} (nm) (range)	Extinction coefficient (ε) (L mol^{-1} cm^{-1}) (range)	Chemical structures
Quinones	205–235 250–295 308–387	7079–70,795 13,183–57,544 10,233–22,387	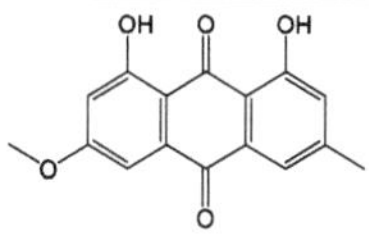 **Parietin** 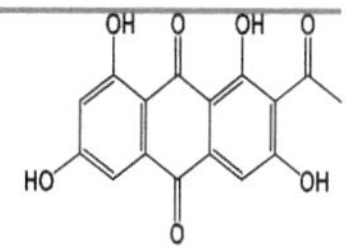 **Haematommone**
Shikimic acid derivatives	236–291 316–389	11,749–54,954 13,183–31,623	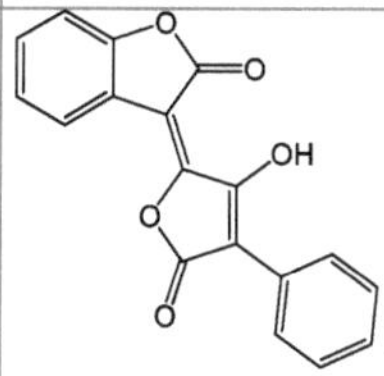 **Calycin** 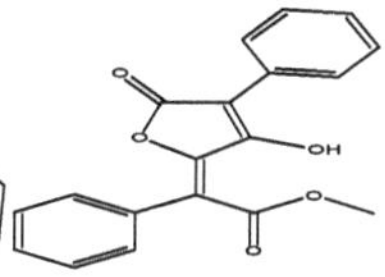 **Vulpinic acid** 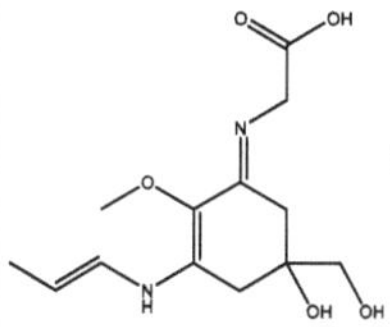 **Mycosporine- glycine** 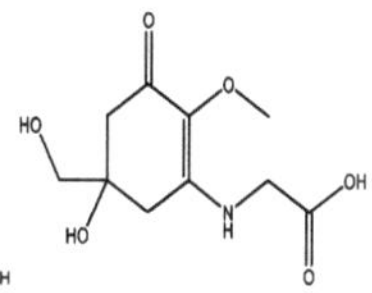 **Palythene** 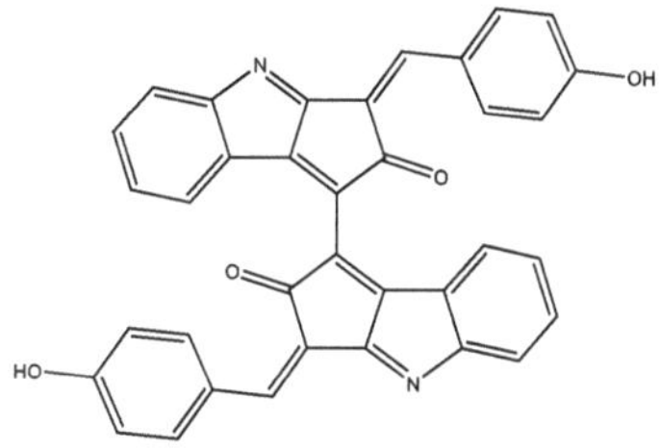 **Scytonemin**

5.4.1.1 Depsides

The majority of lichen aromatic compounds are depsides and depsidones (Elix and Stocker-Wörgötter 2008). Depsides are produced when two depsides or b-orcinol units or three tridepsides are esterified and absorbed in the UV-B zone. In terms of absorption peak, depsides and monoaromatic compounds have comparable spectral features, with the shorter wavelength's intensity generally greater than the longer wavelengths. While methylation and extending the alkyl chain have no influence on the UV spectra, functionalizing the orcinol unit with a formoyl or hydroxyl group (at position 3) results in significant alterations in UV absorption. Atranorin, divaricate, and evernic acids as well as the tridepside gyrophoric acid (Table 5.4) have been found to operate as UV filters among some of the depsides already recognized. Under high light intensities enhanced by UV-A and UV-B, a growing buildup of lichen compounds such as the formation of atranorin in *Cladonia rangiferina* has been reported (Begora and Fahselt 2001). In comparison to the shaded specimens, such metabolite was exclusively found in the sunny specimens, indicating a biological function as a UV screen (Millot et al. 2007). In fact, "atranorin absorbs UV-B with a moderate absorption coefficient (λ_{max} ¼ 252 (3 ¼ 14 454 L mol^{-1} cm^{-1}), 312 nm (3¼ 3715 L mol^{-1} cm^{-1})" (Huneck and Yoshimura 1996) and "dissipates it as fluorescence (λ_{max} emission ¼ 425 nm) within the region in which chlorophyll absorbs" (Rao and LeBlanc 1965). After exposure to high levels of UV-A and UV-B, the photostability of certain depsides such as barbatic acid and atranorin has recently been observed (Begora and Fahselt 2001).

In terms of photoprotection, these depsides performed poorly (or "moderate SPF (Sun Protection Factor)") (Fernández et al. 1996). These findings, together with the fact that atranorin has a mild allergenic potential (Sandberg and Thune 1984), raise concerns about its usage as a sunscreen. The tridepside gyrophoric acid (Begora and Fahselt 2001; Lohézic-Le Dévéhat et al. 2013), a UV-B filter with a high absorption coefficient mainly due to the chromophore interaction of the three orcinol units, seems to have the greatest SPF value. Quilhot et al. (1994) It was noted that although the examined depsides had low fluorescence quantum yields, their large concentration (>5% dry weight) in certain lichen thallus might overcome this inefficiency.

5.4.1.2 Depsidones

Depsidones have spectral properties that are quite similar to depsides. The orsellinic acid derivatives are synthesized through ester and ether bonds between two orsellinic or b-orsellinic acid units, giving them the ability to absorb UV-B and some UV-A rays (Table 5.4). As previously stated, the existence of an ortho may alter the UV spectral properties. The spectral features of depsidones are quite similar to those of depsides. Orsellinic acid derivatives are produced by forming ester and ether linkages between two orsellinic or b-orsellinic acid units, allowing them to absorb UV-B and some UV-A radiation (Table 5.4). The spectral features of depsidones are quite similar to those of depsides. Orsellinic acid derivatives are created by forming ester and ether linkages between two orsellinic or b-orsellinic acid units, allowing them to absorb UV-B and some UV-A radiation (Table 5.4), hydroxy benzaldehyde unit, for example, in pannarin and salazinic acids, and fumar protocetraric, resulting in absorption at shorter wavelengths (308–321 nm) owing to hydrogen bonding. The substitution order on the aromatic molecules has a significant impact on UV characteristics and may be predicted using TD-DFT (time-dependent density-functional theory) calculations, as demonstrated by chlorinated compounds from *Diploicia canescens* (Millot et al. 2012). Alkylation with a lengthy aliphatic chain such as in lobaric acid (Hidalgo et al. 2005) and an extra lactonic cycle in variolaric acid (Lohézic-Le Dévéhat et al. 2013), results in considerable but low-intensity UV-B absorption. This result is consistent with the presence of more variolaric acid in shaded *Odostomia parella* samples than those in direct sunlight (Millot et al. 2007). These findings imply that, by trapping free radicals and singlet oxygen, they can boost their photoprotection characteristics for photobionts, with depsidones having a greater effect than depsides (Hidalgo et al. 1994). Thus, the capacity of salazinic acid to scavenge superoxide radicals was observed, and it also proved to be non-phototoxic.

5.4.2 Dibenzofurans and Derivatives

After depsides and depsidones, dibenzofurans are indeed the third most prevalent class of orsellinic acid derivatives in lichens (Elix and Stocker-Wörgötter 2008). This category is classified into compounds created by the condensation of orsellinic acid and resorcylic acids, molecules formed by the combination of two methyl-phloroacetophenones and derivatives formed by the coupling of two methyl-phloroacetophenones, all of which are derived via the polyketide route. When compared to dibenzofurans, the reduction of aromaticity in the second ring seen in usnic acid and derivative (Table 5.4) implies a reduction in the higher wavelength bands' absorption intensity (Table 5.4). Usnic acid, the most common and widely used chemical in this family, is an excellent UV-B absorber (λ_{max} = 287 nm and ε = 18, 600 L mol^{-1} cm^{-1}) (Lohézic-Le Dévéhat et al. 2013). The authors emphasized that usnic acid retains a good UV-B absorption ability following UV-B radiation that is more than that seen for OMC (octyl-methoxycinnamate), implying its relative photostability. Thus, the fluorescent photoproducts generated following usnic acid irradiation (Rancan et al. 2002) have UV-B absorption capabilities similar to usnic acid, with an extra band at the lower energy bands "(λ_{max} 241, 294 and 344 nm)" (Fernández et al. 2006).

Therefore, usnic acid photodegradation happened at a particular radiation threshold and in the nucleophilic solvents, resulting in the generation of decarboxylated photoproducts with identical photoprotection capacity (similar in vitro SPF values) (Fernández et al. 2006).

As previously stated, one criterion for being a photoprotectant is increased synthesis in the presence of UV light. The association between the rise of usnic acid rates and light radiation augmented with UV-A and UV-B radiation has been revealed (Bjerke et al. 2002) and it is dependent on the species acclimatization to radiation from the sun before UV radiation, environmental factors and lichen structure. Furthermore, photosynthetically active radiation (PAR) in the absence of UV radiation causes the formation of usnic acid in lichens lacking a cortex, with the latter providing direct lichen protection against PAR.

5.4.3 Xanthones and Anthraquinones Derivatives

Lichens may produce other polyfunctionalized aromatic compounds including xanthones and anthraquinones through the polyketide pathway, which can also be found in fungi or plants and have the capacity to absorb UV radiation.

5.4.3.1 Xanthones

Xanthones are non-specific lichen metabolites with intriguing UV spectral properties. While chromones have four absorption bands, they absorb in the UV-B range. UV-A absorbers are xanthones, which already have three absorption maxima with strong absorption coefficients in the reduced energy absorption band (Table 5.4).

In the instance of xanthones, lichexanthone (Table 5.4) was identified in the thallus and thaliana exciple of the Brazilian tropical *H. flourescens* and juvenile mycelia after UV exposure of its grown mycobiont. The presence of these compounds may imply its protective effect as a light filter. Secalonic acids share molecular similarities with avobenzone, which belongs to the phenyl-b-diketo moiety. The enol form absorbs strongly at 340–350 nm, but the keto form absorbs at 260–280 nm (Millot et al. 2012). While avobenzone's photoprotection function is conditional (e.g. formulation), favouring the intramolecular hydrogen-bonded enol "chelated" form (Aspée et al. 2007), secalonic acid B occurs in an enol form that is less conditional.

5.4.3.2 Anthraquinones

The yellow parietin pigment (Table 5.4) is perhaps the most researched cortical lichen anthraquinone, operating as a UV absorber with high UV-B and blue light absorption. Parietin's photoprotection of the photosynthetic machinery is not seen (Solhaug et al. 2003). A strong association between parietin concentration and light site parameters confirms parietin's photoprotective activity in *Xanthoria parietina* (Gauslaa and Ustvedt 2003). Furthermore, both lichen symbionts are engaged in parietin production, and blue light filtering of parietin could be more functionally significant than UV-B screening (Solhaug and Gauslaa 2004). The results demonstrated that screening techniques differ depending on the chemical since parietin absorbs light while atranorin reflects it. Finally, in terms of photoprotective action, parietin has a poor in vitro SPF value when compared to homosalate, a commercial sunscreen chemical, with values of 1.9 and 3.9, respectively (Solhaug et al. 2010). The anthraquinones haematommone and russulone were identified in the red colour apothecia of *Haematomma stevensiae* and were interestingly detected in the mycelia of its grown mycobiont as a reaction to a UV light of 365 nm exposure, like in the earlier lichexanthone investigation. Several anthraquinones, including such perylene quinone pigments, have distinct UV-visible absorption spectra. Their conjugated heterocyclic aromatic quinone components enabled them to emit bright red fluorescence when exposed to UV and blue light. Their own spectroscopic characteristics were employed to determine their spatial distribution in lichens by confocal fluorescence microscopy. They have a major peak fluorescence of about 660 nm with a large range of 250 nm, which may explain their involvement in the lichen's energy dissipation process for light absorbed (Mathey and Lukins 2001).

5.4.4 Shikimic Acid Derivatives

The shikimic acid pathway is another major metabolic route in lichens, giving an alternate approach to create aromatic compounds produced from aromatic amino acids or molecules with UV absorption properties such as mycosporines, MAAs and scytonemin. These latter are widely dispersed in algae and cyanobacteria shikimate aromatic molecules with UV sunscreening properties correlate to pulvinic acid derivatives found only in lichens. Scytonemin, a lipid-soluble phenolic as well as indolic derivative, is formed in the sheaths of cyanobacteria, generally in conjunction with MAAs (Cockell and Knowland 1999) and complex copolymer melanin pigments, which are also seen in lichens. Melanin pigments are a class of complex pigments with biological origins that are found in all organisms, from fungus to humans. They, along with parietin, usnic acid and atranorin, are cortical lichen chemicals that act as sun radiation shields (Boustie et al. 2011). These pigments have phenolic and indolic rings that function as fundamental building blocks for the melanin polymeric matrix. The presence of these pigments suggests that stress-tolerant lichens adopt defensive measures comparable to human skin for protection against the detrimental effects of UV radiation.

5.4.5 Pulvinic Acid Derivatives

These chemicals, which include an oxolane-carbonyl chromophore group, such as calycin, vulpinic acid and rhizocarpon acid (Table 5.4), have UV profiles characterized by two absorption peaks with a strong absorption coefficient (Table 5.4) (Huneck and Yoshimura 1996). The photochemical characteristics of these compounds were investigated, and it was found that they absorb UV-A and UV-B radiations and produce fluorescence at roughly 430 nm (Lohézic-Le Dévéhat et al. 2013; Hidalgo et al. 2002). These compounds are often observed in lichens from high UV radiation habitats (e.g. calycin coexists with usnic acid in alpine *Lecanora somervellei* (Obermayer and Poelt 1992)), and their production rates rise as UV radiation levels increase.

5.5 NPQ and Xanthophyll Cycle in Lower Plants

Photosynthesis is a vital life process carried out by all green plants, which converts light energy into chemical energy. Photosynthesis is a vital life process carried out by all green plants in which light energy is converted into chemical energy. Initially, light is absorbed by the antenna pigment and then finally transferred to the reaction center where charge separation starts. In diverse organisms, reaction center may possess similarities among themselves, but antenna complexes pose great diversity. A building block of the antenna complex is highly dependent on the organism habitat. This is because organisms get to adapt and are highly dependent on the habitat of an organism (Büchel 2015). This is because of the organism's adoption of the quality and quantity of available light. However, the light-harvesting complex must be flexible enough to prevent those conditions that limit the efficiency of the reaction center, such as high light and other stressful condition.

Non-photochemical quenching is the most efficient mechanism through which excess energy can safely dissipate in the form of heat (Demmig-Adams et al. 2014). NPQ has an independent evolutionary origin in different organisms. The molecular processes of NPQ, which depend on species and the kind of antenna system, are still unclear (Ruban 2016). NPQ may have commonalities with certain organisms, which has evolutionary implications (Niyogi and Truong 2013).

5.5.1 *OCP-Mediated Quenching Mechanism in Cyanobacteria*

Cyanobacteria are the most primitive oxygenic prokaryotes characterized by phycobilisomes (PBSs) that act as light-harvesting antenna complex. Phycobiliproteins are the main compositions that form covalent interaction with linear tetrapyrrole pigments. Phycobilins as well as linker proteins are the part of OCP-mediated quenching mechanism (Adir 2005). Phycobilins protein complex has a central core consisting of allophycocyanin (APC) from where rods of phycocyanin (PC) project outs. Apart from PC, some organisms also have phycoerythrin (PE) as a constituent of rods. In cyanobacteria, an orange carotenoid protein (OCP) plays a very important role in NPQ (Wilson et al. 2006), as shown schematically in Fig. 5.1. A ketocarotenoid group is non-covalently attached to OCP and occurs in the inactive dark orange form OCP^0 by crossing both the N-terminal domains and C-terminal domains (NTD and CTD) (Fig. 5.1).

Light induces conformational change and domain rearrangement in OCP protein which results in the conversion of OCP^O into the active red form (OCP^R) (Fig. 5.1). Now the OCP will bind to the phycobilisome protein and quenches the energy from the bilins protein. However, the above-mentioned process is successfully performed under in vitro conditions and the in vivo mechanism of NPQ is still largely unknown. Recent studies based on crystallographic data provide light into the steps involved in OCP activation (Bandara et al. 2017). Because photoactivation causes a keto-enol

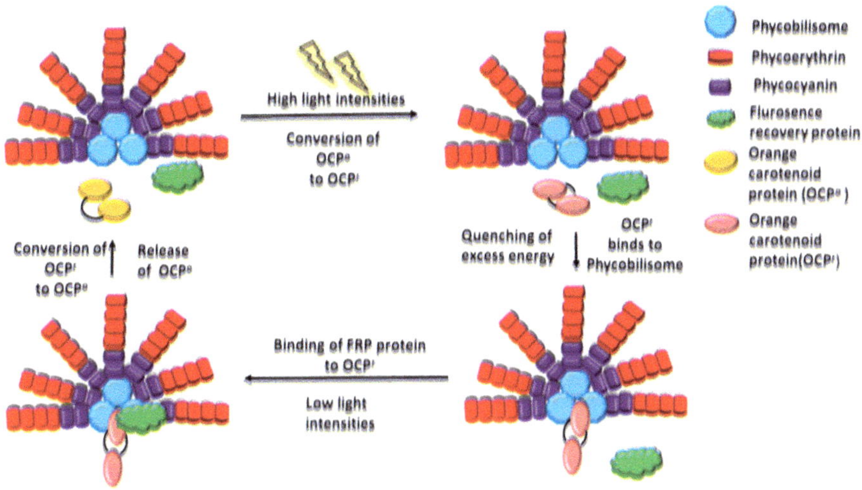

Fig. 5.1 Orange carotenoid protein (OCP)-mediated quenching mechanism in cyanobacteria. OCP^O orange form converts into OCP^R, red form under high light intensities, FRP is a fluorescence recovery protein that recovers the OCP^O from OCP^R

shift in protein, the carotenoid-1 ring's conjugated carbonyl group is no longer connected to the protein. This, in turn, causes separation between the N- and C-terminal domains. OCP has been demonstrated to burrow into the APC core of the PB, bringing the carotenoids near to the exciting bilin for quenching. The binding of OCP and PB is reversible in nature. A fluorescence recovery protein (FRP) plays a crucial role in the removal of the OCP from the PB. Despite previously binding to the CTD of OCP^R, FRP later interacts with NTD. This will form a bridge between both the domains, which results in recovery. However, the site and quenching mechanisms of carotenoids are still unknown, although many studies have been done on OCP. The binding site of the carotenoid may be the APC core, where APC 660 or 680 acts as a site of quenching. Yet there is no research done that can confirm the involvement of carotenoids in the quenching mechanism. But previously, several hypotheses have been made which suggest involvement of carotenoids in quenching. OCP, under its inactivated form, quenches singlet oxygen radicals in the thylakoid membrane in addition to important PB fluorescence quenching (Sedoud et al. 2014).

Recently, single-molecule spectroscopy suggested a novel, OCP-independent PB-quenching mechanism. According to the reports, multiple intrinsic channels are present in the different subunits of PBs, and any of them can participate in the quenching mechanism, but the protein core is the frequent target. It acts as a rapid mechanism of quenching and acts before the OCP-dependent mechanism gets activated. It involves high-light-inducible (Hli) and iron starvation-inducible (IsiA) proteins for photoprotection in cyanobacteria.

An ancestor of the LHC superfamily, that is, Hlips (small Cab-like proteins) is single helix proteins found in cyanobacteria (Engelken et al. 2012). These are important for the survival of cyanobacteria under high light irradiation and other stressful conditions. This protein is known to bind to chlorophyll *a* (Chl *a*). The protein Hlips is thought to play a photoprotective role in chlorophyll biosynthesis and PSII assembly but is not engaged in light harvesting (Komenda and Sobotka 2016).

According to research on transient absorption, energy dissipation in Hlips happens when energy is directly transferred from the excited state of Chl a Qy to the carotenoid (-carotene) S1 state (Staleva et al. 2015). This was the first research to give direct experimental evidence for such a mechanism. HliD/C also validated the energy transfer from chlorophyll a to carotenoid at ambient and cryogenic temperatures, whose conformation is distorted that reduces its S1 energy and may serve as a quencher (Wilson et al. 2007). IsiA's sequence is comparable to CP43, which contains Chl *a* and carotenoids found in PSII's core antenna (Murray et al. 2006). IsiA uncoupled from PSI quenches Chl a fluorescence, with carotenoids previously recognized as potential dissipators via Chl a Qy transfer to the carotenoid (Car) S1 phase (Berera et al. 2010; Chen et al. 2017). Recent studies show that carotenoids are not involved in the quenching but are controlled by the interaction of cysteine residue of IsiA protein and Chl *a* (Chen et al. 2017). This process shows similarities with the redox-dependent quenching mechanism of green sulphur bacteria and oxygenic photosynthetic organisms as well.

5.5.2 OCP-Mediated Quenching Mechanism in Red Algae

Red algae include two light-collecting antenna complexes, phycobilisomes and the LHCI complex, which are linked to the RCs of PSII and PSI, respectively. On the other hand, red algae lack OCP, and nothing is known about their photoprotection mechanisms. According to small-molecule fluorescence data, decoupling of PE from the PB core was recommended as a technique in *Porphyridium cruentum* (Magdaong and Blankenship 2018). State transitions requiring PB mobility are still being debated in cyanobacteria but have been demonstrated to be crucial in the mesophilic red algae *P. cruentum* and *Rhodella violacea*. In the thermophilic red algae (*Cyanidium caldarium* and *Cyanidioschyzon merolae*), the main mechanism for dissipating surplus energy is NPQ; however it is situated in the PSII response centre rather than the antenna (Magdaong and Blankenship 2018).

5.5.3 NPQ in Green Algae, Moss and Diatoms

In the case of green algae, NPQ is inducible in nature. Generally, after a few hours of the high light irradiation or a decrease in CO_2 supply, NPQ comes into effect (Minagawa and Tokutsu 2015). NPQ is the principal mechanism for excessive energy dissipation in halo-tolerant red algae (*Cyanidium caldarium* and *Cyanidioschyzon merolae*), although it is positioned in the PSII reaction centre rather than the antenna (Magdaong and Blankenship 2018). NPQ was regulated by the light-harvesting complex stress-related (LHCSR) protein (Peers et al. 2009). LHCSR binds Chls (*a* and *b*) and xanthophylls (Bonente et al. 2011). The algae *Chlamydomonas reinhardtii* has two forms of LHCSR: one that is produced constitutively and is known as constitutive LHCSR1.

The process of converting violaxanthin to zeaxanthin in the presence of violaxanthin de-epoxidase is known as the violaxanthin (V)-antheraxanthin (A)-zeaxanthin (Z) (VAZ) xanthophyll cycle, is also involved in green algae NPQ. However, zeaxanthin-dependent NPQ in green algae varies depending on the species (Magdaong and Blankenship 2018). A model for NPQ activation in *Chlamydomonas reinhardtii* was proposed where LHCSR3 expression occurs under high light conditions, then this complex is associated with PSII-LHCII and forms PSII–LHCII–LHCSR3 (Fig. 5.2). A low pH of the lumen LHCSR 3 becomes protonated and forms a quenching centre. Charge transfer from Chl to Car is the cause of quenching (Fig. 5.2). According to recent research, blue light receptor protein induces the production of LHCSR3 (Magdaong and Blankenship 2018). As the C terminal is rich in acidic amino acid residues (Asp-177, Glu-221, and Glu-224), this region of the LHCSR3 gets protonated (Ballottari et al. 2016). It shows that although LHCSR interacts with PS-II, it can also move to the PS-I center. LHCSR binds only with chl *a* and is involved in NPQ.

In vivo studies in mutant *Chlamydomonas reinhardtii* having minimal NPQ and without PSII and PS1 cell shows that LHCSR1 is pH-dependent and it will induce LHC II quenching and support NPQ in mutant lacking LHCSR3 (Ballottari et al.

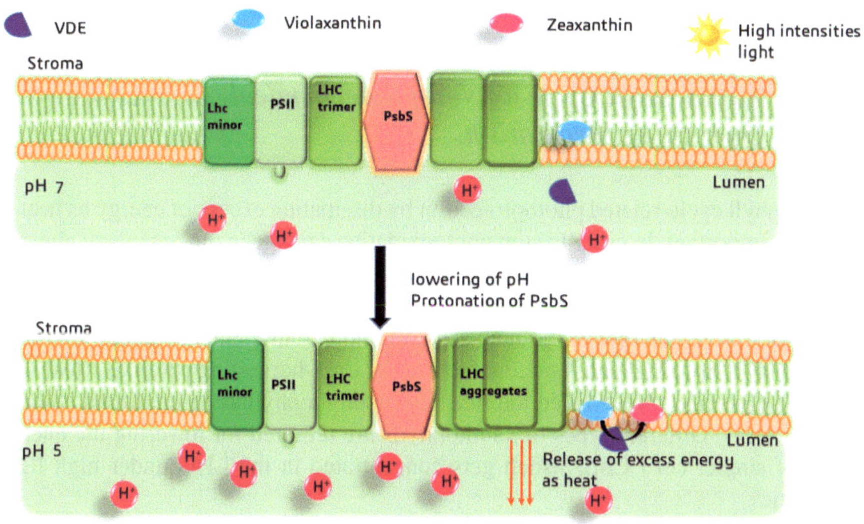

Fig. 5.2 NPQ mechanism in algae and lower plants. Formation of quenching centre upon acidification of lumen cause protonation in components of NPQ including PsbS protein, conversion of violaxanthin (*Viola*) to zeaxanthin (*Zea*) by VDE, cause the creation of quenching centres, which take part in the heat-radiating loss of surplus energy

2016). Single-molecule spectroscopy data shows the presence of two dissipative states, LHCSR1, regulated by carotenoid compositions and by pH (Magdaong and Blankenship 2018). Although gene for photosystem II subunit S (PsbS) is there in *C. reinhardtii*it does not express the same. The mechanism of induction of LHCSR 3 gets affected by PsbS in green algae. However, this one is not enough to carry out LHCSR-dependent NPQ (Magdaong and Blankenship 2018).

Lower plants such as diatoms and moss also possess the gene for LHCSR. Apart from the xanthophyll cycle, LHCSR and PsbS both play an important role in the NPQ of moss *Physcomitrium patens*. When zeaxanthin binds to LHCSR, the quenching mechanism regulated by LHCSR gets enhanced. But the same case cannot be found in green algae; diatom *Phaeodactylum tricornutum* possesses LHCX1, which regulates NPQ; LHCX1 is ancestrally related to LHCSR with some dissimilarities. Such as LHCX1 expresses in a constitutive manner. It is not sensitive to pH because it does not have residues that can undergo protonation. In *Cyclotella meneghiniana*, LHCX1 has recently been linked to antenna aggregation and altered pigment linkages (Magdaong and Blankenship 2018).

NPQ also depends upon the pH-dependent xanthophyll cycle, which involves the conversion of diadinoxanthin to diatoxanthin (DT) through de-epoxidation. After the formation of DT, it will interact with LHCX and lead to the aggregation and formation of a quenching canter. Till now, two quenching centers have been reported according to the fluorescence data: Q_1 is located in detached LHC and Q_2 is found in LHCX–DT–PSII (Magdaong and Blankenship 2018). Because diatoms live in dynamically variable light conditions, the availability of flexible and quick quenching mechanisms is critical.

5.5.4 Xanthophyll Cycle as Non-photochemical Quenching Mechanism

Xanthophyll cycle-related photoprotection by dissipating excess of energy as heat is as much important. In order to stop photooxidative damage in the organism, alternative photoprotective processes may be used, such as the production of UV-absorbing compounds (Sinha et al. 1998) or changing the orientation of the chloroplast.

In photosynthetic organisms such as algae and vascular plants, xanthophyll pigments play a structural and functional role in the light-harvesting complex. Research in algae *C. reinhardtii* revealed the function of xanthophyll as an important component of NPQ. Different types of xanthophylls are required for lowering the energy level of singlet chl (^{1}Chl), which gets concentrated in the LHC under high light intensities.

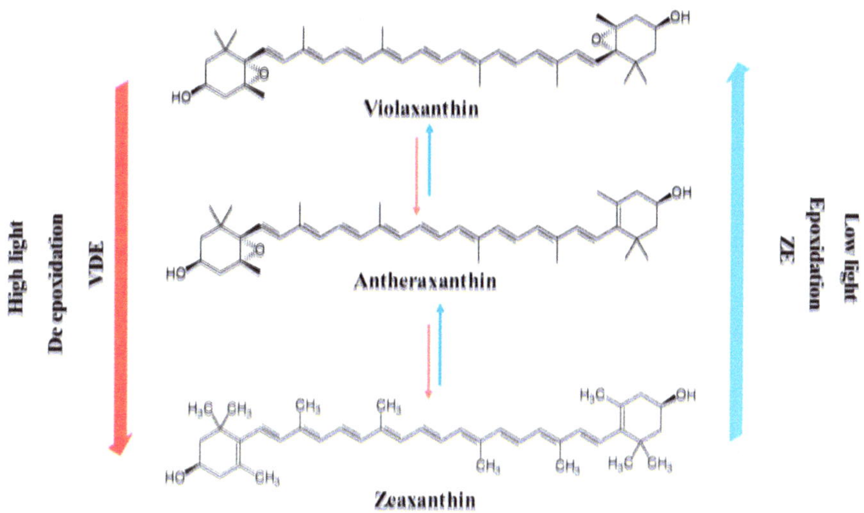

Fig. 5.3 Diagrammatic representation of violaxanthin cycle

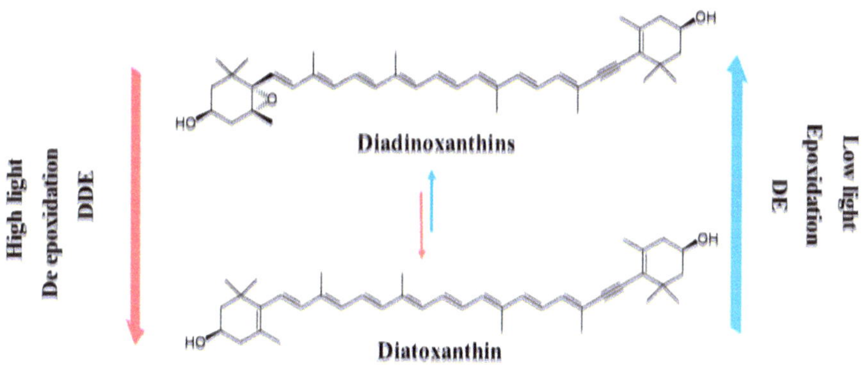

Fig. 5.4 Diagrammatic representation of diadinoxanthin cycle

5.5.4.1 The Violaxanthin Cycle

Green plants, green algae (Chlorophyta) and brown algae (Phaeophyceae) possess violaxanthin (Vx) cycle (Fig. 5.3) (Stransky and Hager 1970). There are also some algal groups known to possess both the Vx cycle and diadinoxanthin (Ddx) cycle (Fig. 5.4) (Lohr and Wilhelm 2001). However, the Ddx cycle is the major xanthophyll cycle in those species and pigments from the Vx cycle act as intermediate in the production of Ddx cycle pigments.

5.5.4.2 Mechanism of the Violaxanthin Cycle

Under high light intensities, the Vx cycle undergoes a forward reaction. This consists of two de-epoxidation stages in which di-epoxy xanthophyll Vx is first transformed to antheraxanthin (Ax) with one epoxy group and then to zeaxanthin (Zx) with no epoxy group (Fig. 5.3). But under darkness and low light intensities, the above-mentioned mechanism gets reversed where Zx again converts back to Vx through antheraxanthin (Ax) (backward reaction) (Fig. 5.3). The Vx de-epoxidase enzyme (VDE) catalyses the forward process, and the backward reaction is catalyzed by the enzyme Zx epoxidase (ZEP) located in the thylakoid lumen (Hager and Holocher 1994). Not much is known about the biochemistry of the Vx cycle in Phaeophyceae. However, the epoxidation and de-epoxidation steps of the Vx cycle are comparable to that of green algae and vascular plants. It shows the conversion of Vx to Zx induced by high intensities of light, and the reverse process is induced by low light intensities in case of brown algae (García-Mendoza and Colombo-Pallotta 2007).

5.5.4.3 Role of Vx Cycle in Photoprotection

In the case of green algae, brown algae and vascular plants, the Vx cycle is critical for photoprotection. When Vx converts to Zx, this will lead to the increased dissipation of excess energy in the antenna complex of PS II. Excess excitation energy is released as heat. This results in the prevention of damage and inactivation of a photosynthetic pigment-protein complex (Horton and Ruban 1992). The direct quenching mechanism Zx directly interacts with excited Chl *a* (Frank et al. 1994). Zx will accept excitation energy from a single state Chl *a* molecule. According to this process, Chl, a molecule in its singlet excited state, is given excitation energy by Zx, which then transfers that energy to the ground state and releases it as heat (Goss and Jakob 2010). A lower energy level of the Zx molecule's initial singlet excited state than Chl a is required for the direct quenching process, which depends on energy transfer between Chl a and the Zx molecule (Goss and Jakob 2010). The structural difference in Zx and Vx are crucial factors for the indirect quenching mechanism. Conversion of Vx to Zx brings an accumulation of LHC II (Horton et al. 2008). Protonation of apoprotein residues of the LHC II, low luminal pH and different steric hindrance structures of Zx cause aggregation. As per the LHC II aggregation model, Zx enhances aggregation of LHC II, whereas Vx inhibits the aggregation (Horton et al. 2008). LHC II accumulation converts the antenna system of PS II into a state that can dissipate excess energy expeditiously.

According to recent research, the real quenching sites in LHC II are indeed a Chl *a* dimer or a Chl *a*/lutein heterodimer. Zx binds to the V_1 site of LHC II, which is in close proximity with the quenching site. The binding of Zx regulates the magnitude of NPQ through the conformational change of the quenching center. Vx in the V_1 site of LHC II will reduce the extent of NPQ. The de-epoxide structure, that is, Zx of the xanthophyll cycle, can also impact the fluidity of the thylakoid membrane

(Gruszecki and Strzałka 1991). When Vx is removed from the binding site of LHC II, it will interact with the lipid of the thylakoid membrane. Formation of a temporary pool of unbound Vx pigments occurs during the removal of Vx and re-binding of Zx into the LCH II. Unbound Vx is able to move in the lipid phase of the thylakoid membrane. The Zx molecule makes the thylakoid membrane stiffer, which boosts the protein pigment complex's thermal stability when exposed to strong light (Havaux and Tardy 1996). Furthermore, Zx and Ax stop lipid peroxidation.

5.5.4.4 The Diadinoxanthin Cycle

Diadinoxanthin cycle (Ddx cycle) (Fig. 5.4) is another xanthophyll cycle consisting of one de-epoxidation step. Ddx cycle is found mostly in algae groups such as Bacillariophyceae, Xanthophyceae, Haptophyceae and Dinophyceae (Hager 1980; Goss and Jakob 2010). As mentioned above, algae with Ddx cycle also possess pigments involved in Vx cycle. However, pigments of the Vx cycle only come into action when algae are exposed to high light intensities for a longer period of time (Lohr and Wilhelm 2001). The Vx cycle pigments act as a precursor for Ddx and Dtx, that is, pigments of the Ddx cycle and fucoxanthin (Goss and Jakob 2010).

5.5.4.5 Mechanism of Diadinoxanthin Cycle

As with Vx cycle, Ddx cycle involves the conversion of epoxy-xanthophyll (Ddx) to carotenoid, which is epoxy-free (Dtx) (Fig. 5.4). High light intensities induce de-epoxidation, and darkness induces reverse reaction. Apart from high light intensities, prolonged darkness also induces the accumulation of Dtx (Jakob et al. 1999; Brunet et al. 2007). Ddx de-epoxidase (DDE) and Dtx epoxidase (DEP) are the enzymes that catalyze de-epoxidation and epoxidation reactions in Ddx cycle (Goss and Jakob 2010).

5.5.4.6 Role of Ddx Cycle in Photoprotection

In diatoms, the Ddx cycle represents most important photoprotective mechanisms. This cycle serves as NPQ and helps in the disposal of extra energy in the form of heat (Goss and Jakob 2010). Those algae which have Ddx cycle have a simple NPQ mechanism and are not like that in vascular plants, which possesses heterogenous NPQ (consisting of three components qE high-energy state quenching, qT a state-transition and qI a photo inhibitory state (qI). NPQ is closely related to the Ddx cycle in the case of the diatoms. In diatoms, quenching components qT is missing as diatoms do not have state transitions. Increased concentration of Ddx cycle pigment in diatoms' thylakoid membranes also indicates the role of Ddx cycle pigments in NPQ. Diatoms cultivated under high light intensities also show an increase in the concentration of Ddx cycle pigments (Goss and Jakob 2010). Apart from Ddx cycle-dependent NPQ, reaction center-type NPQ is reported in *Phaeodactylum tricornutum* in recent studies (Eisenstadt et al. 2008).

5.5.4.7 Xanthophyll Cycle in Red Algae

The presence of the xanthophyll cycle in red algae is unclear at the present time but there are species that are known to possess a high concentration of Zx. Reports also show the presence of Ax and Vx derived from Zx. There are also some studies that do not find any evidence of the xanthophyll cycle in red algae irradiated with high light and low light intensities (Carnicas et al. 1999). For example, *Gracilaria domingensis,* which has a high concentration of Ax did not convert to Zx under high light illumination for a very brief period of time. In the case of the red algae *Porphyra aumbilicalis*, the carotenoid acts as an antioxidant, protecting the photosynthetic system from damaging free radicals (Sampath-Wiley et al. 2008).

It is also unlikely that the red algae's light-harvesting complexes could undergo xanthophyll cycle-dependent quenching, known as phycobilisomes. Cyanobacteria which have phycobilisome as the main antenna system also need a special structure for efficient energy dissipation. The special structure is a carotenoid binding protein that interacts with phycobilisomes and IsiA protein which is produced under stress conditions in the case of *Synechocystis* PCC 6803. It was proposed that quenching occurs as a result of a singlet-singlet excitation energy transfer between Chl *a* and xanthophyll on IsiA aggregates (Berera et al. 2010). Here we can observe that *Synechocystis* PCC 6803 does not depend upon the xanthophyll cycle for quenching. However, it is caused by the carotenoid echinenon, which interacts with Chl *a*.

5.5.5 Evolutionary Significance of NPQ

NPQ intensities and kinetics are variable in algae belonging to the same taxa such as *Zygnema*, *Cosmarium* and *Mesotaenium* of Zygnematales, according to some reports. It has also been shown that in some species NPQ response constitutively expresses; however, in some species like *Mesotaenium* and *Chlamydomonas*, NPQ responds only under high light intensities (Bonente et al. 2011). From such reports, it was concluded that NPQ intensities and kinetics are not related to the taxonomic position, but it was more closely related to the ecological niche and adaptation of a particular organism. If NPQ response will constitutively express or will be induced by high light intensities, it may be favoured by natural selection as per the environmental condition around the organism (Gerotto and Morosinotto 2013). NPQ mechanism is important for the fitness of the photosynthetic organisms, but according to different environmental conditions, other photoprotective mechanisms are also developed by the organisms (Peers et al. 2009). Species belonging to different surroundings may have different impacts on different photoprotective mechanisms. *Klebsormidium, Cosmarium, Zygnema, Interfilum,* all have similar abilities of NPQ, but under different light conditions, they show different abilities of their growth (Gerotto and Morosinotto 2013).

When we consider the proteins which are involved in the activation of NPQ, a different picture comes into existence. Data taken from different sources depict that PSBS protein is important for plant NPQ, but algae depend on LHCSR. In comparison to green algae, diatoms that do not belong to viridiplantae more exclusively depend on LHCSR for NPQ activation as their genome does not have genes for PSBS protein. The above-mentioned data suggest that LHCSR protein plays a general role in algae NPQ; however, clear information in the case of red algae are still missing (Gerotto and Morosinotto 2013).

5.5.6 *Evolutionary Significance of Xanthophyll Cycle*

Xanthophyll cycle pigments are present in almost all species belonging to viridiplantae and mostly related to photoprotective function. Zeaxanthin modulates the rate of thermal energy dissipation and also acts as membrane stabilizers and acts as an antioxidant. These pigments get easily modified under different environmental conditions and also show high degree of plasticity and impact their evolutionary studies of them (Esteban et al. 2009).

But still, it has been shown that the xanthophyll pigments show a retrogression, and α-tocopherol shows the progression from green algae to vascular plants. It has been shown that both the xanthophyll pigments and α-tocopherol share complementary effects on photoprotective mechanisms (Esteban et al. 2009). Studies suggest that the pigments of xanthophyll and α-tocopherol appeared early in the evolution and were widespread among the viridiplantae (Esteban et al. 2009). But tocopherol is absent in unicellular algae. This shows that photoprotective pathways which are present now in higher plants are cyanobacterial in nature. Xanthophylls are the photoprotective pigments present in the ancient group of organisms such as algae, bryophytes. However, vascular plants α-tocopherol and Xanthophyll cycle both operated simultaneously. Xanthophyll cycle takes part in the energy dissipation as heat in NPQ and helps to maintain the balance of photosynthetic pigments under changing low light and high light conditions. Species with a long lifespan and high growth rate tocopherol acts as a membrane stabilizer and protects from lipid peroxidation (Esteban et al. 2009).

5.6 Photoprotective Effects of Flavonoids

All plant species contain a category of polyphenolic substances known as flavonoids, which have a benzo-pyrone structure. They are synthesized via the phenylpropanoid pathway as a 15-carbon phenylpropanoid chain. Functional hydroxyl groups present in flavonoids have antioxidant properties that are mediated by either

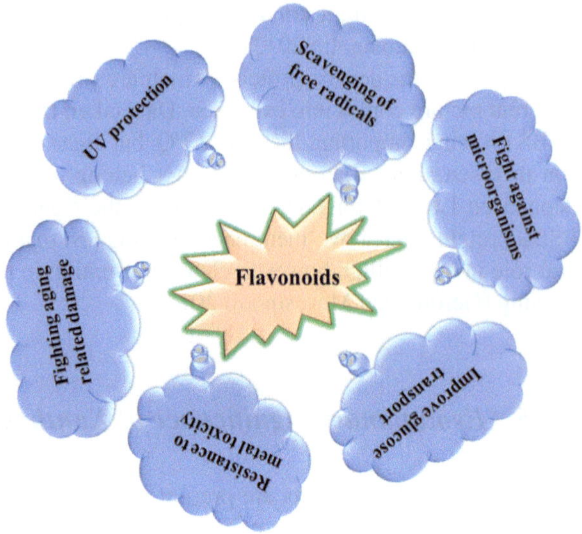

Fig. 5.5 Biological function of flavonoids

scavenging free radicals or chelating metal ions (Kumar et al. 2013). Chalcone synthase (CHS) facilitates the condensation of one p-coumaroyl-CoA molecule with three molecules of malonyl-CoA to form the first step in the production of flavonoids. Chalcone is isomerized into flavanone in the presence of chalcone isomerase (CHI). From this step, the biosynthetic pathway branches to several different flavonoid classes such as flavones, flavonols, flavanones, flavanol, isoflavone, flavanonol, anthocyanins and proanthocyanidins. The UV spectral properties of flavanones define a saturated heterocyclic C ring, with no conjugation between the A and B rings (Rice-Evans et al. 1996). They assist plants in protecting themselves from biotic and abiotic stressors because of their unique chemical composition and variability. Flavonoids have antioxidative activity due to the presence of conjugated double bonds and functional groups in the rings (Rice-Evans et al. 1996). Flavonoids reduce the production of reactive oxygen species (ROS) through elimination of singlet oxygen, quenching cascades of free-radical reactions in lipid peroxidation, inhibition of enzymes that generate ROS, chelating ions of transition metals, recycling of other antioxidants (Jovanovic et al. 1994).

Flavonoids have long been thought to have numerous photoprotective effects (Agati et al. 2012) (Fig. 5.5). Because of their strong antioxidant activity in both in vivo and in vitro conditions, flavonoids are regarded to have health-promoting characteristics (Cook and Samman 1996). Flavonoid production is almost entirely boosted by oxidative stress. They can absorb harmful UV radiations such as UV-A and UV-B and inhibit or quench the generation of ROS (Agati et al. 2012). The type of substitution occurring on various flavonoid rings has a significant impact on flavonoids' capacity to absorb UV radiation (Agati et al. 2012). It was shown that dihydroxy flavonoids that respond to light have a substantially better capacity to prevent ROS production than their monohydroxy counterparts (Rice-Evans et al. 1997; Pourcel et al. 2007). The reduction potential of flavonoids is of key significance in plants under stress conditions. Oxidative stress causes lipid peroxidation, which affects the cell membrane's integrity. Rutin (quercetin 3-*O*-rutinoside) may interact with the polar head of phospholipids, enhancing membrane rigidity; therefore, it protects membranes from oxidative damage (Erlejman et al. 2004).

5.7 Evolution of Flavonoid Genes in Lower Plants

Flavonoids may have played crucial role during the establishment of plants on the land that goes beyond their ability to block UV-B rays (Pollastri and Tattini 2011). During the evolution of aquatic to land habitat, carbon-based flavonoids replaced MAAs as UV-B screening pigments. This shift in the metabolism of early land plants suggests low nutrient availability in soils (Bonfante and Genre 2008). Flavonoids such as flavonols have a lower ability to absorb wavelengths between 290 and 320 nm than MAAs in liverworts and mosses. As a result of the evolution of distinct branches of the general phenylpropanoid pathways, the early plants colonized the land to perform UV-B screening tasks (Ferrer et al. 2008). Gene duplication's significance in the evolution of new characteristics and the relative relevance of structural and regulatory genes in the development of ecologically significant characters have both been studied using the structural and regulatory flavonoid pathway genes as a model system. The distribution of different types of flavonoids and flavonoid enzymes provides a crucial clue to the gradual evolution of the flavonoid pathway. Land plants are believed to be evolved from organisms like green algae (Charales). Flavonoids are widespread in algae, and the concentration of flavonoids increases during stress conditions. Bryophytes (mosses, liverworts, and hornworts) represent the earliest plants to colonize the land.

Chalcone synthase (CHS), chalcone-flavanone isomerase (CHI), and flavanone 3-hydroxylase are the first three enzymes in the flavonoid pathway, and they are responsible for producing the three forms of flavonoids that these plants produce: chalcones, flavonols, and flavones (F3H) (Fig. 5.6). All three enzymes appear to have been produced from genes coding for primary metabolic enzymes through gene duplication. Strong sequence similarity between bacterial genes coding for polyketide synthases involved in fatty acid production and chalcone synthase may be seen (Verwoert et al. 1992).

The oxoglutarate-dependent dioxygenase family of enzymes includes flavanone 3-hydroxylase, which is produced by the duplication of one of its members (Winkel-Shirley 2001). However, several bacteria and fungi have been found to have enzymes with identical sequences and secondary structures (Gensheimer and Mushegian 2004). As plants began to colonize land, flavonoids emerged as an excellent sunscreen to defend against UV radiation. This finding implies that even basic flavonoids like chalcones, aurones, and flavanones substantially absorb UV rays (Lois and Buchanan 1994). Chalcone synthase evolved first, followed by flavanone 3-hydroxylase and then chalcone-flavanone isomerase as chalcone synthase seems likely to be the first enzyme in the biosynthetic pathway. *Physcomitrella patens*

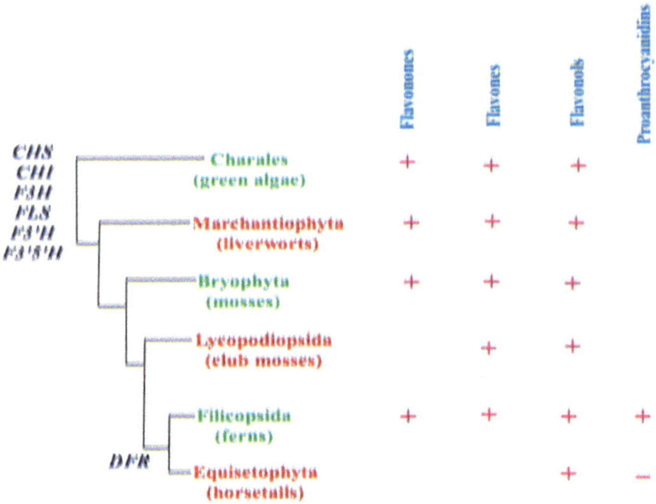

Fig. 5.6 Land plant phylogeny showing the evolution of flavonoid enzymes. +, Documented presence of flavonoid; —, possible evolutionary loss of flavonoid. Enzymes are listed in bold italics. *CHS* chalcone synthase, *CHI* chalcone-flavanone isomerase, *F3H* flavanone-3-hydroxylase, *FLS* flavonol synthase, *F3′ H* flavonoid 3′ hydroxylase, *F3′5′H* flavonoid 3′5′ hydroxylase, *DFR* dihydroflavonols. (Modified from Savolainen and Chase 2003)

(moss) contain sequences like flavonoid 3′ hydroxylase (F3′H) and flavonoid 3′ 5′ hydroxylase (F3′5′H), which are similar to the gene present in higher plants. The emergence of flavonol synthase appears to be another enzymatic innovation that occurred at this time. Flavonol synthase (*FLS*) enzymes synthesize flavonols from dihydroflavonols which are present in all bryophytes. Flavonol synthase is derived from the 2-oxoglutarate-dependent dioxygenase gene family (Holton et al. 1993). The ferns are believed to be derived from ancestors like bryophytes. They are the oldest plants group known to produce proanthocyanidins. In ferns, proanthocyanidins are synthesized by the enzyme dihydroflavonol-4-reductase (DFR) using dihydroflavonols as substrates. This is one of the main benefits that accelerated the evolution of proanthocyanidin synthesis in the fern's lineage. Moreover, the ferns continue to manufacture the flavonols kaempferol, quercetin, and myricetin as well as procyanidin and prodelphinidin, preserving the three branches of the flavonoid pathway that were developed in the bryophytes.

5.8 Conclusion

The solar UVR has a deleterious impact on the majority of sun-exposed organisms. This might be one of the driving forces for the creation of photoprotectants such as phenylpropanoids, flavonoids, mycosporines, scytonemin, mycosporine-like amino acids (MAAs), parietin, photopigments (xanthophyll and phycobilins) and other chemical substances. Natural molecules derived from non-flowering lower plants can be utilized efficiently in skincare and cosmeceuticals since they are less damaging to the skin than manufactured compounds. The evolution in the photoprotecting mechanism was seen in the case of non-flowering lower plants. The use of unique biologically produced sunscreen molecules in cosmetics has increased significantly as a result of a breakthrough in biotechnology, genetic modification of the organism and a diverse microbiological diversity.

Acknowledgements A. Gupta (09/013(0912)/2019-EMR-I), N. Kumari (09/013(0819)/2018-EMR-I), and P.R. Singh (09/013(0795)/2018-EMR-I) are thankful to CSIR, New Delhi, India, for providing financial support in the form of Senior Research Fellowship. N. Sahu (09/013(0927)/2020-EMR-I) is thankful to CSIR, New Delhi, India, for providing financial support in the form of a Junior Research Fellowship. A.P. Singh (NTA Ref. No. 191620014505) and J. Jaiswal (926/CSIR-UGC-JRF DEC, 2018) are thankful to the University Grants Commission (UGC) New Delhi, India, for the financial assistance in the form of fellowship as Senior Research Fellow. Incentive grant received from IoE (Scheme No. 6031), Banaras Hindu University, Varanasi, India, to Rajeshwar P. Sinha is highly acknowledged.

References

Adir N (2005) Elucidation of the molecular structures of components of the phycobilisome: reconstructing a giant. Photosynth Res 85(1):15–32

Agati G, Azzarello E, Pollastri S, Tattini M (2012) Flavonoids as antioxidants in plants: location and functional significance. Plant Sci 196:67–76

Almeida-Paes R, Nosanchuk JD, Zancope-Oliveira RM (2012) Fungal melanins: biosynthesis and biological functions. In: Ma X-P, Sun X-X (eds) Melanin: biosynthesis, functions, and health effects. Nova Science Publishers, New York, pp 77–107

Aspée A, Aliaga C, Scaiano JC (2007) Transient enol isomers of dibenzoylmethane and avobenzone as efficient hydrogen donors toward a nitroxide pre-fluorescent probe. Photochem Photobiol 83(3):481–485

Avalos J, Carmen Limón M (2015) Biological roles of fungal carotenoids. Curr Genet 61(3):309–324

Ballaré CL, Caldwell MM, Flint SD, Robinson SA, Bornman JF (2011) Effects of solar ultraviolet radiation on terrestrial ecosystems. Patterns, mechanisms, and interactions with climate change. Photochem Photobiol Sci 10(2):226–241

Ballottari M, Truong TB, De Re E, Erickson E, Stella GR, Fleming GR, Bassi R, Niyogi KK (2016) Identification of pH-sensing sites in the light harvesting complex stress-related 3 protein essential for triggering non-photochemical quenching in *Chlamydomonas reinhardtii*. J Biol Chem 291(14):7334–7346

Bandara S, Ren Z, Lu L, Zeng X, Shin H, Zhao KH, Yang X (2017) Photoactivation mechanism of a carotenoid-based photoreceptor. Proc Natl Acad Sci 114(24):6286–6291

Bayry J, Beaussart A, Dufrêne YF, Sharma M, Bansal K, Kniemeyer O, Aimanianda V, Brakhage AA, Kaveri SV, Kwon-Chung KJ, Latgé JP (2014) Surface structure characterization of *Aspergillus fumigatus* conidia mutated in the melanin synthesis pathway and their human cellular immune response. Infect Immun 82(8):3141–3153

Begora MD, Fahselt D (2001) Usnic acid and atranorin concentrations in lichens in relation to bands of UV irradiance. Bryologist 104(1):134–140

Belozerskaya TA, Gessler NN, Aver'Yanov AA (2017) Melanin pigments of fungi. In: Fungal metabolites, pp 263–291

Berera R, van Stokkum IH, Kennis JT, van Grondelle R, Dekker JP (2010) The light-harvesting function of carotenoids in the cyanobacterial stress-inducible IsiA complex. Chem Phys 373(1–2):65–70

Bhatia S, Garg A, Sharma K, Kumar S, Sharma A, Purohit AP (2011) Mycosporine and mycosporine-like amino acids: a paramount tool against ultra violet irradiation. Pharmacogn Rev 5(10):138

Bjerke JW, Lerfall K, Elvebakk A (2002) Effects of ultraviolet radiation and PAR on the content of usnic and divaricatic acids in two arctic-alpine lichens. Photochem Photobiol Sci 1(9):678–685

Bonente G, Ballottari M, Truong TB, Morosinotto T, Ahn TK, Fleming GR, Niyogi KK, Bassi R (2011) Analysis of LhcSR3, a protein essential for feedback de-excitation in the green alga *Chlamydomonas reinhardtii*. PLoS Biol 9(1):e1000577

Bonfante P, Genre A (2008) Plants and arbuscular mycorrhizal fungi: an evolutionary-developmental perspective. Trends Plant Sci 13(9):492–498

Boustie J, Tomasi S, Grube M (2011) Bioactive lichen metabolites: alpine habitats as an untapped source. Phytochem Rev 10(3):287–307

Braga GU, Rangel DE, Fernandes EK, Flint SD, Roberts DW (2015) Molecular and physiological effects of environmental UV radiation on fungal conidia. Curr Genet 61(3):405–425

Brunet C, Casotti R, Vantrepotte V, Conversano F (2007) Vertical variability and diel dynamics of picophytoplankton in the Strait of Sicily, Mediterranean Sea, in summer. Mar Ecol Prog Ser 346:15–26

Bryan R, Jiang Z, Friedman M, Dadachova E (2011) The effects of gamma radiation, UV and visible light on ATP levels in yeast cells depend on cellular melanization. Fungal Biol 115(10):945–949

Büchel C (2015) Evolution and function of light harvesting proteins. J Plant Physiol 172:62–75

Bultel-Poncé V, Félix-Théodose F, Sarthou C, Ponge JF, Bodo B (2004) New pigments from the terrestrial cyanobacterium *Scytonema* sp. collected on the Mitaraka Inselberg, French Guyana. J Nat Prod 67(4):678–681

Cadet J, Douki T, Ravanat JL (2015) Oxidatively generated damage to cellular DNA by UVB and UVA radiation. Photochem Photobiol B Biol 91(1):140–155

Carnicas E, Jiménez C, Niell FX (1999) Effects of changes of irradiance on the pigment composition of *Gracilaria tenuistipitata* var. liui Zhang et Xia. J Photochem Photobiol B Biol 50(2–3):149–158

Carreto JI, Carignan MO, Montoya NG (2005) A high-resolution reverse-phase liquid chromatography method for the analysis of mycosporine-like amino acids (MAAs) in marine organisms. Mar Biol 146(2):237–252

Castenholz RW (1997) Multiple strategies for UV tolerance in cyanobacteria. Spectrum 10:10–16

Chen HYS, Liberton M, Pakrasi HB, Niedzwiedzki DM (2017) Reevaluating the mechanism of excitation energy regulation in iron-starved cyanobacteria. Biochim Biophys Acta Bioenerg 1858(3):249–258

Cockell CS, Knowland J (1999) Ultraviolet radiation screening compounds. Biol Rev 74(3):311–345

Conde FR, Churio MS, Previtali CM (2000) The photoprotector mechanism of mycosporine-like amino acids. Excited-state properties and photostability of porphyra-334 in aqueous solution. J Photochem Photobiol B Biol 56(2–3):139–144

Conde FR, Churio MS, Previtali CM (2004) The deactivation pathways of the excited-states of the mycosporine-like amino acids shinorine and porphyra-334 in aqueous solution. Photochem Photobiol Sci 3(10):960–967

Cook NC, Samman S (1996) Flavonoids—chemistry, metabolism, cardioprotective effects, and dietary sources. J Nutr Biochem 7(2):66–76

Cordero RJ, Casadevall A (2017) Functions of fungal melanin beyond virulence. Fungal Biol Rev 31(2):99–112

Crutzen PJ (1992) Ultraviolet on the increase. Nature 356(6365):104–105

Daniel S, Cornelia S, Fred Z (2004) UV-A sunscreen from red algae for protection against premature skin aging. In: Cosmet toiletries manufacture worldwide, vol 2004, pp 139–143
De la Coba F, Aguilera J, Figueroa FL, De Gálvez MV, Herrera E (2009) Antioxidant activity of mycosporine-like amino acids isolated from three red macroalgae and one marine lichen. J Appl Phycol 21(2):161–169
Demmig-Adams B, Garab G, Govindjee U, Adams W (2014) Non-photochemical quenching and energy dissipation in plants, algae and cyanobacteria. Springer, Dordrecht, pp 1–33
Eisenstadt D, Ohad I, Keren N, Kaplan A (2008) Changes in the photosynthetic reaction centre II in the diatom *Phaeodactylum tricornutum* result in non-photochemical fluorescence quenching. Environ Microbiol 10(8):1997–2007
Elix JA, Stocker-Wörgötter E (2008) Biochemistry and secondary metabolites. In: Lichen biology, pp 104–133
Engelken J, Funk C, Adamska I (2012) The extended light harvesting complex (LHC) protein superfamily: classification and evolutionary dynamics. In: Burnap R, Vermaas W (eds) Functional genomics and evolution of photosynthetic systems. Springer, Dordrecht, pp 265–284
Erlejman AG, Verstraeten SV, Fraga CG, Oteiza PI (2004) The interaction of flavonoids with membranes: potential determinant of flavonoid antioxidant effects. Free Radic Res 38(12):1311–1320
Esteban R, Olano JM, Castresana J, Fernández-Marín B, Hernández A, Becerril JM, García-Plazaola JI (2009) Distribution and evolutionary trends of photoprotective isoprenoids (xanthophylls and tocopherols) within the plant kingdom. Physiol Plant 135(4):379–389
Fernández E, Quilhot W, González I, Hidalgo ME, Molina X (1996) Lichen metabolites as UVB filters: lichen metabolites show photoprotector capacity. Cosmet Toilet 111(12):69–74
Fernández E, Quilhot W, Rubio C, Hidalgo ME, Diaz R, Ojeda J (2006) Effects of UV radiation on usnic acid in *Xanthoparmelia microspora* (Müll. Arg. Hale). Photochem Photobiol B Biol 82(4):1065–1068
Ferrer JL, Austin MB, Stewart C Jr, Noel JP (2008) Structure and function of enzymes involved in the biosynthesis of phenylpropanoids. Plant Physiol Biochem 46(3):356–370
Fleming ED, Castenholz RW (2007) Effects of periodic desiccation on the synthesis of the UV-screening compound, scytonemin, in cyanobacteria. Environ Microbiol 9(6):1448–1455
Frank HA, Cua A, Chynwat V, Young A, Gosztola D, Wasielewski MR (1994) Photophysics of the carotenoids associated with the xanthophyll cycle in photosynthesis. Photosynth Res 41(3):389–395
Fuentes-Tristan S, Parra-Saldivar R, Iqbal HM, Carrillo-Nieves D (2019) Bioinspired biomolecules: Mycosporine-like amino acids and scytonemin from *Lyngbya* sp. with UV-protection potentialities. J Photochem Photobiol B Biol 201:111684
Fuller KK, Loros JJ, Dunlap JC (2015) Fungal photobiology: visible light as a signal for stress, space and time. Curr Genet 61(3):275–288
Gao Q, Garcia-Pichel F (2011) Microbial ultraviolet sunscreens. Nat Rev Microbiol 9(11):791–802
García-Mendoza E, Colombo-Pallotta MF (2007) The giant kelp *Macrocystis pyrifera* presents a different nonphotochemical quenching control than higher plants. New Phytol 173(3):526–536
Garcia-Pichel F, Castenholz RW (1991) Characterization and biological implications of scytonemin, a cyanobacterial sheath pigment. J Phycol 27(3):395–409
Garcia-Pichel F, Wingard CE, Castenholz RW (1993) Evidence regarding the UV sunscreen role of a mycosporine-like compound in the cyanobacterium *Gloeocapsa* sp. Appl Environ Microbiol 59(1):170–176
Gauslaa Y, Ustvedt EM (2003) Is parietin a UV-B or a blue-light screening pigment in the lichen *Xanthoria parietina*? Photochem Photobiol Sci 2(4):424–432
Gensheimer M, Mushegian A (2004) Chalcone isomerase family and fold: no longer unique to plants. Protein Sci 13(2):540–544
Geraldes V, Jacinavicius FR, Genuário DB, Pinto E (2020) Identification and distribution of mycosporine-like amino acids in Brazilian cyanobacteria using ultrahigh-performance liquid chromatography with diode array detection coupled to quadrupole time-of-flight mass spectrometry. Rapid Commun Mass Spectrom 34:e8634

Gerotto C, Morosinotto T (2013) Evolution of photoprotection mechanisms upon land colonization: evidence of PSBS-dependent NPQ in late Streptophyte algae. Physiol Plant 149(4):583–598
Goss R, Jakob T (2010) Regulation and function of xanthophyll cycle-dependent photoprotection in algae. Photosynth Res 106(1):103–122
Grant CS, Louda JW (2013) Scytonemin-imine, a mahogany-colored UV/Vis sunscreen of cyanobacteria exposed to intense solar radiation. Org Geochem 65:29–36
Grégoire G, Lucas B, Barat M, Fayeton JA, Dedonder-Lardeux C, Jouvet C (2009) UV photoinduced dynamics in protonated aromatic amino acid. Eur Phys J D 51(1):109–116
Gruszecki WI, Strzałka K (1991) Does the xanthophyll cycle take part in the regulation of fluidity of the thylakoid membrane? Biochim Biophys Acta Bioenerg 1060(3):310–314
Häder DP, Helbling EW, Williamson CE, Worrest RC (2011) Effects of UV radiation on aquatic ecosystems and interactions with climate change. Photochem Photobiol Sci 10(2):242–260
Hager A (1980) The reversible, light-induced conversions of xanthophylls in the chloroplast. In: Pigments in plants, pp 57–79
Hager A, Holocher K (1994) Localization of the xanthophyll-cycle enzyme violaxanthin de-epoxidase within the thylakoid lumen and abolition of its mobility by a (light-dependent) pH decrease. Planta 192(4):581–589
Hall RSB, Bornman JF, Björn LO (2002) UV-induced changes in pigment content and light penetration in the fruticose lichen *Cladonia arbuscula* sp. mitis. J Photochem Photobiol B Biol 66(1):13–20
Havaux M, Tardy F (1996) Temperature-dependent adjustment of the thermal stability of photosystem II in vivo: possible involvement of xanthophyll-cycle pigments. Planta 198(3):324–333
Hidalgo ME, Fernandez E, Quilhot W, Lissi E (1994) Antioxidant activity of depsides and depsidones. Phytochemistry 37(6):1585–1587
Hidalgo ME, Fernández E, Ponce M, Rubio C, Quilhot W (2002) Photophysical, photochemical, and thermodynamic properties of shikimic acid derivatives: calycin and rhizocarpic acid (lichens). J Photochem Photobiol B Biol 66(3):213–217
Hidalgo ME, Bascuñan L, Quilhot W, Fernández E, Rubio C (2005) Spectroscopic and photochemical properties of the lichen compound lobaric acid. J Photochem Photobiol B Biol 81(6):1447–1449
Hinnebusch AG, Natarajan K (2002) Gcn4p, a master regulator of gene expression, is controlled at multiple levels by diverse signals of starvation and stress. Eukaryot Cell 1(1):22–32
Holton TA, Brugliera F, Tanaka Y (1993) Cloning and expression of flavanol synthase from *Petunia hybrida*. Plant J 4(6):1003–1010
Horton P, Ruban AV (1992) Regulation of photosystem II. Photosynth Res 34(3):375–385
Horton P, Johnson MP, Perez-Bueno ML, Kiss AZ, Ruban AV (2008) Photosynthetic acclimation: does the dynamic structure and macro-organisation of photosystem II in higher plant grana membranes regulate light harvesting states? FEBS J 275(6):1069–1079
Huijser A, Pezzella A, Sundström V (2011) Functionality of epidermal melanin pigments: current knowledge on UV-dissipative mechanisms and research perspectives. Phys Chem Chem Phys 13(20):9119–9127
Huneck S, Yoshimura I (1996) Identification of lichen substances. Springer, Berlin, pp 11–123
Jacobson ES, Ikeda R (2005) Effect of melanization upon porosity of the cryptococcal cell wall. Med Mycol 43(4):327–333
Jakob T, Goss R, Wilhelm C (1999) Activation of diadinoxanthin de-epoxidase due to a chiororespiratory proton gradient in the dark in the diatom *Phaeodactylum tricornutum*. Plant Biol 1(1):76–82
Jovanovic SV, Steenken S, Tosic M, Marjanovic B, Simic MG (1994) Flavonoids as antioxidants. J Am Chem Soc 116(11):4846–4851
Karentz D, McEuen FS, Land MC, Dunlap WC (1991) Survey of mycosporine-like amino acid compounds in Antarctic marine organisms: potential protection from ultraviolet exposure. Mar Biol 108(1):157–166

Karsten U, Garcia-Pichel F (1996) Carotenoids and mycosporine-like amino acid compounds in members of the genus *Microcoleus* (Cyanobacteria): a chemosystematic study. Syst Appl Microbiol 19(3):285–294

Keyhani NO (2018) Lipid biology in fungal stress and virulence: Entomopathogenic fungi. Fungal Biol 122(6):420–429

Komenda J, Sobotka R (2016) Cyanobacterial high-light-inducible proteins-protectors of chlorophyll–protein synthesis and assembly. Biochim Biophys Acta Bioenerg 1857(3):288–295

Kumar S, Mishra A, Pandey AK (2013) Antioxidant mediated protective effect of *Parthenium hysterophorus* against oxidative damage using in vitro models. BMC Complement Altern Med 13(1):1–9

Kumari N, Pathak J, Dwivedy AK, Sinha RP (2021) Bioprospection of UV-screening compounds from lichens inhabiting the Indian state of Sikkim. Plant Arch 21:1168–1177

Kumari N, Pandey A, Gupta A, Mishra S, Sinha RP (2022) Characterization of UV-screening pigment scytonemin from cyanobacteria inhabiting diverse habitats of Varanasi, India. Biologia 78:1–12

Lee TM, Shiu CT (2009) Implications of mycosporine-like amino acid and antioxidant defenses in UV-B radiation tolerance for the algae species *Pterocladiella capillacea* and *Gelidium amansii.* Mar Environ Res 67(1):8–16

Leite B, Nicholson RL (1992) Mycosporine-alanine: a self-inhibitor of germination from the conidial mucilage of *Colletotrichum graminicola*. Exp Mycol 16(1):76–86

Li Z, Guo M (2018) Healthy efficacy of Nostoc commune Vaucher. Oncotarget 9(18):14669

Libkind D, Moline M, van Broock M (2011) Production of the UVB-absorbing compound mycosporine–glutaminol–glucoside by *Xanthophyllomyces dendrorhous* (*Phaffia rhodozyma*). FEMS Yeast Res 11(1):52–59

Llewellyn CA, Airs RL (2010) Distribution and abundance of MAAs in 33 species of microalgae across 13 classes. Mar Drugs 8(4):1273–1291

Lohézic-Le Dévéhat F, Legouin B, Couteau C, Boustie J, Coiffard L (2013) Lichenic extracts and metabolites as UV filters. J Photochem Photobiol B Biol 120:17–28

Lohr M, Wilhelm C (2001) Xanthophyll synthesis in diatoms: quantification of putative intermediates and comparison of pigment conversion kinetics with rate constants derived from a model. Planta 212(3):382–391

Lois R, Buchanan BB (1994) Severe sensitivity to ultraviolet radiation in an *Arabidopsis* mutant deficient in flavonoid accumulation. Planta 194(4):504–509

Lushchak VI (2011) Adaptive response to oxidative stress: bacteria, fungi, plants and animals. Comp Biochem Physiol C Toxicol Pharmacol 153(2):175–190

Magdaong NCM, Blankenship RE (2018) Photoprotective, excited-state quenching mechanisms in diverse photosynthetic organisms. J Biol Chem 293(14):5018–5025

Mathews-Roth MM, Pathak MA (1975) Phytoene as a protective agent against sunburn (> 280nm) radiation in Guinea pigs. Photochem Photobiol 21(4):261–263

Mathey A, Lukins PB (2001) Spatial distribution of perylenequinones in lichens and extended quinones in quincyte using confocal fluorescence microscopy. Micron 32(2):107–113

Meléndez-Martínez AJ, Mapelli-Brahm P, Benítez-González A, Stinco CM (2015) A comprehensive review on the colorless carotenoids phytoene and phytofluene. Arch Biochem Biophys 572:188–200

Millot M, Tomasi S, Articus K, Rouaud I, Bernard A, Boustie J (2007) Metabolites from the lichen *Ochrolechiaparella* growing under two different heliotropic conditions. J Nat Prod 70(2):316–318

Millot M, Di Meo F, Tomasi S, Boustie J, Trouillas P (2012) Photoprotective capacities of lichen metabolites: a joint theoretical and experimental study. J Photochem Photobiol B Biol 111:17–26

Minagawa J, Tokutsu R (2015) Dynamic regulation of photosynthesis in *Chlamydomonas reinhardtii*. Plant J 82(3):413–428

Misonou T, Saitoh J, Oshiba S, Tokitomo Y, Maegawa M, Inoue Y, Hori H, Sakurai T (2003) UV-absorbing substance in the red alga *Porphyra yezoensis* (Bangiales, Rhodophyta) block thymine photodimer production. Mar Biotechnol 5(2):194–200

Moliné M, Flores MR, Libkind D, Carmen Diéguez MDC, Farías ME, van Broock M (2010) Photoprotection by carotenoid pigments in the yeast *Rhodotorula mucilaginosa*: the role of torularhodin. Photochem Photobiol Sci 9(8):1145–1151

Murray JW, Duncan J, Barber J (2006) CP43-like chlorophyll binding proteins: structural and evolutionary implications. Trends Plant Sci 11(3):152–158

Nägeli C, Schwendener S (1877) Das Mikroskop, 2nd edn. Wilhelm Engelmann, Leipzig

Nakamura H, Kobayashi JI, Hirata Y (1982) Separation of mycosporine-like amino acids in marine organisms using reversed-phase high-performance liquid chromatography. J Chromatogr A 250:113–118

Nakayama R, Tamura Y, Kikuzaki H, Nakatani N (1999) Antioxidant effect of the constituents of Susabinori (Porphyrayezoensis). J Am Oil Chem Soc 76(5):649–653

Nazifi E, Wada N, Asano T, Nishiuchi T, Iwamuro Y, Chinaka S, Matsugo S, Sakamoto T (2015) Characterization of the chemical diversity of glycosylated mycosporine-like amino acids in the terrestrial cyanobacterium *Nostoc commune*. J Photochem Photobiol B Biol 142:154–168

Ngoennet S, Nishikawa Y, Hibino T, Waditee-Sirisattha R, Kageyama H (2018) A method for the isolation and characterization of mycosporine-like amino acids from cyanobacteria. Methods Protoc 1(4):46

Niyogi KK, Truong TB (2013) Evolution of flexible non-photochemical quenching mechanisms that regulate light harvesting in oxygenic photosynthesis. Curr Opin Plant Biol 16(3):307–314

Obermayer W, Poelt J (1992) Contributions to the knowledge of the lichen flora of the Himalayas III. On *Lecanora somervellii* Paulson (lichenized Ascomycotina, Lecanoraceae). Lichenologist 24(2):111–117

Oren A, Gunde-Cimerman N (2007) Mycosporines and mycosporine-like amino acids: UV protectants or multipurpose secondary metabolites? FEMS Microbiol Lett 269(1):1–10

Oyamada C, Kaneniwa M, Ebitani K, Murata M, Ishihara K (2008) Mycosporine-like amino acids extracted from scallop (*Patinopecten yessoensis*) ovaries: UV protection and growth stimulation activities on human cells. Mar Biotechnol 10(2):141–150

Peers G, Truong TB, Ostendorf E, Busch A, Elrad D, Grossman AR, Hippler M, Niyogi KK (2009) An ancient light-harvesting protein is critical for the regulation of algal photosynthesis. Nature 462(7272):518–521

Pollastri S, Tattini M (2011) Flavonols: old compounds for old roles. Ann Bot 108(7):1225–1233

Pourcel L, Routaboul JM, Cheynier V, Lepiniec L, Debeaujon I (2007) Flavonoid oxidation in plants: from biochemical properties to physiological functions. Trends Plant Sci 12(1):29–36

Proteau PJ, Gerwick WH, Garcia-Pichel F, Castenholz R (1993) The structure of scytonemin, an ultraviolet sunscreen pigment from the sheaths of cyanobacteria. Experientia 49(9):825–829

Quilhot W, Fernández E, Hidalgo ME (1994) British Lichen Soc. Bull 75:1–5

Raj S, Kuniyil AM, Sreenikethanam A, Gugulothu P, Jeyakumar RB, Bajhaiya AK (2021) Microalgae as a source of mycosporine-like amino acids (MAAs); advances and future prospects. Int J Environ Res Public Health 18(23):12402

Rancan F, Rosan S, Boehm K, Fernández E, Hidalgo ME, Quihot W, Rubio C, Boehm F, Piazena H, Oltmanns U (2002) Protection against UVB irradiation by natural filters extracted from lichens. J Photochem Photobiol B Biol 68(2–3):133–139

Rangel DE, Fernandes ÉK, Braga GU, Roberts DW (2011) Visible light during mycelial growth and conidiation of *Metarhizium robertsii* produces conidia with increased stress tolerance. FEMS Microbiol Lett 315(2):81–86

Rao DN, LeBlanc F (1965) A possible role of atranorin in the lichen thallus. Bryologist 68:284–289

Rastogi RP, Sinha RP, Singh SP, Häder DP (2010a) Photoprotective compounds from marine organisms. J Ind Microbiol Biotechnol 37(6):537–558

Rastogi RP, Kumar A, Tyagi MB, Sinha RP (2010b) Molecular mechanisms of ultraviolet radiation-induced DNA damage and repair. J Nucleic Acids 2010:1–32

Rastogi RP, Singh SP, Incharoensakdi A, Häder DP, Sinha RP (2014) Ultraviolet radiation-induced generation of reactive oxygen species, DNA damage and induction of UV-absorbing compounds in the cyanobacterium *Rivularia* sp. HKAR-4. S Afr J Bot 90:163–169
Rastogi RP, Madamwar D, Incharoensakdi A (2015) Multiple defence mechanisms in cyanobacteria in response to solar UV radiation. In: Cyanobacteria: ecological importance, biotechnological uses and risk management, pp 125–158
Rice-Evans CA, Miller NJ, Paganga G (1996) Structure-antioxidant activity relationships of flavonoids and phenolic acids. Free Radic Biol Med 20(7):933–956
Rice-Evans C, Miller N, Paganga G (1997) Antioxidant properties of phenolic compounds. Trends Plant Sci 2(4):152–159
Ruban AV (2016) Nonphotochemical chlorophyll fluorescence quenching: mechanism and effectiveness in protecting plants from photodamage. Plant Physiol 170(4):1903–1916
Saelices L, Youssar L, Holdermann I, Al-Babili S, Avalos J (2007) Identification of the gene responsible for torulene cleavage in the *Neurospora* carotenoid pathway. Mol Gen Genomics 278(5):527–537
Sampath-Wiley P, Neefus CD, Jahnke LS (2008) Seasonal effects of sun exposure and emersion on intertidal seaweed physiology: fluctuations in antioxidant contents, photosynthetic pigments and photosynthetic efficiency in the red alga *Porphyra umbilicalis* Kützing (Rhodophyta, Bangiales). J Exp Mar Biol Ecol 361(2):83–91
Sandberg M, and Thune P, (1984) The sensitizing capacity of atranorin. Contact Dermatitis, 11(3), pp. 168–173
Savolainen V, Chase MW (2003) A decade of progress in plant molecular phylogenetics. Trends Genet 19(12):717–724
Sedoud A, López-Igual R, ur Rehman A, Wilson A, Perreau F, Boulay C, Vass I, Krieger-Liszkay A, Kirilovsky D (2014) The cyanobacterial photoactive orange carotenoid protein is an excellent singlet oxygen quencher. Plant Cell 26(4):1781–1791
Shick JM, Dunlap WC (2002) Mycosporine-like amino acids and related gadusols: biosynthesis, accumulation, and UV-protective functions in aquatic organisms. Annu Rev Physiol 64(1):223–262
Shuryak I, Bryan RA, Nosanchuk JD, Dadachova E (2014) Mathematical modelling predicts enhanced growth of X-ray irradiated pigmented fungi. PLoS One 9(1):e85561
Singh S, Malhotra AG, Pandey A, Pandey KM (2013) Computational model for pathway reconstruction to unravel the evolutionary significance of melanin synthesis. Bioinformation 9(2):94
Sinha RP, Häder DP (2008) UV-protectants in cyanobacteria. Plant Sci 174(3):278–289
Sinha RP, Klisch M, Gröniger A, Häder DP (1998) Ultraviolet-absorbing/screening substances in cyanobacteria, phytoplankton and macroalgae. J Photochem Photobiol B Biol 47(2–3):83–94
Sinha RP, Singh SP, Häder DP (2007) Database on mycosporines and mycosporine-like amino acids (MAAs) in fungi, cyanobacteria, macroalgae, phytoplankton and animals. J Photochem Photobiol B Biol 89(1):29–35
Solhaug KA, Gauslaa Y (2004) Photosynthates stimulate the UV-B induced fungal anthraquinone synthesis in the foliose lichen *Xanthoria parietina*. Plant Cell Environ 27(2):167–176
Solhaug KA, Gauslaa Y, Nybakken L, Bilger W (2003) UV-induction of sun-screening pigments in lichens. New Phytol 158(1):91–100
Solhaug KA, Larsson P, Gauslaa Y (2010) Light screening in lichen cortices can be quantified by chlorophyll fluorescence techniques for both reflecting and absorbing pigments. Planta 231(5):1003–1011
Staleva H, Komenda J, Shukla MK, Šlouf V, Kaňa R, Polívka T, Sobotka R (2015) Mechanism of photoprotection in the cyanobacterial ancestor of plant antenna proteins. Nat Chem Biol 11(4):287–291
Stevenson CS, Capper EA, Roshak AK, Marquez B, Eichman C, Jackson JR, Mattern M, Gerwick WH, Jacobs RS, Marshall LA (2002) The identification and characterization of the marine natural product scytonemin as a novel antiproliferative pharmacophore. J Pharmacol Exp Ther 303(2):858–866

Stransky HT, Hager A (1970) Das Carotinoidmuster und die Verbreitung des lichtinduzierten Xanthophyllcyclus in verschiedenen Algenklassen. Arch Fuer Mikrobiol 71(2):164–190

Suh SS, Lee SG, Youn UJ, Han SJ, Kim IC, Kim S (2017) Comprehensive expression profiling and functional network analysis of porphyra-334, one mycosporine-like amino acid (MAA), in human keratinocyte exposed with UV radiation. Mar Drugs 15(7):196

Swanson A, Fahselt D (1997) Effects of ultraviolet on polyphenolics of *Umbilicaria americana*. Can J Bot 75(2):284–289

Upadhyay S, Xu X, Lowry D, Jackson JC, Roberson RW, Lin X (2016) Subcellular compartmentalization and trafficking of the biosynthetic machinery for fungal melanin. Cell Rep 14(11):2511–2518

Varnali T, Edwards HG (2014) Reduced and oxidised scytonemin: theoretical protocol for Raman spectroscopic identification of potential key biomolecules for astrobiology. Spectrochim Acta A Mol Biomol Spectrosc 117:72–77

Verwoert II, Verbree EC, Van der Linden KH, Nijkamp HJ, Stuitje AR (1992) Cloning, nucleotide sequence, and expression of the *Escherichia coli fabD* gene, encoding malonyl coenzyme A-acyl carrier protein transacylase. J Bacteriol 174(9):2851–2857

Wada N, Sakamoto T, Matsugo S (2013) Multiple roles of photosynthetic and sunscreen pigments in cyanobacteria focusing on the oxidative stress. Metabolites 3(2):463–483

Wilson A, Ajlani G, Verbavatz JM, Vass I, Kerfeld CA, Kirilovsky D (2006) A soluble carotenoid protein involved in phycobilisome-related energy dissipation in cyanobacteria. Plant Cell 18(4):992–1007

Wilson A, Boulay C, Wilde A, Kerfeld CA, Kirilovsky D (2007) Light-induced energy dissipation in iron-starved cyanobacteria: roles of OCP and IsiA proteins. Plant Cell 19(2):656–672

Winkel-Shirley B (2001) Flavonoid biosynthesis. A colorful model for genetics, biochemistry, cell biology, and biotechnology. Plant Physiol 126(2):485–493

Yagura T, Schuch AP, Garcia CCM, Rocha CRR, Moreno NC, Angeli JPF, Mendes D, Severino D, Sanchez AB, Di Mascio P, de Medeiros MHG (2017) Direct participation of DNA in the formation of singlet oxygen and base damage under UVA irradiation. Free Radic Biol Med 108:86–93

Yan GL, Liang HY, Wang ZQ, Yang XF, Liu D, Liu JF, Duan CQ (2011) Important role of catalase in the production of β-carotene by recombinant *Saccharomyces cerevisiae* under H_2O_2 stress. Curr Microbiol 62(3):1056–1061

Zhang L, Li L, Wu Q (2007) Protective effects of mycosporine-like amino acids of *Synechocystis* sp. PCC 6803 and their partial characterization. J Photochem Photobiol B Biol 86(3):240–245

Chapter 6
Impacts of Climate Alterations on the Biosynthesis of Defensive Natural Products

Pooja Singh and Krishna Kumar Choudhary

Abstract Climate change fluctuations, specifically CO_2 concentration, temperature, rainfall patterns, droughts, and soil salinity, are increasing due to anthropogenic activities. These variations are identified as major constraints to plant survival and therefore limit plant growth and productivity. Photosynthesis inhibition, excessive ROS (reactive oxygen species) production, biomass reduction, increased pathogen infestation, and ultimately lower yields are the major limiting attributes that have attracted a lot of attention from researchers worldwide. Since climate change predictions indicate that ecological damage will be more frequent and severe in the upcoming futuristic scenarios, the question of fulfilling the food requirement of the ever-growing population becomes imperative. Plants are sensitive to the effects of climate change. Alterations in photosynthesis and carbon assimilation mechanisms are attributed to reduced productivity. To cope with these stresses, secondary metabolite production elicits defensive responses in plants. These natural by-products are synthesized from primary metabolites and protect against various abiotic and biotic stresses. Synthesis and accumulation of secondary metabolites differ among plant species growing in different environmental conditions. Phenolics, flavonoids, alkaloids, terpenoids, tannins, glucosinolates, and so on are a useful array of natural products that increase plant resistance against various stresses. Although these are synthesized in minimal concentrations, they display a crucial role in the scavenging of ROS molecules.

Keywords Abiotic stress · Climate change · Photosynthesis · Reactive oxygen species · Secondary metabolites · Yield

P. Singh · K. K. Choudhary (✉)
Department of Botany, MMV, Banaras Hindu University, Varanasi, India
e-mail: kkc@bhu.ac.in

V. K. Kannaujiya et al. (eds.), *Photoprotective Green Pharmacology: Challenges, Sources and Future Applications*,
https://doi.org/10.1007/978-981-99-0749-6_6

6.1 Introduction

Anthropogenic activities including the burning of fossil fuels, urbanization, and a rise in the concentration of greenhouse gases (GHGs) are the major factors responsible for global climate change (Dutta et al. 2020). Elevated concentration of GHGs in the atmosphere since the industrial revolution has increased the concentration of CO_2 from 280 ppm to >410 ppm and is expected to rise further to 730–1000 ppm by 2100 (IPCC 2014). Recently, IPCC (2021) revealed that this enhancement in atmospheric GHGs further raises the global temperature approximately by 0.84–1.10 °C, and consequently disturbing the rainfall patterns and prevailing drought conditions in arid regions of the world (IPCC 2014). Such variability in climate is influencing crop production with each successive year and somewhere evokes an uncertainty in terms of food production (Reddy and Hodges 2000). It is predicted that agricultural outputs will be declined (10–20%) by the end of 2080 in developing countries (Thompson and Cohen 2012). Consequently, the subject of achieving food security worldwide becomes a daunting task with an ever-growing population (Barnett 2011; Funk and Brown 2009; Rice and Garcia 2011). In the current scenario, about 1 billion people are food-deprived, 150 million children are chronically undernourished, 50 million children are acutely malnourished with a higher mortality rate, and another 38 million children are overweight (Misselhorn et al. 2012; Fanzo 2018). The situation in India is similar to the global scenario: whether the increasing population and demand for food supply will continue to rise with climate change. Temperature increases of 1–2 °C have a negative influence on the productivity of major cereal crops, which in turn affects the nutritional status of the population (Easterling et al. 2007; Rao et al. 2016).

Climate change unavoidably disturbs plants by hampering the physiological and biochemical processes such as altered photosynthesis, plant–water interactions, and CO_2 assimilations, which severely affects their growth and yield (Fig. 6.1) (Anjum et al. 2011). These variations induced oxidative stress in plants via increased generation of reactive oxygen species (ROS), leading to lipid peroxidation, DNA damage, and inactivation of important enzymes (Akula and Ravishankar 2011). In addition to this, overproduction of ROS also inhibits CO_2 fixation in chloroplasts, as they are the primary source of ROS generation (Asada 2006). In response to such constraints, plants have acquired alternative strategies such as increased antioxidative response, phytohormones, osmotic adjustment, and enhanced production of secondary metabolites (Yadav et al. 2021; Jogawat et al. 2021; Zandalinas et al. 2022). Secondary metabolites play a vital role in plant defense against herbivory, insect attack, and environmental stress (Chomel et al. 2016). Several biotic and abiotic stresses act as an elicitor for the stimulation of secondary metabolites (Radman et al. 2003; Ghorbanpour et al. 2014). Their synthesis and accumulation differ among plant species grown under different environmental conditions (Radušienė et al. 2012). Shikimate pathway, acetate–malonate pathway, and side reactions involving glycolysis and TCA cycle are different metabolic routes through which biosynthesis of secondary metabolites takes place in plants (Geilfus 2019;

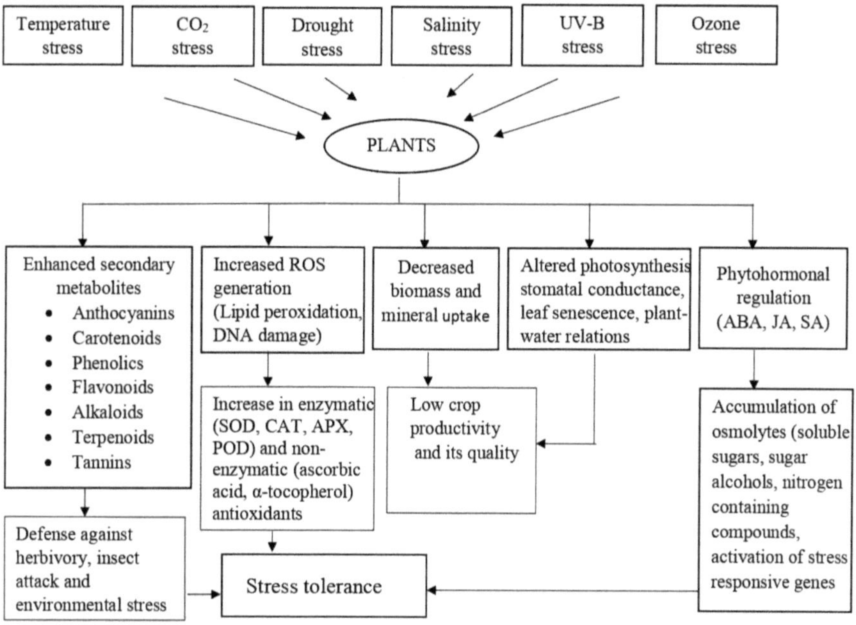

Fig. 6.1 Effect of different abiotic stress responses in plants due to climate change

Nabavi et al. 2020). Phenolics, flavonoids, and terpenes synthesized in very low concentrations facilitate antioxidative defense mechanisms, thus increasing their acclimatization to oxidative stress in plants (Edreva et al. 2008). Phytohormones, particularly ABA and jasmonic acid (JA), are positively correlated with the production of secondary metabolites, as they work in a synergistic manner. For example, ABA and JA were responsible for the increase in phenolics and flavonoid contents in *Castanea sativa* (Camisón et al. 2019). This could protect the plants against increased oxidative stress through the activation of NAC transcription factors (Choudhary et al. 2021).

6.2 Elevated CO_2 and Temperature Stress

Increased anthropogenic activities have accelerated the level of CO_2 concentrations in the atmosphere. At the time of pre-industrialization, the CO_2 levels were 280 ppm initially, but with increasing trends, it has been reported to be nearly 410 ppm (September 2019) (IPCC 2019). In view of this, a two-fold increase in CO_2 concentration has been expected (IPCC 2013). These elevated levels of CO_2 not only causes global warming but also reduces ecosystem productivity. According to NOAA (2020), the surface temperature of land and oceans is 0.98 °C warmer than the twentieth-century average (13.9 °C). Kimball (2016) found that increasing CO_2

concentration by 200 ppm will increase canopy temperature (ET) by 0.7 °C. Increased ET becomes a problem, particularly for developing countries, as it has reduced crop productivity and grain yield (Chaturvedi et al. 2017; Wang et al. 2017). CO_2 levels play a significant role in plant metabolic processes. Short-term exposure to elevated CO_2 (~400 ppm) reported enhanced photosynthesis, biomass, and decreased oxidative stress. Various plants, such as *Solanum lycopersicum* L., *Stevia rebaudiana* L., and *Parthenium hysterophorus* L., have demonstrated the beneficial impact of elevated CO_2 (Hussin et al. 2017; Bajwa et al. 2019; Pan et al. 2020). However, prolonged exposure to higher CO_2 levels (~800 ppm), promoted negative effects on plant growth, i.e., reduced photosynthesis, and altered biomass that ultimately affected crop yield and its quality (Wang et al. 2013). For instance, decreased photosynthesis and fruit yield has been observed under high CO_2 concentrations in strawberry (Balasooriya et al. 2018). This significant reduction is due to the low availability of RuBisCO content and nitrogen concentration (Gamage et al. 2018; Rosa et al. 2019). Many crops, including *Lactuca sativa* and *Spinacia oleracea*, had lower nutritional quality (Mg, N, Fe, Zn, and S) under increased CO_2 concentrations (Giri et al. 2016; Dong et al. 2018).

Variable environmental factors influence secondary metabolite biosynthesis in plants. Phenolics such as flavonoids, condensed tannins, and alkaloids in response to elevated CO_2 concentration have significantly modulated secondary metabolism in plants (Levine et al. 2008; Jia et al. 2014). CO_2 enrichment in the atmosphere increases the susceptibility of plants to insect attack by boosting photosynthesis and higher production of carbohydrates (Ainsworth and Rogers 2007; Bernacchi et al. 2007). To avoid insect damage, plants allocate the primary metabolites to secondary metabolites grown under high CO_2 levels. In woody plants, phenolic compounds and terpenoids provide defense against herbivory at higher CO_2 concentrations (Feeny 1976; Rhoades and Cates 1976). Robinson et al. (2012) reported increased total phenolics (19%), flavonoids (27%), and tannins (22%) in plants grown under elevated CO_2. On the other hand, flavonoids such as quercetin, fisetin, and kaempferol were enhanced in the leaves and rhizomes of ginger, hence exhibiting higher antioxidative defense responses (Ghasemzadeh et al. 2010). Similarly, soybean plants mediate anti-herbivory by increasing the ratios of quercetin and kaempferol while decreasing the genistein concentration (Piubelli et al. 2005). Higher phenylalanine ammonia-lyase (PAL) enzyme activity is linked with the upregulation of secondary metabolites in elevated CO_2. For example, a significant increase in phenolics and flavonoid concentration was observed in *Eleais guneensis* L. due to increased PAL enzyme activity (Ibrahim and Jaafar 2012). Similarly, *Triticum aestivum* L. exhibited higher PAL activity along with an accumulation of phenolic compounds (Mishra et al. 2013). Generally, warming conditions are associated with phenolic contents in leaves and increased terpenoid concentrations in foliage (Peñuelas and Staudt 2010; Zvereva and Kozlov 2006). However, under elevated CO_2, phenolic concentrations were increased in the foliage while decreased in woody tissues (Zvereva and Kozlov 2006). On the other hand, terpenoid concentration was significantly lowered as CO_2 concentration increased in conifers. Similarly, the emission of phenolics and flavonoid content was significantly intensified by elevated temperature in *Zingiber officinale* L. (Ghasemzadeh et al. 2011). Several plant species, such as *Thymus hyemalis* L., *Thymus vulgaris* L., *Valeriana jatamansi* L., and *Camellia sinensis* L., have been reported with an increased concentration of secondary metabolites in plants (Biel et al. 2005; Vurro et al. 2009; Li et al. 2017; Kaundal et al. 2018). Sobuj et al. (2018) observed the differential response of flavonoid concentration in male and female plants. Under elevated CO_2 conditions, female plants had a significantly higher concentration of flavonoids as compared to male plants. The anti-carcinogenic and anti-inflammatory activities of glucoraphanin and sulforaphane have been linked to increased hydrolysis of glucosinolates (GSs) in response to elevated CO_2 (Table 6.1) (Almuhayawi et al. 2020). Jasmonic acid (JA) plays an integral role in plant defense mechanisms through the elicitation of different secondary metabolites such as alkaloids, flavonoids, phenylpropanoids, and terpenoids (Tamogami et al. 1997). For example, higher ascorbic acid and carotenoid content have been observed in *Origanum majorana* L. (Złotek 2017). Similarly, various plants have been reported to produce differential secondary metabolites being elicited by JA (Thakur et al. 2019).

Temperature stress also affects plant ontology and metabolic processes, i.e., physiological and biochemical changes such as chlorophyll pigment breakdown, leaf senescence, membrane damage, and protein denaturation (Waraich et al. 2012). Other effects of higher temperature (heat stress) can be identified by decreased quantum efficiency of Photosystem II (PSII), stomatal conductance, CO_2 fixation, altered secondary metabolites, and ROS generation (Hasanuzzaman et al. 2013; Verma and Shukla 2015). However, low temperature is responsible for disturbing the plant–water interactions and metabolic activities, ultimately hampering the plant growth and productivity (Chinnusamy et al. 2007). Plant growing under low temperature synthesizes cryoprotective substances such as soluble sugars (trehalose, raffinose, stachyose, and saccharose), sugar alcohols (inositol, ribitol, and sorbitol), and nitrogen-containing compounds (glycine betaine, proline) to maximize cold stress tolerance (Janská et al. 2010). In *Arnica montana*, enhanced ratios of quercetin–kaempferol have been reported under low temperature (Albert et al. 2009). Increased artemisinin content has been observed after exposure of *Artemisia annua* to cold stress (Yin et al. 2008; Vashisth et al. 2018).

Variations in temperature influence the biosynthesis and accumulation of alkaloids in plants. For instance, in *Papaver somniferum* L., the accumulation of morphinanc, benzylisoquinoline, and phthalisoquinoline becomes restricted at low temperature (Bernáth and Tetenyi 1979). Contrary to this, the concentration of isoflavonoids (genistein, genistin, and daidzein) is significantly enhanced in the roots of *Glycine max* L. at low temperature (Janas et al. 2002). Similarly, several studies have been reported with increased alkaloid contents in plants incubated at a higher temperature. For example, *Lupinus angustifolius* has been reported with higher alkaloids concentration when grown under elevated temperature (Jansen et al. 2009). In *Catharanthus roseus* L., increased concentrations of catharanthine, vindoline, and vinblastine were observed at a higher temperature, while incubation at low temperature resulted in a two- to four-fold reduction of catharanthine and vindoline contents (Dutta et al. 2007). These findings suggest that higher temperature

Table 6.1 Impact of CO_2 and temperature stress on different secondary metabolites in plants

Stress	Dose	Plants	Endogenous response of secondary metabolites	References
CO_2	(400–1000 ppm)	*Triticum aestivum* L.	Flavonoid contents (homoorientin and rhamnoside) increased by 28–64%; a significant increase in tyrosine and trans-caffeic acid	Levine et al. (2008)
	(350–700 μmol mol^{-1})		Significant reduction in total phenolic acids (21.4%), condensed tannins (22.2%), and indole alkaloids (48.1%)	Jia et al. (2014)
	(700 ppm)		Increased PAL activity by 39.2% along with a significant increase in total phenolics by 11.7%	Mishra et al. (2013)
	(400–800 μmol mol^{-1})	*Zingiber officinale* L.	Elevated flavonoids contents (kaempferol, fisetin, and naringenin) in leaves by 44.9% and rhizomes by 86.3%; phenolic compounds (gallic acid, vanillic acid, and ferulic acid) increased in leaves by 112.2% and rhizomes by 109.2%	Ghasemzadeh et al. (2010)
	(400–1200 μmol mol^{-1})	*Eleais guineensis* L.	Enhanced total flavonoids by 132%; total phenolics by 91%	Ibrahim and Jaafar (2012)
	(550 μmol mol^{-1})	*Valeriana jatamansi*	Elevated essential oil content by 17.7%; sesquiterpenes by 17.2%	Kaundal et al. (2018)
	(800 μmol mol^{-1})	*Camellia sinensis* L.	Upregulated expression of catechins and theanine biosynthetic genes while caffeine synthetic genes were downregulated	Li et al. (2017)
	(400 μmol m^{-2} s^{-1})	*Brassica oleracea* L.	Slight increase in myrosinase activity accounts for the effective production of sulforaphene	Almuhayawi et al. (2020)

(continued)

Table 6.1 (continued)

Stress	Dose	Plants	Endogenous response of secondary metabolites	References
Temperature	(7.5–12.5 °C)	*Arnica montana* L.	Ratios of quercetin–kaempferol were significantly enhanced	Albert et al. (2009)
	(4 °C)	*Artemisia annua* L.	Artemisinin content significantly elevated by 27.16%	Vashisth et al. (2018)
	(10 °C)	*Glycine max* L.	Increased concentration of isoflavonoids consisting of daidzein (200%), genistein (240%), and genistin (310%)	Janas et al. (2002)
	(15.5 °C)	*Lupinus angustifolius* L.	Estimated increase in alkaloid contents by 0.11%	Jansen et al. (2008)
	(4 °C)	*Catharanthus roseus* L.	Significant reductions in catharanthine content by two-fold along with two- to four-fold reduction in vindoline content	Dutta et al. (2007)
	(20 °C)	*Chamomilla recutita* L.	Increased concentrations of α-bisabolol were detected	Fahlén et al. (1997)
	(40 °C)	*Pinus ponderosa* L.	Significant increase in sesquiterpenes (α-bergamotene, α-farnesene, β-caryophyllene, and β-farnesene) concentrations	Helmig et al. (2007)
	(6–22 °C)	*Betula pendula* L. *Populus tremula* L.	DMNT (homoterpene) increased consistently; SQTs (β-bourbonene, γ-cadinene) were significantly enhanced	Ibrahim et al. (2010)
	(26.8–31.2 °C)	*Panax quinquefolius* L.	Ginsenoside concentrations in roots significantly increased by 49%	Jochum et al. (2007)

DMNT 4,8-dimethylnona-1,3,7-triene, *PAL* phenylalanineammonia-lyase, *SQTs* sesquiterpenes

enhanced the concentration of alkaloids in plants and low temperature significantly hinders their biosynthetic pathway genes (Dutta et al. 2007). The antioxidative properties of terpenes provide stability to the thylakoid membrane of the chloroplast. In *Chamomilla recutita*, the combination of photoperiod (21-3h) and temperature (20 ± 2 °C) resulted in the highest concentration of α-bisabolol (Fahlén et al. 1997). Temperature dependency is correlated with the yield of terpenoids. For example, pine species have been reported with increased emissions of sesquiterpene compounds (α-bergamotene, α-farnesene, β-caryophyllene, and β-farnesene) at elevated temperature (Table 6.1) (Helmig et al. 2007). Differential response of temperature on volatile organic compounds (VOCs) has been studied in *Betula pendula* and *Populus tremula*, resulting in an exponential increase in DMNT (4,8-dimethyl-nona-1,3,7-triene) concentration (Ibrahim et al. 2010). In the roots of *Panax ginseng* and *Panax quinquefolius*, ginsenoside content was significantly enhanced under elevated temperature, while photosynthesis and biomass were considerably reduced (Yu et al. 2005; Jochum et al. 2007).

6.3 Drought Stress

High temperature and solar radiations are accompanied by water deficit conditions that induce drought (Xu et al. 2010). Among abiotic stressors, drought hampers agricultural productivity by upto 50–70% (Verma and Deepti 2016). It has been estimated that drought affects 40% of the global population and now has been predicted to pose a risk of displacement to 700 million populations by 2030 (WHO 2020). Drought stress severely alters plant growth through photosynthesis inhibition, decreased stomatal conductance, CO_2 assimilation, and leaf senescence (Nezhadahmadi et al. 2013; Wang et al. 2018; Zargar et al. 2017). Plant defense responses, including secondary metabolites production, are triggered by decreased water potential and turgor pressure caused by increased transpiration rate (Ashraf et al. 2018). Drought stress induces ROS production through oxidative stress, resulting in enhanced production of flavonoids and phenolic acids (Larson and Weber 2018). Through transcriptomics, Morales et al. (2017) identified pathways as well as genes involved in drought-tolerant quinoa. Upregulation of drought-tolerant genes such as *GmbZIP44*, *GmbZIP46*, *GmbZIP62*, and *GmbZIP78* has been known to provide tolerance against drought (Xie et al. 2009). In addition, enhanced expression of the *GmbZIP1* gene in wheat has been reported as an excellent resource for overcoming drought stress (Gao et al. 2011). Activation of the PAL gene resulted in enhanced phenolic and flavonoid contents in *Lactuca sativa* L. (Rajabbeigi et al. 2013). Various plants such as *Artemisia*, *Hypericum brasiliense*, *Hypericum perforatum*, and *Trachyspermum ammi* have been reported with increased secondary metabolites such as artemisinin, betulinic, ruetin, hyperforin, and quercitin (Azhar et al. 2011; Zobayed et al. 2007; Verma and Shukla 2015). Similarly, water-deficit conditions (80–85% field capacity) decreased the number of total flavonoids in *Glechoma longituba* (Zhang et al. 2012).

The major enzymes responsible for the biosynthesis of flavonoids include chalcone synthase (CHS), chalcone isomerase (CHI), flavone synthase (FNS), flavanone 3-hydroxylase (F3H), flavonol synthase (FLS), dihydroflavonol 4--reductase (DFR), and anthocyanidin synthase (ANS) (Shih et al. 2008). The antioxidant property of flavonoids lies in the position of hydroxyl groups and carbon modifications such as glycosylation, methylation, and prenylation (Rice-Evans et al. 1997). Flavonoids under drought stress act as an antioxidant and protect plants from severe damage induced under water-deficit conditions (Nichols et al. 2015). For instance, in *Pisum sativum* L., flavonoid concentration was significantly increased by 45% in response to drought stress (Nogués et al. 1998). Similarly, roots of *Scutellaria baicalensis* were reported with elevated concentrations of flavonoids (Yuan et al. 2012). This increased accumulation of flavonoids represents effective detoxification of H_2O_2 molecules induced via drought stress (Hernández et al. 2009). In addition, drought stress also influenced phenolic concentration in plants, which was mediated via alteration in the phenylpropanoid pathway (Table 6.2) (Gharibi et al. 2019; Rezayian et al. 2018; Li et al. 2018). *Salvia dolomitica* and *Salvia officinalis* showed an increase in flavonoids (101%) and phenolics (139%) content under drought conditions (Caser et al. 2018, 2019). Drought stress reduced oil, sesamin, and quercetin concentration, however, a significant increment was noticed in flavonoids and phenolics contents in *Sesamum indicum* L " (Kermani et al. 2019). Biosynthesis of glycine betaine via enhanced expressions of glycine betaine hydrogenase was responsible for the alleviation of drought stress in *C. roseus* (Jaleel et al. 2007). Similarly, the artificial introduction of mannitol in seedlings elevated the concentrations of carbohydrates, proline, thymol, and γ-terpinene (Razavizadeh and Komatsu 2018). Water stress altered essential oil content (geraniol and citral) in *Cymbopogon citratus* L. (Singh-Sangwan et al. 1994). However, moderate drought conditions exhibited a higher concentration of β-thujone and camphor in *Salvia officinalis* (Bettaieb et al. 2009). Additionally, Nowak et al. (2010) reported higher concentrations of monoterpenes (33%) in the same plant. Essential oil contents do not always increase; however, it depends on the plant species and the severity of the stress. Paulsen and Selmar (2016) reported a considerable increase in terpene content, whereas the total amount of terpene was markedly reduced due to biomass reduction.

6.4 Salinity Stress

Salinity stress is one of the major limiting factors in plant growth and development. Due to increased anthropogenic activities and global climate change, it is projected to worsen in the near future (Rengasamy 2010). For instance, salinity stress significantly decreased crop yield by 10–50% in most salt-sensitive plant species (Panta et al. 2014). Globally, salinization has recorded an estimated economic loss of US$ 27.3 billion/year (Qadir et al. 2014). This significant increasing trend in salinity becomes a subject of great concern for national as well as global food security. Keeping this in view, the Indian government has planned to restore 26 million ha of salt-affected lands by 2030 (Kumar and Sharma 2020). Photosynthesis inhibition, ROS production, and reduced germination are some of the negative impacts commonly observed under salt stress. The generation of ROS mediated via salt stress

Table 6.2 Impact of drought stress on different secondary metabolites in plants

Stress	Dose	Plants	Endogenous response of secondary metabolites	References
Drought	Field capacity (60–100%)	*Trachyspermum ammi* L.	Increased total phenolic contents (4.44 mg/g) was estimated	Azhar et al. (2011)
	9–12 days	*Hypericum perforatum* L.	70-fold higher hyperforin concentration; antioxidants increased by 2.5-fold	Zobayed et al. (2007)
	Field capacity (80–85%)	*Glechoma longituba* L.	Significant increase in yield of total flavonoids	Zhang et al. (2012)
	30–70 days	*Scutellaria baicalensis* L.	Increased total flavonoid contents both in roots and leaves; baicalin and baicalein (major active compounds) contents remained unchanged	Yuan et al. (2012)
	7–28 days	*Achillea pachycephala* L.	Phenolic acids and flavones such as chlorogenic acid (7.23 mg/100 g DW) and luteolin (5.1 mg/100 mg DW) were markedly elevated, major flavonoid (apigenin-7-*O*-glucoside) present in abundant concentration (10.41 mg/100 g)	Gharibi et al. (2019)
	PEG (0, 5, 10, 15%)	*Brassica napus* L.	Influence of increased PAL enzyme activity on total phenols, flavonoids, and flavonols concentration accompanied with increased tocopherol content and decreased anthocyanin contents significantly	Rezayian et al. (2018)
	PEG (5–10%)	*Cucumis sativus* L.	Upregulated expression of phenolic compounds (vanillic acid and 4-hydroxycinnamic acid)	Li et al. (2018)
	0–34 days	*Salvia sinaloensis* L.	Altered chemical profiles of BVOC and EO; significant increase in phenolics, flavonoids, and monoterpenes (camphor) while sesquiterpene (Germacrene D) contents decreased significantly	Caser et al. (2018)
		Salvia dolomitica L.	Substantial reduction in total phenols and flavonoid contents; sesquiterpene hydrocarbons (66.32%) were significantly increased; monoterpene hydrocarbons (29.41%) and oxygenated hydrocarbons (2.19%) were considerably reduced	Caser et al. (2019)
	–	*Sesamum indicum* L.	Elevated phenolics (caffeic, *p*-coumaric and ferulic acids) and flavonoids (rutin and apigenin) levels; oil contents, sesamin, and quercetin decreased significantly	Kermani et al. (2019)
	Field capacity (25, 50, 100%)	*Salvia officinalis* L.	Enhanced essential oil constituents (β-thujone, camphor, and 1,8-cineole)	Bettaieb et al. (2009)

BVOC biogenic volatile organic compounds, *EO* essential oils, *PAL* phenylalanineammonia-lyase alters plant metabolic activities such as the disruption of membrane and ion toxicity (Ashraf et al. 2015; Chaudhary and Choudhary 2021). Secondary metabolites can scavenge ROS through the enhanced accumulation of phenolic compounds. Polyphenol concentrations significantly increased in *Cakile maritime* after exposure to different concentrations of NaCl (0, 100, 400 mM), indicating a protective role against salt stress (Ksouri et al. 2007). Similarly, *Cynara cardunculus* were reported to have increased phenolic contents on exposure to moderate levels of NaCl (>75 mM) (Hanen et al. 2008). *Fagopyrum esculentum* L. under variable salt concentrations (10–200 mM) showed a remarkable increase in phenolic contents (isoorientin, rutin, orientin, and vitexin) compared to control (Lim et al. 2012). Exposure to increased salinity levels (0–200 mM) significantly enhanced the total non-flavonoids (30%), total phenolics (135%), and total tannins (72%) content in *Brassica napus* L. (Falcinelli et al. 2017). In contrast, *Brassica oleracea* L. showed a decrease in phenolic compounds (chlorogenic and derivatives of sinapic acid), indicating the accumulation of phenolic acids in a plant-specific manner (Lopez-Berenguer et al. 2009). Similarly, *Salvia macrosiphon* L. has been reported with a remarkable decrease in total phenolics (2.6 times) after exposure to 8 dS ms^{-1} salinity level (Valifard et al. 2017). Furthermore, this lack of correlation, however, depends on the synergistic interactions of different antioxidant molecules (Tarchoune et al. 2012a, b). The effects of different salt concentrations on various plant species are shown in Table 6.3.

Salt stress stimulates the production of tropane alkaloids in *Datura innoxia* L. (Brachet and Cosson 1986). In *C. roseus*, vincristine content was significantly enhanced in response to 150 mM NaCl but gradually declined with increasing salinity levels (Osman et al. 2007). Ali et al. (2008) reported altered ricinine content in *Ricinus communis* L. Similarly, reserpine and vincristine (alkaloids) contents significantly increased in *C. roseus* and *R. tetraphylla*, respectively (Ahl and Omer 2011). *Rosmarinus officinalis* governs increased concentrations of camphor and cineole on account of salt stress, whereas borneol, camphene, nopol, and α-terpineol concentrations were decreased significantly (Tounekti et al. 2011). Furthermore, roots of *Zea mays* L. drastically improved the zealexins levels by five-fold at higher levels of NaCl (500 mM); however, kauralexins contents increased upto two-fold at lower levels (100 mM) (Vaughan et al. 2015). Different concentrations of salt (0–150 mM) significantly enhanced the expression of flavonoid biosynthetic genes (CHS, FS, and PAL) and resulted in increased production of lutein and quercetin in *Solanum nigrum* L. (Ben Abdallah et al. 2016).

6.5 UV-B Stress

Depletion of the ozone layer raises its concern over increased exposure to UV-B radiation on plants and animals. UV-B radiation, which comprises 0.5% of total solar radiation, possesses a significant impact on terrestrial life forms (Rozema et al. 2009; Verdaguer et al. 2012; Correia et al. 2012). Equatorial regions receive

Table 6.3 Impact of salinity stress on different secondary metabolites in plants

Stress	Dose	Plants	Endogenous response of secondary metabolites	References
Salinity	(0, 100, 400 mM)	*Cakile maritima* L.	Significant increase in polyphenol concentration (56%) along with higher MDA contents (1.6–2.6-fold)	Ksouri et al. (2007)
	(25–150 mM)	*Cynara cardunculus* L.	Two-fold increase in polyphenol concentrations at moderate salinity levels, 50-fold higher (IC_{50}) antioxidant activity	Hanen et al. (2008)
	(10–200 mM)	*Fagopyrum esculentum* L.	Two-fold increase in phenolic contents (isoorientin, orientin, rutin, and vitexin) with subsequent increase in antioxidant activity (60%); carotenoids contents elevated upto 40%	Lim et al. (2012)
	(0–200 mM)	*Brassica napus* L.	Total phenolics increased upto 135%, non-flavonoids upto 30%, and total tannins upto 72%; higher DPPH exhibited positive correlation with total phenolic contents	Falcinelli et al. (2017)
	(4–80 mM)	*Brassica oleracea* L.	Enhanced glucosinolate and flavonoid levels; significant decrease in phenolic contents, i.e., chlorogenic acid and sinapic acid (63.8%)	Lopez-Berenguer et al. (2009)
	2.3–6.8 dS m^{-1}	*Salvia macrosiphon* L.	2.6 times decrease in phenolic compounds accompanied with significant increase in antioxidant activity; negative correlation is established	Valifard et al. (2017)
	(100–150 mM)	*Catharanthus roseus* L.	Remarkable increase in alkaloid (vincristine) content (2-peak) attributed to increased levels of arginine	Osman et al. (2007)
	(50–150 mM)	*Solanum nigrum* L.	Quercetin levels increased about 2.6-fold; amount of carotenoids (lutein and β-carotene) was substantially high	Ben Abdallah et al. (2016)

DPPH 2,2-diphenyl-1-piacrylhydrazyl, *MDA* malondialdehyde content

about 12 kJ m^{-2} d^{-1} of solar UV-B radiation (Forster 2011). During the pre-1980s, about 6–14% of increment was detected. However, current scenarios reflect this percentage remaining elevated for the next decades (WMO 2010). UV-B influenced plants by reducing photosynthesis, biomass, deformities in chloroplast structure, and increased ROS generation (Pandey and Chaplot 2007; Yang et al. 2007; Kataria et al. 2014; Yao and Liu 2006; Kakani et al. 2003; Choudhary et al. 2017). Elevated UV-B levels significantly altered the concentrations of secondary metabolites such

Table 6.4 Impact of UV-B stress on different secondary metabolites in plants

Stress	Dose	Plants	Endogenous response of secondary metabolites	References
UV-B radiation	5.4–31 kJ m^{-2} d^{-1}	*Rosamatinus officinalis* L.	Predominant rosmarinic acid and carnosic concentrations followed by naringin and carnosol, while vanillic acid and hispidulin are considerably reduced	Luis et al. (2007)
	5.8–7.2 kJ m^{-2} d^{-1}	*Pisum sativum* L.	Significant increase in quercetin (114%) and kaempferol (72%) contents	Choudhary and Agrawal (2014b)
	5.8–7.2 kJ m^{-2} d^{-1}	*Vigna radiata* L.	Total flavonoids increased significantly by 36% along with maximum induction of PAL activity (105%)	Choudhary and Agrawal (2014a)
		Glycine max L.	Increased PAL activity (45%) correlates with increased kaempferol (83%) and quercetin (95%) contents; higher contents of lignin (60%) and wax (88%)	Choudhary and Agrawal (2016)
	0.43–1.13 W m^{-2}	*Glycyrrhiza uralensis* L.	Stimulated glycyrrhizin concentration in root tissues; melatonin present in roots and leaves	Afreen et al. (2005)
	ambient+1.8 kJ m^{-2} d^{-1}	*Acorus calamus* L.	Increased percentage of aristolene, *p*-cymene, caryophyllene oxide, and carvacrol; reduced contents of β-asarone	Kumari et al. (2009a, b)
	35 μmol s^{-1} m^{-2}	*Fagopyrum tataricum* L.	9.35-fold increase of rutin concentration in leaves with substantial increase in quercetin content; 30–40-fold higher abundance of FtCHI and FtCHS transcripts	Huang et al. (2016)
	8.64–9.50 kJ m^{-2} d^{-1}	*Glycine max* L.	Considerable increase in flavonoids concentration (quercetin, rutin, ferulic acid); no significant difference in phenolic compounds	Mao et al. (2017)
	6.5–12.4 kJ m^{-2} d^{-1}	*Olea europaea* L.	Significant decrease in phenolic contents (seciridoids, oleuropein (54%), and 2″-methoxy oleuropein) increased significantly by 68%; flavonoids (4′-methoxy luteolin) decreased; ouercetin-3-*O*-rutinoside, luteolin-7-*O*-glucoside, luteolin-7,4′-diglucoside, and apigenin 7-*O*-glucoside contents remains unchanged; HCAds and β-hydroxyverbacoside increased significantly by 75%	Dias et al. (2020)
	0.5–2.0 kJ m^{-2} d^{-1}	*Brassica oleracea* L.	Accumulation of kaempferol-3-*O*-disinapoyl-triglucoside-7-*O*-glucoside, kaempferol, and quercetin derivatives significantly decreased	Neugart et al. (2012)
	ambient +9.6 kJ m^{-2} d^{-1}	*Curcuma caesia* L.	Total flavonoid content increased by 62%, anthocyanin content by 44%; reduction in D-camphor, eucalyptol, curcumenol, isocurcumenol compounds; increment in 1,8-cineole, epicurzerenone, and elemene compounds; stimulation of anti-cancerous compounds (caryophllene, furanodiene, curzerene, epicurzerenone, and verrucarol)	Jaiswal et al. (2020)
	54 kJ m^{-2} d^{-1}	*Cuminum cyminum* L.	Increased trends in flavonoid and alkaloid contents with more pronounced effect on expression levels of PAL and DAHPs	Ghasemi et al. (2019)
	2.8 W m^{-2}	*Withania coagulans* L.	Enhanced contents of both withanolide A and withaferin by 3.42- and 1.38-folds; upregulated expression of terpenoid biosynthetic genes (FPPS, SQS, and $CYP_{51}G_1$)	Tripathi et al. (2021)

CHI chalcone isomerase, *CHS* chalcone synthase, *$CYP_{51}G_1$* Cytochrome P45051G1, *DAHPs* deoxyriboninoheptulosinate-7-phosphate synthase, *FPPS* farnesyl pyrophosphate synthase, *HCAds* hydroxycinnamic acid, *PAL* phenylalanineammonia-lyase, *SQS* squalene synthase.

as alkaloids, anthocyanins, cyanogenic glycosides, flavonoids, and tannins in plants (Table 6.4) (Hirata et al. 1993; Morales et al. 2010; Gouvea et al. 2012). For instance, in *C. roseus*, the amount of catharanthine and vindoline production was significantly enhanced after supplemental UV-B radiation (Ramani and Jayabaskaran 2008). In another study, increased kaempferol and quercetin contents have been reported in *Populus trichocarpa* (Warren et al. 2003). Similarly, different rice cultivars have been observed with increased C-glycosyl flavones content under high UV-B intensity. Enhanced UV-B radiation leads to a more pronounced effect on flavanols accumulation in *Trifolium repens*, resulting in increased quercetin levels by 200% (Hofmann et al. 2000). These flavanols protect by acting as UV-B filters and further help to scavenge ROS (Agati et al. 2009, 2011).

UV-B-absorbing compounds such as flavonoids and hydroxycinnamic acids (derivatives of phenolic acids) confer protection at elevated UV-B levels (Agati and Tattini 2010; Jansen et al. 2008; Qian et al. 2020). For example, an increased concentration of flavonoids was reported in *Lactuca sativa* and *Gynura bicolor*, which were grown under ambient and high UV-B radiation (García-Macías et al. 2007; Schirrmacher et al. 2004). Accumulation of flavonoids in leaf epidermis confers resistance to the detrimental effects of UV-B radiation. Higher flavonoid contents illustrate increased PAL activity, a key enzyme involved in the phenylpropanoid pathway (Liu et al. 2002). Quercetin and kaempferol levels are certainly beneficial for plants to quench free radicals generated at the initial stage of UV-B exposure (Harborne and Williams 2000). Important crop plants, mainly *Vigna radiata* L., *Pisum sativum* L., and *Glycine max* L., demonstrated enhanced concentrations of quercetin and kaempferol contents induced via elevated UV-B exposure (Choudhary and Agrawal 2014a, b, 2016). In *Fagopyrum tataricum*, UV-B treatment resulted in a dramatic increase in concentrations of rutin (4.82 mg/g) DW and quercetin (0.04 mg/g) DW, respectively (Huang et al. 2016). Similarly, Mao et al. (2017) reported enhanced concentrations of rutin and quertein (flavonoids) in soyabean. Prolonged exposure to UV-B resulted in upregulation of flavonoid synthetic genes, i.e., FLS and F3′H in *Gingko biloba* (Zhao et al. 2020). The highest estimated flavonoid concentration was recorded in *Alternanthera sessilis* (Klein et al. 2018). In *Olea europaea* L. leaves, abundant concentrations of luteolin-7-*O*-glucoside account for the species' high tolerance to UV-B stress (Dias et al. 2020).

UV-B elicitation greatly influences the biosynthesis of phenolic compounds in plants. Increased ROS production initially triggered by UV-B resulted in enhanced phenolic contents that acts direct scavenger of ROS (Solovchenko and Merzlyak 2008). Ambient UV-B doses significantly enhanced flavonoid concentration in root and leaves of *Tropaeolum majus* L. and *Brassica oleracea*, suggesting UV-B as a systemic inducer of phenolic compounds in plants (Schreiner et al. 2009; Neugart et al. 2012). Phenolic compounds under UV-B exposure become elevated in post-harvested fruits and crops, including apples, peaches, onions, and strawberries (Marais et al. 2001; Kataoka and Beppu 2004; Higashio et al. 2004). The upregulation of phenylpropanoid enzymes by UV-B causes an increase in phenolic concentration (Tomás-Barberán and Espín 2001; Treutter 2005). Moderate UV-B exposure increased catharanthine concentration in *C. roseus* (Ramani and Chelliah 2007).

Various plants such as *Acorus calamus*, *Cyambopogon citratus*, *Mentha piperata*, and *Ocimum basilicum* have been reported with important pharmacological compounds, induced via UV-B (Kumari et al. 2009a, b; Dolzhenko et al. 2010; Maffei and Scannerini 2000). Sesquiterpenes such as artemisinin and Germacrene-D concentrations were found to be elevated by 11.6% and 10.5% under UV-B exposure (Kumari and Agrawal 2011; Rai et al. 2011). Another plant, *Glycyrrhiza uralensis*, exhibited a 1.5-fold increase in Glycyrrhizin content on exposure to UV-B dose (0.43 W m^{-2}) (Afreen et al. 2005). One of the major pharmacological important diterpenes, carnosic acid, present in *R. officinalis*, becomes elevated with UV-B dose (31 kJ m^{-2} d^{-1}) (Luis et al. 2007).

Exposure to UV-B radiation induced the production of terpenoids in various medicinal plants such as *Artemisia annua*, *Curcuma caesia*, *Cuminum cyminum* L., and *Vitis vignifera* L. (Li et al. 2021; Jaiswal et al. 2020; Ghasemi et al. 2019; Gil et al. 2012). These terpenoids protect the plant leaves from heat stress induced via UV-B (Liu et al. 2017). Withaferin A and withanolide A contents are increased by 1.38- and 3.42-fold in *Withania coagulans* L. (Tripathi et al. 2021). Thus, it can be concluded that UV-B can be used as a potential elicitor in increasing the contents of pharmacologically important compounds (Tripathi et al. 2021; Takshak and Agrawal 2014, 2015; Choudhary et al. 2021).

6.6 Tropospheric Ozone Stress

Ozone (O_3) is a potent air pollutant and greenhouse gas that may influence vegetation and human health directly or indirectly (DeLang et al. 2021; Wedow et al. 2021). Consumption of fossil fuel increases the concentration of precursor gases such as nitrogen oxide, carbon monoxide, and volatile organic compounds (VOCs), including methane and CO_2 that drive increased O_3 concentrations (Bhatia et al. 2012). Currently, tropospheric O_3 has reached 35–40 ppm globally and is expected to rise further to 70 ppm or more by 2050 (Frei 2015; Sicard et al. 2017; Pfister et al. 2014). Being a strong antioxidant, O_3 incorporates into plant tissues through stomata and induces ROS production that ultimately causes lipid peroxidation, DNA and RNA degradation, and programmed cell death (Mishra and Agrawal 2015; Picchi et al. 2017; Choudhury et al. 2017). Likewise, a variety of responses marked by elevated O_3, i.e., foliar injury, reduced chlorophyll and RuBisCO content, stomatal conductance, photosynthesis inhibition, and alteration in carbon allocation, cause a reduction in biomass, yield, and its quality (Emberson 2020). The activation of the PAL enzyme corresponds to increased production of flavonoids, phenolic acids, and monolignols, which improves the tolerance ability of plants by acting as scavengers against O_3 stress (Iriti and Faoro 2009). Long-term exposure to elevated O_3 concentrations concerning accumulation of phenolic compounds has been extensively studied (Richet et al. 2012). In *Linum usitatissimum* L., various secondary metabolites (flavonoids, anthocyanins, lignin, and wax) were enhanced under elevated O_3 (27.7–59.0 ppb) (Tripathi and Agrawal 2013). This enhancement reflects more utilization of assimilate in the production of secondary metabolites and less availability for reproductive organs that ultimately contribute to less yield (Singh et al. 2014). Fatima et al. (2018) investigated the effects of treatment of higher O_3 concentration (ambient + 30 ppb) on different wheat cultivars. These findings state that higher induction of flavonoids and total phenols subsequently declined reproductive structures and final yield. Differential responses in the accumulation of total phenolic contents in early and late sown cultivars of wheat indicated a correlation with higher ascorbic acid involved in the production of polyphenols (Yadav et al. 2019). Furthermore, various plants such as wheat, caster, groundnut, and cotton elucidate the sensitivity to ozone pollution (Chaudhary et al. 2021; Rathore and Chaudhary 2019; Ghosh et al. 2020a, b; Chaudhary and Rathore 2021a, b).

Weed invasion delineates the struggle of crop plants for their healthy survival under progressive climate change (Clements et al. 2014). It has been reported that weed interference causes an annual yield loss of 34% in some agronomically important crop species (Oerke 2006). The reason behind the aggressiveness of weeds lies in their higher content of phenolics and alkaloids, which alters the nutrient uptake in the soil (Majeed et al. 2012). A recent study was performed on *Chenopodium album* L. and *Triticum aestivum* L. plants to investigate the allelopathic interaction with a concomitant elevation in the concentration of O_3. The study revealed that O_3 raised the concentrations of ferulic acid (FA) and *p*-coumaric acid (CA) in the roots of the former and attributed a negative change in the root length of the latter (Ghosh et al. 2020a, b). Greater tolerance to O_3 stress is determined by increased PAL activity (Di Baccio et al. 2008). Elevated O_3 concentrations significantly increased total phenolic contents and PAL activity in *Vigna radiata* L. (Mishra and Agrawal 2015). Accumulated phenolic compounds triggered by higher O_3 levels during the initial days of exposure were later observed with a slight decrement in *Salvia officinalis* L. This suggested that higher doses of O_3 displayed a priming effect, and later, these plants failed to invest in their response strategy, indicating a slow production of secondary metabolites (Marchica et al. 2021). *Brassica campestris* L., a rich source of glucosinolate (GLS), exhibited an alteration in the amount of indole, aliphatic, and aromatic GLS (Han et al. 2021).

Exposure to higher levels of O_3 also affects isoprene emissions. Isoprene biosynthesis in plants maintains photochemical efficiency and ROS levels induced via excess O_3 (Pollastri et al. 2019; Loreto and Velikova 2001). O_3-induced emission of isoprene has been documented in several studies (Hewitt et al. 2009; Arab et al. 2016). Taking this into account, date palm has a high potential to resist photochemical changes induced by short-term exposure to O_3 (Du et al. 2018). A more realistic Free-air CO_2 enrichment (FACE) study demonstrated that emission of isoprene declined significantly with higher O_3 concentrations, but the number of total monoterpenes stimulated in date palm leaves was attributed to increased emission of aldehyde volatiles (Table 6.5) (Paoletti et al. 2021).

Table 6.5 Impact of tropospheric ozone stress on different secondary metabolites in plants

Stress	Dose	Plants	Endogenous responses of secondary metabolites	References
Ozone	27.7–59.0 ppb	*Linum usitatissimum* L.	Significant increment in flavonoids (32.8%) and anthocyanins (34.4%); increased lignin content (14.1%) and epicuticular wax	Tripathi and Agrawal (2013)
	15–30 ppb	*Zea mays* L.	Higher carotenoid and flavonoids levels; more induction of phenols accompanied with higher PAL activity	Singh et al. (2014)
	30 ppb	*Triticum aestivum* L.	Quercetin and kaempferol content responded differently; phenylpropanoid enzymes (CAD, 4CL) showed enhanced activities	Fatima et al. (2018)
	58.3 ppb	*Triticum aestivum* L. *Chenopodium album* L.	Strong stimulation of flavonols (kaempferol and Quercetin) and total phenolics in roots and leaves	Ghosh et al. (2020a, b)
	68.9 ppb	*Vigna radiata* L.	Total phenols elevated significantly by 34.2% along with increased PAL activity by 37%	Mishra and Agrawal (2015)
	45–90 ppb	*Phoenix dactylifera* L.	Isoprene emission declined significantly at elevated O_3 (−58% and −50%); stimulation of monoterpenes (α-pinene, β-octanal, nonanal, camphor, iso-bornrol)	Paoletti et al. (2021)
	60 ppb	*Brassica campestris* L.	Significant increase in lycopene, total carotenoids, and lutein content; negative correlation between total aromatic GLS and total aliphatic GLS	Han et al. (2021)

CAD cinnamyl alcohol dehydrogenase, *4CL* 4-coumarate CoA ligase, *GLS* glucosinolate synthase, *PAL* phenylalanineammonia-lyase

6.7 Conclusion

Climate change caused by increased anthropogenic activities has significantly altered CO_2 concentrations, temperature fluctuations, water-deficit conditions, salinity stress, UV-B intensity, and tropospheric ozone concentrations on Earth's surface. This is accompanied by a parallel decrease in physiological processes in plants. Elevated CO_2 concentrations induce photosynthetic processes, but plants become more susceptible to insect attack at the same time. To avoid insect damage, plants allocate photo-assimilates to secondary metabolite production. Phenolics and terpenoids decreased as CO_2 concentrations increased; however, these were significantly intensified by elevated temperature. Higher temperature prevails drought conditions and, with a concomitant increase in salt levels, severely impacts plant growth and yield via ROS production and reduced osmotic potential, which mediates biochemical changes. To confer resistance, plants facilitate antioxidative defense mechanisms through enhanced production of phenolics and flavonoids. Similarly, increased UV-B exposure and ozone stress induce morphological, physiological, and biochemical alterations in plants. Despite this, it needs to further investigate the synergistic role of different abiotic stresses responsible for actual synthesis and modulation at the same time. More importantly, scientists mimic the climate change perspectives through experimental studies, which raises concern about achieving food security and nutritional status worldwide.

Acknowledgments P.S. is thankful to CSIR, New Delhi, India, for the Junior Research Fellowship (F. No.: 09/0013(12956)/2021-EMR-I). K.K.C. is grateful to UGC-BSR Start-up Grant (No. F. 30-432/2018(BSR)), New Delhi, India, and to the seed grant, Institute of Eminence (IoE—Dev. Scheme no. 6031), Banaras Hindu University, Varanasi, India.

References

Afreen F, Zobayed SMA, Kozai T (2005) Spectral quality and UV-B stress stimulate glycyrrhizin concentration of *Glycyrrhiza uralensis* in hydroponic and pot system. Plant Physiol Biochem 43(12):1074–1081

Agati G, Tattini M (2010) Multiple functional roles of flavonoids in photoprotection. New Phytol 186(4):786–793

Agati G, Stefano G, Biricolti S, Tattini M (2009) Mesophyll distribution of 'antioxidant' flavonoid glycosides in *Ligustrum vulgare* leaves under contrasting sunlight irradiance. Ann Bot 104(5):853–861

Agati G, Biricolti S, Guidi L, Ferrini F, Fini A, Tattini M (2011) The biosynthesis of flavonoids is enhanced similarly by UV radiation and root zone salinity in *L. vulgare* leaves. J Plant Physiol 168(3):204–212

Ahl SA, Omer EA (2011) Medicinal and aromatic plants production under salt stress. A review. Herba Polonica 57(2):72–87

Ainsworth EA, Rogers A (2007) The response of photosynthesis and stomatal conductance to rising [CO_2]: mechanisms and environmental interactions. Plant Cell Environ 30(3):258–270

Akula R, Ravishankar GA (2011) Influence of abiotic stress signals on secondary metabolites in plants. Plant Signal Behav 6(11):1720–1731

Albert A, Sareedenchai V, Heller W, Seidlitz HK, Zidorn C (2009) Temperature is the key to altitudinal variation of phenolics in *Arnica montana* L. cv. ARBO. Oecologia 160(1):1–8

Ali RM, Elfeky SS, Abbas H (2008) Response of salt stressed *Ricinus communis* L. to exogenous application of glycerol and/or aspartic acid. J Biol Sci 8(1):171–175

Almuhayawi MS, AbdElgawad H, Al Jaouni SK, Selim S, Hassan AH, Khamis G (2020) Elevated CO_2 improves glucosinolate metabolism and stimulates anticancer and anti-inflammatory properties of broccoli sprouts. Food Chem 328:127102

Anjum SA, Xie XY, Wang LC, Saleem MF, Man C, Lei W (2011) Morphological, physiological and biochemical responses of plants to drought stress. Afr J Agric Res 6(9):2026–2032

Arab L, Kreuzwieser J, Kruse J, Zimmer I, Ache P, Alfarraj S, Al-Rasheid KA, Schnitzler JP, Hedrich R, Rennenberg H (2016) Acclimation to heat and drought—lessons to learn from the date palm (*Phoenix dactylifera*). Environ Exp Bot 125:20–30

Asada K (2006) Production and scavenging of reactive oxygen species in chloroplasts and their functions. Plant Physiol 141(2):391–396

Ashraf M, Iqbal M, Hussain I, Rasheed R (2015) Physiological and biochemical approaches for salinity tolerance. In: Managing salt tolerance in plants: molecular and genomic perspectives, p 79

Ashraf MA, Iqbal M, Rasheed R, Hussain I, Riaz M, Arif MS (2018) Environmental stress and secondary metabolites in plants: an overview. In: Plant metabolites and regulation under environmental stress, pp 153–167

Azhar N, Hussain B, Ashraf MY, Abbasi KY (2011) Water stress mediated changes in growth, physiology and secondary metabolites of desi ajwain (*Trachyspermum ammi* L.). Pak J Bot 43(1):15–19

Bajwa AA, Wang H, Chauhan BS, Adkins SW (2019) Effect of elevated carbon dioxide concentration on growth, productivity and glyphosate response of parthenium weed (*Parthenium hysterophorus* L.). Pest Manag Sci 75(11):2934–2941

Balasooriya HN, Dassanayake KB, Seneweera S, Ajlouni S (2018) Interaction of elevated carbon dioxide and temperature on strawberry (*Fragaria*× *ananassa*) growth and fruit yield. Int J Biol Biomol Agric Food Biotechnol Eng 12:279–287

Barnett J (2011) Dangerous climate change in the Pacific Islands: food production and food security. Reg Environ Chang 11(1):229–237

Ben Abdallah S, Aung B, Amyot L, Lalin I, Lachâal M, Karray-Bouraoui N, Hannoufa A (2016) Salt stress (NaCl) affects plant growth and branch pathways of carotenoid and flavonoid biosynthesis in *Solanum nigrum*. Acta Physiol Plant 38(3):1–13

Bernacchi CJ, Kimball BA, Quarles DR, Long SP, Ort DR (2007) Decreases in stomatal conductance of soybean under open-air elevation of [CO_2] are closely coupled with decreases in ecosystem evapotranspiration. Plant Physiol 143(1):134–144

Bernáth J, Tetenyi P (1979) The Effect of environmental factors on growth. Development and alkaloid production of Poppy (*Papaver somniferum L.*): I. Responses to day-length and light intensity. Biochem Physiol Pflanz 174(5–6):468–478

Bettaieb I, Zakhama N, Wannes WA, Kchouk ME, Marzouk B (2009) Water deficit effects on *Salvia officinalis* fatty acids and essential oils composition. Sci Hortic 120(2):271–275

Bhatia A, Tomer R, Kumar V, Singh SD, Pathak H (2012) Impact of tropospheric ozone on crop growth and productivity—a review. J Sci Ind Res 71(2):97–112

Biel C, Save R, Cristobal R, Cases MA, del Medio Natural ADC (2005) Effects of atmospheric carbon dioxide concentrations on *Thymus vulgaris*, *Thymus zygis* and *Thymus hyemalis*. Acta Hortic 676:61–65

Brachet J, Cosson L (1986) Changes in the total alkaloid content of *Datura innoxia* Mill. subjected to salt stress. J Exp Bot 37(5):650–656

Camisón Á, Martín MÁ, Sánchez-Bel P, Flors V, Alcaide F, Morcuende D, Pinto G, Solla A (2019) Hormone and secondary metabolite profiling in chestnut during susceptible and resistant interactions with *Phytophthora cinnamomi*. J Plant Physiol 241:153030

Caser M, D'Angiolillo F, Chitarra W, Lovisolo C, Ruffoni B, Pistelli L, Pistelli L, Scariot V (2018) Ecophysiological and phytochemical responses of *Salvia sinaloensis* Fern. to drought stress. Plant Growth Regul 84(2):383–394

Caser M, Chitarra W, D'Angiolillo F, Perrone I, Demasi S, Lovisolo C, Pistelli L, Scariot V (2019) Drought stress adaptation modulates plant secondary metabolite production in *Salvia dolomitica* Codd. Ind Crop Prod 129:85–96

Chaturvedi AK, Bahuguna RN, Pal M, Shah D, Maurya S, Jagadish KS (2017) Elevated CO_2 and heat stress interactions affect grain yield, quality and mineral nutrient composition in rice under field conditions. Field Crop Res 206:149–157

Chaudhary N, Choudhary KK (2021) Photosynthetic responses and tolerance mechanism of crop plants against salt stress. In: Srivastava PK, Kumar J, Prasad SM (eds) Salt stress responses in plants: perception, signaling, omics and tolerance mechanisms. Nova Science Publications, pp 65–84

Chaudhary IJ, Rathore D (2021a) Assessment of dose–response relationship between ozone dose and groundnut (*Arachis hypogaea* L.) cultivars using Open Top Chamber (OTC) and Ethylenediurea (EDU). Environ Technol Innov 22:101494

Chaudhary IJ, Rathore D (2021b) Assessment of ozone toxicity on cotton (*Gossypium hirsutum* L.) cultivars: its defensive system and intraspecific sensitivity. Plant Physiol Biochem 166:912–927

Chaudhary N, Bonfil DJ, Tas E (2021) Physiological and yield responses of spring wheat cultivars under realistic and acute levels of ozone. Atmosphere 12(11):1392

Chinnusamy V, Zhu J, Zhu JK (2007) Cold stress regulation of gene expression in plants. Trends Plant Sci 12(10):444–451

Chomel M, Guittonny-Larchevêque M, Fernandez C, Gallet C, DesRochers A, Paré D, Jackson BG, Baldy V (2016) Plant secondary metabolites: a key driver of litter decomposition and soil nutrient cycling. J Ecol 104(6):1527–1541

Choudhary KK, Agrawal SB (2014a) Cultivar specificity of tropical mung bean (*Vigna radiata* L.) to elevated ultraviolet-B: changes in antioxidative defense system, nitrogen metabolism and accumulation of jasmonic and salicylic acids. Environ Exp Bot 99:122–132

Choudhary KK, Agrawal SB (2014b) Ultraviolet-B induced changes in morphological, physiological, and biochemical parameters of two cultivars of pea (*Pisum sativum* L.). Ecotoxicol Environ Saf 100:178–187

Choudhary KK, Agrawal SB (2016) Assessment of fatty acid profile and seed mineral nutrients of two soybean (*Glycine max* L.) cultivars under elevated UV-B: role of ROS pigments and antioxidants. Photochem Photobiol 92:134–143

Choudhary KK, Chaudhary N, Agrawal SB, Agrawal M (2017) Reactive oxygen species: generation, damage, and quenching in plants during stress. In: Singh VP, Singh S, Tripathi DK, Prasad SM, Chauhan DK (eds) Reactive oxygen species in plants: boon or bane—revisiting the role of ROS. Wiley, pp 89–115

Choudhary KK, Singh S, Agrawal M, Agrawal SB (2021) Role of jasmonic and salicylic acid signaling in plants under UV-B stress. In: Aftab T, Yusuf M (eds) Jasmonates and salicylates signaling in plants. Signaling and communication in plants. Springer, Cham, pp 45–63

Choudhury FK, Rivero RM, Blumwald E, Mittler R (2017) Reactive oxygen species, abiotic stress and stress combination. Plant J 90(5):856–867

Clements DR, DiTommaso A, Hyvönen T (2014) Ecology and management of weeds in a changing climate. In: Chauhan B, Mahajan G (eds) Recent advances in weed management. Springer, New York, pp 13–37

Correia CM, Coutinho JF, Bacelar EA, Gonçalves BM, Björn LO, Moutinho Pereira J (2012) Ultraviolet-B radiation and nitrogen affect nutrient concentrations and the amount of nutrients acquired by above-ground organs of maize. Sci World J 2012:11

DeLang MN, Becker JS, Chang KL, Serre ML, Cooper OR, Schultz MG, Schröder S, Lu X, Zhang L, Deushi M, Josse B (2021) Mapping yearly fine resolution global surface ozone through the Bayesian maximum entropy data fusion of observations and model output for 1990–2017. Environ Sci Technol 55(8):4389–4398

Di Baccio D, Castagna A, Paoletti E, Sebastiani L, Ranieri A (2008) Could the differences in O_3 sensitivity between two poplar clones be related to a difference in antioxidant defense and secondary metabolic response to O_3 influx? Tree Physiol 28(12):1761–1772

Dias MC, Pinto DC, Freitas H, Santos C, Silva AM (2020) The antioxidant system in Olea europaea to enhanced UV-B radiation also depends on flavonoids and secoiridoids. Phytochemistry 170:112199

Dolzhenko Y, Bertea CM, Occhipinti A, Bossi S, Maffei ME (2010) UV-B modulates the interplay between terpenoids and flavonoids in peppermint (*Mentha*× *piperita* L.). J Photochem Photobiol B Biol 100(2):67–75

Dong J, Gruda N, Lam SK, Li X, Duan Z (2018) Effects of elevated CO_2 on nutritional quality of vegetables: a review. Front Plant Sci 9:924

Du B, Kreuzwieser J, Winkler JB, Ghirardo A, Schnitzler JP, Ache P, Alfarraj S, Hedrich R, White P, Rennenberg H (2018) Physiological responses of date palm (*Phoenix dactylifera*) seedlings to acute ozone exposure at high temperature. Environ Pollut 242:905–913

Dutta A, Sen J, Deswal R (2007) Downregulation of terpenoid indole alkaloid biosynthetic pathway by low temperature and cloning of a AP2 type C-repeat binding factor (CBF) from *Catharanthus roseus* (L). G Don. Plant Cell Rep 26(10):1869–1878

Dutta P, Chakraborti S, Chaudhuri KM, Mondal S (2020) Physiological responses and resilience of plants to climate change. In: Rakshit A, Singh H, Singh A, Singh U, Fraceto L (eds) New frontiers in stress management for durable agriculture. Springer, Singapore, pp 3–20
Easterling WE, Aggarwal PK, Batima P, Brander KM, Erda L, Howden SM, Kirilenko A, Morton J, Soussana JF, Schmidhuber J, Tubiello FN (2007) Food, fibre and forest products. Climate Change 2007:273–313
Edreva A, Velikova V, Tsonev T, Dagnon S, Gürel A, Aktaş L, Gesheva E (2008) Stress-protective role of secondary metabolites: diversity of functions and mechanisms. Gen Appl Plant Physiol 34(1-2):67–78
Emberson L (2020) Effects of ozone on agriculture, forests and grasslands. Phil Trans R Soc A 378(2183):20190327
Fahlén A, Welander M, Wennersten R (1997) Effects of light–temperature regimes on plant growth and essential oil yield of selected aromatic plants. J Sci Food Agric 73(1):111–119
Falcinelli B, Sileoni V, Marconi O, Perretti G, Quinet M, Lutts S, Benincasa P (2017) Germination under moderate salinity increases phenolic content and antioxidant activity in rapeseed (*Brassica napus* var oleifera Del.) sprouts. Molecules 22(8):1377
Fanzo J (2018) Challenges and impacts of poor nutrition (No. 2175-2019-152)
Fatima A, Singh AA, Mukherjee A, Agrawal M, Agrawal SB (2018) Variability in defence mechanism operating in three wheat cultivars having different levels of sensitivity against elevated ozone. Environ Exp Bot 155:66–78
Feeny P (1976) Plant apparency and chemical defense. In: Wallace JW, Mansell RL (eds) Biochemical interaction between plants and insects. Recent advances in phytochemistry. Springer, Boston, MA, pp 1–40
Forster PM (2011) Stratospheric changes and climate. Scientific assessment of ozone depletion: 2010. Global Ozone research and monitoring project—report no. 52. World Meteorological Organization, Geneva, Switzerland, pp 1–60
Frei M (2015) Breeding of ozone resistant rice: relevance, approaches and challenges. Environ Pollut 197:144–155
Funk CC, Brown ME (2009) Declining global per capita agricultural production and warming oceans threaten food security. Food Secur 1(3):271–289
Gamage D, Thompson M, Sutherland M, Hirotsu N, Makino A, Seneweera S (2018) New insights into the cellular mechanisms of plant growth at elevated atmospheric carbon dioxide concentrations. Plant Cell Environ 41(6):1233–1246
Gao SQ, Chen M, Xu ZS, Zhao CP, Li L, Xu HJ, Tang YM, Zhao X, Ma YZ (2011) The soybean GmbZIP1 transcription factor enhances multiple abiotic stress tolerances in transgenic plants. Plant Mol Biol 75(6):537–553
García-Macías P, Ordidge M, Vysini E, Waroonphan S, Battey NH, Gordon MH, Hadley P, John P, Lovegrove JA, Wagstaffe A (2007) Changes in the flavonoid and phenolic acid contents and antioxidant activity of red leaf lettuce (Lollo Rosso) due to cultivation under plastic films varying in ultraviolet transparency. J Agric Food Chem 55(25):10168–10172
Geilfus CM (2019) Plant secondary compounds. In: Controlled environment horticulture. Springer, Cham, pp 19–33
Gharibi S, Tabatabaei BES, Saeidi G, Talebi M, Matkowski A (2019) The effect of drought stress on polyphenolic compounds and expression of flavonoid biosynthesis related genes in *Achillea pachycephala* Rech. f. Phytochemistry 162:90–98
Ghasemi S, Kumleh HH, Kordrostami M (2019) Changes in the expression of some genes involved in the biosynthesis of secondary metabolites in *Cuminum cyminum* L. under UV stress. Protoplasma 256(1):279–290
Ghasemzadeh A, Jaafar HZ, Rahmat A (2010) Elevated carbon dioxide increases contents of flavonoids and phenolic compounds, and antioxidant activities in Malaysian young ginger (Zingiber officinale Roscoe.) varieties. Molecules 15(11):7907–7922

Ghasemzadeh A, Jaafar HZ, Rahmat A (2011) Effects of solvent type on phenolics and flavonoids content and antioxidant activities in two varieties of young ginger (*Zingiber officinale Roscoe*) extracts. J Med Plants Res 5(7):1147–1154

Ghorbanpour M, Khavazi K, Hatami M (2014) Chemical compositions and antimicrobial activity of *Salvia officinalis* L. essential oil under rhizobacteria (*Pseudomonas fluorescens* and Putida) inoculation. Eur J Soil Biol

Ghosh A, Agrawal M, Agrawal SB (2020a) Effect of water deficit stress on an Indian wheat cultivar (*Triticum aestivum* L. HD 2967) under ambient and elevated level of ozone. Sci Total Environ 714:136837

Ghosh A, Pandey B, Agrawal M, Agrawal SB (2020b) Interactive effects and competitive shift between *Triticum aestivum* L.(wheat) and *Chenopodium album* L.(fat-hen) under ambient and elevated ozone. Environ Pollut 265:114764

Gil M, Pontin M, Berli F, Bottini R, Piccoli P (2012) Metabolism of terpenes in the response of grape (*Vitis vinifera* L.) leaf tissues to UV-B radiation. Phytochemistry 77:89–98

Giri A, Armstrong B, Rajashekar CB (2016) Elevated carbon dioxide level suppresses nutritional quality of lettuce and spinach. Am J Plant Sci 7(01):246

Gouvea DR, Gobbo-Neto L, Lopes NP (2012) The influence of biotic and abiotic factors on the production of secondary metabolites in medicinal plants. In: Plant bioactives and drug discovery: principles, practice, and perspectives, vol 17, p 419

Han YJ, Gharibeshghi A, Mewis I, Förster N, Beck W, Ulrichs C (2021) Effect of different durations of moderate ozone exposure on secondary metabolites of *Brassica campestris* L. ssp. chinensis. J Hortic Sci Biotechnol 96(1):110–120

Hanen F, Ksouri R, Megdiche W, Trabelsi N, Boulaaba M, Abdelly C (2008) Effect of salinity on growth, leaf-phenolic content and antioxidant scavenging activity in *Cynara cardunculus* L. In: Öztürk M, Ashraf M, Grignon C (eds) Abdelly C. Biosaline agriculture and high salinity tolerance, Birkhäuser, Basel, pp 335–343

Harborne JB, Williams CA (2000) Advances in flavonoid research since 1992. Phytochemistry 55(6):481–504

Hasanuzzaman M, Nahar K, Alam M, Roychowdhury R, Fujita M (2013) Physiological, biochemical, and molecular mechanisms of heat stress tolerance in plants. Int J Mol Sci 14(5):9643–9684

Helmig D, Ortega J, Duhl T, Tanner D, Guenther A, Harley P, Wiedinmyer C, Milford J, Sakulyanontvittaya T (2007) Sesquiterpene emissions from pine trees—identifications, emission rates and flux estimates for the contiguous United States. Environ Sci Technol 41(5):1545–1553

Hernández I, Alegre L, Van Breusegem F, Munné-Bosch S (2009) How relevant are flavonoids as antioxidants in plants? Trends Plant Sci 14(3):125–132

Hewitt CN, MacKenzie AR, Di Carlo P, Di Marco CF, Dorsey JR, Evans M, Fowler D, Gallagher MW, Hopkins JR, Jones CE, Langford B (2009) Nitrogen management is essential to prevent tropical oil palm plantations from causing ground-level ozone pollution. Proc Natl Acad Sci U S A 106(44):18447–18451

Higashio H, Hirokane H, Sato F, Tokuda S, Uragami A (2004) Effect of UV irradiation after the harvest on the content of flavonoid in vegetables. In: V International postharvest symposium, vol 682, pp 1007–1012

Hirata K, Asada M, Yatani E, Miyamoto K, Miura Y (1993) Effects of near-ultraviolet light on alkaloid production in *Catharanthus roseus* plants. Planta Med 59(01):46–50

Hofmann RW, Swinny EE, Bloor SJ, Markham KR, Ryan KG, Campbell BD, Jordan BR, Fountain DW (2000) Responses of nine *Trifolium repens* L. populations to ultraviolet-B radiation: differential flavonol glycoside accumulation and biomass production. Ann Bot 86(3):527–537

Huang X, Yao J, Zhao Y, Xie D, Jiang X, Xu Z (2016) Efficient rutin and quercetin biosynthesis through flavonoids-related gene expression in *Fagopyrum tataricum* Gaertn. hairy root cultures with UV-B irradiation. Front Plant Sci 7:63

Hussin S, Geissler N, El-Far MM, Koyro HW (2017) Effects of salinity and short-term elevated atmospheric CO_2 on the chemical equilibrium between CO_2 fixation and photosynthetic electron transport of *Stevia rebaudiana* Bertoni. Plant Physiol Biochem 118:178–186

Ibrahim MH, Jaafar HZ (2012) Impact of elevated carbon dioxide on primary, secondary metabolites and antioxidant responses of *Eleais guineensis* Jacq. (Oil Palm). Molecules 17(5):5195–5211

Ibrahim MA, Mäenpää M, Hassinen V, Kontunen-Soppela S, Malec L, Rousi M, Pietikäinen L, Tervahauta A, Kärenlampi S, Holopainen JK, Oksanen EJ (2010) Elevation of night-time temperature increases terpenoid emissions from *Betula pendula* and *Populus tremula*. J Exp Bot 61(6):1583–1595

IPCC (2013) Climate change 2013: the physical science basis. In: Stocker TF, Qin D, Plattner GK, Tignor MMHL, Allen SK, Boschung J, Nauels A, Xia Y, Bex B, Midgley PM (eds) Contribution of working group I to the fifth assessment report of the intergovernmental panel on climate change, vol 2. Cambridge University Press, pp 1–27

IPCC (2014) Climate change 2014: synthesis report. In: Pachauri RK, Allen MR, Barros VR, Broome J, Cramer W, Christ R, Church JA, Clarke L, Dahe Q, Dasgupta P, Dubash NK et al (eds) Proceedings of the contribution of working groups I, II and III to the fifth assessment report of the Intergovernmental Panel on Climate Change. IPCC, Geneva, p 151

IPCC (2019) In: Shukla PR, Skea J, Calvo Buendia E, Masson-Delmotte V, Pörtner HO, Roberts DC, Zhai P, Slade R, Connors S, Van Diemen R, Ferrat M et al (eds) Climate change and land: an IPCC special report on climate change, desertification, land degradation, sustainable land management, food security, and greenhouse gas fluxes in terrestrial ecosystems. IPCC, Geneva, p 865

IPCC (2021) Summary for policymakers. In: Masson Delmotte V, Zhai P, Pirani A, Connors SL, Péan C, Berger S, Caud N et al (eds) Climate change 2021: the physical science basis. Contribution of working group I to the sixth assessment report of the Intergovernmental Panel on Climate Change. Cambridge University Press

Iriti M, Faoro F (2009) Chemical diversity and defense metabolism: how plants cope with pathogens and ozone pollution. Int J Mol Sci 10(8):3371–3399

Jaiswal D, Pandey A, Mukherjee A, Agrawal M, Agrawal SB (2020) Alterations in growth, antioxidative defense and medicinally important compounds of *Curcuma caesia* Roxb. under elevated ultraviolet-B radiation. Environ Exp Bot 177:104152

Jaleel CA, Manivannan P, Sankar B, Kishorekumar A, Gopi R, Somasundaram R, Panneerselvam R (2007) Water deficit stress mitigation by calcium chloride in *Catharanthus roseus*: Effects on oxidative stress, proline metabolism and indole alkaloid accumulation. Colloids Surf B: Biointerfaces 60(1):110–116

Janas KM, Cvikrová M, Pałagiewicz A, Szafranska K, Posmyk MM (2002) Constitutive elevated accumulation of phenylpropanoids in soybean roots at low temperature. Plant Sci 163(2):369–373

Jansen MA, Hectors K, O'Brien NM, Guisez Y, Potters G (2008) Plant stress and human health: do human consumers benefit from UV-B acclimated crops? Plant Sci 175(4):449–458

Jansen G, Jürgens HU, Ordon F (2009) Effects of temperature on the alkaloid content of seeds of *Lupinus angustifolius* cultivars. J Agron Crop Sci 195(3):172–177

Janská A, Maršík P, Zelenková S, Ovesná J (2010) Cold stress and acclimation–what is important for metabolic adjustment? Plant Biol 12(3):395–405

Jia X, Wang W, Chen Z, He Y, Liu J (2014) Concentrations of secondary metabolites in tissues and root exudates of wheat seedlings changed under elevated atmospheric CO_2 and cadmium-contaminated soils. Environ Exp Bot 107:134–143

Jochum GM, Mudge KW, Thomas RB (2007) Elevated temperatures increase leaf senescence and root secondary metabolite concentrations in the understory herb *Panax quinquefolius* (Araliaceae). Am J Bot 94(5):819–826

Jogawat A, Yadav B, Lakra N, Singh AK, Narayan OP (2021) Crosstalk between phytohormones and secondary metabolites in the drought stress tolerance of crop plants: a review. Physiol Plant 172(2):1106–1132

Kakani VG, Reddy KR, Zhao D, Sailaja K (2003) Field crop responses to ultraviolet-B radiation: a review. Agric For Meteorol 120(1–4):191–218

Kataoka I, Beppu K (2004) UV irradiance increases development of red skin color and anthocyanins in Hakuho'Peach. HortScience *39*(6):1234–1237

Kataria S, Jajoo A, Guruprasad KN (2014) Impact of increasing ultraviolet-B (UV-B) radiation on photosynthetic processes. J Photochem Photobiol B Biol 137:55–66

Kaundal M, Bhatt V, Kumar R (2018) Elevated CO_2 and temperature effect on essential oil content and composition of *Valeriana jatamansi* Jones. with organic manure application in a Western Himalayan region. J Essent Oil-Bear Plants 21(4):1041–1050

Kermani SG, Saeidi G, Sabzalian MR, Gianinetti A (2019) Drought stress influenced sesamin and sesamolin content and polyphenolic components in sesame (*Sesamum indicum* L.) populations with contrasting seed coat colors. Food Chem 289:360–368

Kimball BA (2016) Crop responses to elevated CO_2 and interactions with H_2O, N, and temperature. Curr Opin Plant Biol 31:36–43

Klein FRS, Reis A, Kleinowski AM, Telles RT, Amarante LD, Peters JA, Braga EJB (2018) UV-B radiation as an elicitor of secondary metabolite production in plants of the genus *Alternanthera*. Acta Bot Bras 32:615–623

Ksouri R, Megdiche W, Debez A, Falleh H, Grignon C, Abdelly C (2007) Salinity effects on polyphenol content and antioxidant activities in leaves of the halophyte *Cakile maritima*. Plant Physiol Biochem 45(3–4):244–249

Kumar P, Sharma PK (2020) Soil salinity and food security in India. Front Sustain Food Syst 4:174

Kumari R, Agrawal SB (2011) Comparative analysis of essential oil composition and oil containing glands in *Ocimum sanctum* L. (Holy basil) under ambient and supplemental level of UV-B through gas chromatography–mass spectrometry and scanning electron microscopy. Acta Physiol Plant 33(4):1093–1101

Kumari R, Agrawal SB, Sarkar A (2009a) Evaluation of changes in oil cells and composition of essential oil in lemongrass (*Cymbopogon citratus* (DC) Stapf.) due to supplemental ultraviolet-B irradiation. Curr Sci 97:1137–1142

Kumari R, Singh S, Agrawal SB (2009b) Effects of supplemental ultraviolet-B radiation on growth and physiology of *Acorus calamus* L. (sweet flag). Acta Biol Cracov Ser Bot 51:19–27

Larson RA, Weber EJ (2018) Reaction mechanisms in environmental organic chemistry. Routledge

Levine LH, Kasahara H, Kopka J, Erban A, Fehrl I, Kaplan F, Zhao W, Littell RC, Guy C, Wheeler R, Sager J (2008) Physiologic and metabolic responses of wheat seedlings to elevated and super-elevated carbon dioxide. Adv Space Res 42(12):1917–1928

Li X, Zhang L, Ahammed GJ, Li ZX, Wei JP, Shen C, Yan P, Zhang LP, Han WY (2017) Stimulation in primary and secondary metabolism by elevated carbon dioxide alters green tea quality in *Camellia sinensis* L. Sci Rep 7(1):1–12

Li M, Li Y, Zhang W, Li S, Gao Y, Ai X, Zhang D, Liu B, Li Q (2018) Metabolomics analysis reveals that elevated atmospheric CO_2 alleviates drought stress in cucumber seedling leaves. Anal Biochem 559:71–85

Li Y, Qin W, Fu X, Zhang Y, Hassani D, Kayani SI, Xie L, Liu H, Chen T, Yan X, Peng B (2021) Transcriptomic analysis reveals the parallel transcriptional regulation of UV-B-induced artemisinin and flavonoid accumulation in *Artemisia annua* L. Plant Physiol Biochem 163:189–200

Lim JH, Park KJ, Kim BK, Jeong JW, Kim HJ (2012) Effect of salinity stress on phenolic compounds and carotenoids in buckwheat (*Fagopyrum esculentum* M.) sprout. Food Chem 135(3):1065–1070

Liu C, Blount JW, Steele CL, Dixon RA (2002) Bottlenecks for metabolic engineering of isoflavone glycoconjugates in *Arabidopsis*. Proc Natl Acad Sci U S A 99:14578–14583

Liu H, Cao X, Liu X, Xin R, Wang J, Gao J, Wu B, Gao L, Xu C, Zhang B, Grierson D (2017) UV-B irradiation differentially regulates terpene synthases and terpene content of peach. Plant Cell Environ 40(10):2261–2275

Lopez-Berenguer C, Martínez-Ballesta MDC, Moreno DA, Carvajal M, Garcia-Viguera C (2009) Growing hardier crops for better health: salinity tolerance and the nutritional value of broccoli. J Agric Food Chem 57(2):572–578
Loreto F, Velikova V (2001) Isoprene produced by leaves protects the photosynthetic apparatus against ozone damage, quenches ozone products, and reduces lipid peroxidation of cellular membranes. Plant Physiol 127(4):1781–1787
Luis JC, Pérez RM, González FV (2007) UV-B radiation effects on foliar concentrations of rosmarinic and carnosic acids in rosemary plants. Food Chem 101(3):1211–1215
Maffei M, Scannerini S (2000) UV-B effect on photomorphogenesis and essential oil composition in peppermint (*Mentha piperita* L.). J Essent Oil Res 12(5):523–529
Majeed A, Chaudhry Z, Muhammad Z (2012) Allelopathic assessment of fresh aqueous extracts of *Chenopodium album* L. for growth and yield of wheat (*Triticum aestivum* L.). Pak J Bot 44(1):165–167
Mao B, Wang Y, Zhao TH, Tian RR, Wang W, Ye JS (2017) Combined effects of elevated O_3 concentrations and enhanced UV-B radiation of the biometric and biochemical properties of soybean roots. Front Plant Sci 8:1568
Marais E, Jacobs G, Holcroft DM (2001) Postharvest irradiation enhances anthocyanin synthesis in apples but not in pears. HortScience 36(4):738–740
Marchica A, Ascrizzi R, Flamini G, Cotrozzi L, Tonelli M, Lorenzini G, Nali C, Pellegrini E (2021) Ozone as eustress for enhancing secondary metabolites and bioactive properties in *Salvia officinalis*. Ind Crop Prod 170:113730
Mishra AK, Agrawal SB (2015) Biochemical and physiological characteristics of tropical mung bean (*Vigna radiata* L.) cultivars against chronic ozone stress: an insight to cultivar-specific response. Protoplasma 252(3):797–811
Mishra AK, Rai R, Agrawal SB (2013) Individual and interactive effects of elevated carbon dioxide and ozone on tropical wheat (*Triticum aestivum* L.) cultivars with special emphasis on ROS generation and activation of antioxidant defence system. Indian J Biochem Biophys 50:139–149
Misselhorn A, Aggarwal P, Ericksen P, Gregory P, Horn-Phathanothai L, Ingram J, Wiebe K (2012) A vision for attaining food security. Curr Opin Environ Sustain 4(1):7–17
Morales LO, Tegelberg R, Brosche M, Keinänen M, Lindfors A, Aphalo PJ (2010) Effects of solar UV-A and UV-B radiation on gene expression and phenolic accumulation in *Betula pendula* leaves. Tree Physiol 30(7):923–934
Morales A, Zurita-Silva A, Maldonado J, Silva H (2017) Transcriptional responses of Chilean quinoa (*Chenopodium quinoa* Wild.) under water deficit conditions uncovers ABA-independent expression patterns. Front Plant Sci 8:216
Nabavi SM, Šamec D, Tomczyk M, Milella L, Russo D, Habtemariam S, Suntar I, Rastrelli L, Daglia M, Xiao J, Giampieri F (2020) Flavonoid biosynthetic pathways in plants: versatile targets for metabolic engineering. Biotechnol Adv 38:107316
Neugart S, Zietz M, Schreiner M, Rohn S, Kroh LW, Krumbein A (2012) Structurally different flavonol glycosides and hydroxycinnamic acid derivatives respond differently to moderate UV-B radiation exposure. Physiol Plant 145(4):582–593
Nezhadahmadi A, Prodhan ZH, Faruq G (2013) Drought tolerance in wheat. Sci World J 2013
Nichols SN, Hofmann RW, Williams WM (2015) Physiological drought resistance and accumulation of leaf phenolics in white clover interspecific hybrids. Environ Exp Bot 119:40–47
NOAA (National Centers for Environmental Information) (2020). State of the climate: global climate report for annual 2020. https://www.ncdc.noaa.gov/sotc/global/202013. Accessed 15 Mar 2021
Nogués S, Allen DJ, Morison JI, Baker NR (1998) Ultraviolet-B radiation effects on water relations, leaf development, and photosynthesis in droughted pea plants. Plant Physiol 117(1):173–181
Nowak M, Kleinwaechter M, Manderscheid R, Weigel HJ, Selmar D (2010) Drought stress increases the accumulation of monoterpenes in sage (*Salvia officinalis*), an effect that is compensated by elevated carbon dioxide concentration. J Appl Bot Food Qual 83(2):133–136

Oerke EC (2006) Crop losses to pests. J Agric Sci 144(1):31–43
Osman ME, Elfeky SS, El-Soud KA, Hasan AM (2007) Response of *Catharanthus roseus* shoots to salinity and drought in relation to vincristine alkaloid content. Asian J Plant Sci
Pan T, Wang Y, Wang L, Ding J, Cao Y, Qin G, Yan L, Xi L, Zhang J, Zou Z (2020) Increased CO_2 and light intensity regulate growth and leaf gas exchange in tomato. Physiol Plant 168(3):694–708
Pandey J, Chaplot K (2007) Effects of enhanced UV-B radiation on physiological and biochemical characteristics of wheat [*Triticum aestivum* (L.) var. Raj 3077]. Res Crops 8(2):401
Panta S, Flowers T, Lane P, Doyle R, Haros G, Shabala S (2014) Halophyte agriculture: success stories. Environ Exp Bot 107:71–83
Paoletti E, Hoshika Y, Arab L, Martini S, Cotrozzi L, Weber D, Ache P, Neri L, Baraldi R, Pellegrini E, Müller HM (2021) Date palm responses to a chronic, realistic ozone exposure in a FACE experiment. Environ Res 195:110868
Paulsen J, Selmar D (2016) Case study: the difficulty of correct reference values when evaluating the effects of drought stress: a case study with *Thymus vulgaris*. J Appl Bot Food Qual 89:287–289
Peñuelas J, Staudt M (2010) BVOCs and global change. Trends Plant Sci 15(3):133–144
Pfister GG, Walters S, Lamarque JF, Fast J, Barth MC, Wong J, Done J, Holland G, Bruyère CL (2014) Projections of future summertime ozone over the US. J Geophys Res Atmos 119(9):5559–5582
Picchi V, Monga R, Marzuoli R, Gerosa G, Faoro F (2017) The ozone-like syndrome in durum wheat (*Triticum durum* Desf.): mechanisms underlying the different symptomatic responses of two sensitive cultivars. Plant Physiol Biochem 112:261–269
Piubelli GC, Hoffmann-Campo CB, Moscardi F, Miyakubo SH, Neves De Oliveira MC (2005) Are chemical compounds important for soybean resistance to *Anticarsia gemmatalis*? J Chem Ecol 31(7):1509–1525
Pollastri S, Jorba I, Hawkins TJ, Llusià J, Michelozzi M, Navajas D, Peñuelas J, Hussey PJ, Knight MR, Loreto F (2019) Leaves of isoprene-emitting tobacco plants maintain PSII stability at high temperatures. New Phytol 223(3):1307–1318
Qadir M, Quillérou E, Nangia V, Murtaza G, Singh M, Thomas RJ, Drechsel P, Noble AD (2014) Economics of salt-induced land degradation and restoration. Nat Resour Forum 38(4):282–295
Qian C, Chen Z, Liu Q, Mao W, Chen Y, Tian W, Liu Y, Han J, Ouyang X, Huang X (2020) Coordinated transcriptional regulation by the UV-B photoreceptor and multiple transcription factors for plant UV-B responses. Mol Plant 13(5):777–792
Radman R, Saez T, Bucke C, Keshavarz T (2003) Elicitation of plants and microbial cell systems. Biotechnol Appl Biochem 37(1):91–102
Radušienė J, Karpavičienė B, Stanius Ž (2012) Effect of external and internal factors on secondary metabolites accumulation in St. John's worth. Bot Lith 18(2):101–108
Rai R, Meena RP, Smita SS, Shukla A, Rai SK, Pandey-Rai S (2011) UV-B and UV-C pre-treatments induce physiological changes and artemisinin biosynthesis in *Artemisia annua* L.—an antimalarial plant. J Photochem Photobiol B Biol 105(3):216–225
Rajabbeigi E, Eichholz I, Beesk N, Ulrichs C, Kroh LW, Rohn S, Huyskens-Keil S (2013) Interaction of drought stress and UV-B radiation-impact on biomass production and flavonoid metabolism in lettuce (*Lactuca sativa* L.). J Appl Bot Food Qual 86(1):190–197
Ramani S, Chelliah J (2007) UV-B-induced signaling events leading to enhanced-production of catharanthine in *Catharanthus roseus* cell suspension cultures. BMC Plant Biol 7(1):1–17
Ramani S, Jayabaskaran C (2008) Enhanced catharanthine and vindoline production in suspension cultures of *Catharanthus roseus* by ultraviolet-B light. J Mol Signal 3(1):1–6
Rao CR, Raju BMK, Rao AS, Rao KV, Rao VUM, Ramachandran K, Venkateswarlu B, Sikka AK, Rao MS, Maheswari M, Rao CS (2016) A district level assessment of vulnerability of Indian agriculture to climate change. Curr Sci 110:1939–1946
Rathore D, Chaudhary IJ (2019) Ozone risk assessment of castor (*Ricinus communis* L.) cultivars using open top chamber and ethylenediurea (EDU). Environ Pollut 244:257–269

Razavizadeh R, Komatsu S (2018) Changes in essential oil and physiological parameters of callus and seedlings of *Carum copticum* L. under in vitro drought stress. J Food Meas Charact 12(3):1581–1592

Reddy KR, Hodges HF (eds) (2000) Climate change and global crop productivity. CABI

Rengasamy P (2010) Soil processes affecting crop production in salt-affected soils. Funct Plant Biol 37(7):613–620

Rezayian M, Niknam V, Ebrahimzadeh H (2018) Differential responses of phenolic compounds of *Brassica napus* under drought stress. Iran J f Plant Physiol 8(3):2417–2425

Rhoades DF, Cates RG (1976) Toward a general theory of plant anti-herbivore chemistry. In: Wallace JW, Mansell RL (eds) Biochemical interaction between plants and insects. Recent advances in phytochemistry. Springer, Boston, MA, pp 168–213

Rice JC, Garcia SM (2011) Fisheries, food security, climate change, and biodiversity: characteristics of the sector and perspectives on emerging issues. ICES J Mar Sci 68(6):1343–1353

Rice-Evans C, Miller N, Paganga G (1997) Antioxidant properties of phenolic compounds. Trends Plant Sci 2(4):152–159

Richet N, Tozo K, Afif D, Banvoy J, Legay S, Dizengremel P, Cabané M (2012) The response to daylight or continuous ozone of phenylpropanoid and lignin biosynthesis pathways in poplar differs between leaves and wood. Planta 236(2):727–737

Robinson EA, Ryan GD, Newman JA (2012) A meta-analytical review of the effects of elevated CO_2 on plant–arthropod interactions highlights the importance of interacting environmental and biological variables. New Phytol 194(2):321–336

Rosa BL, Souza JP, Pereira EG (2019) Increased atmospheric CO_2 changes the photosynthetic responses of *Acrocomia aculeata* (Arecaceae) to drought. Acta Bot Bras 33:486–497

Rozema J, Blokker P, Fuertes M, Broekman R (2009) UV-B absorbing compounds in present-day and fossil pollen, spores, cuticles, seed coats and wood: evaluation of a proxy for solar UV radiation. Photochem Photobiol Sci 8(9):1233–1243

Schirrmacher G, Schnitzler WH, Grassmann J (2004) Determination of secondary plant metabolites and antioxidative capacity as new parameter for quality evaluation: Indicated by the new Asia salad *Gynura bicolor*. J Appl Bot 78(2):133–134

Schreiner M, Krumbein A, Mewis I, Ulrichs C, Huyskens-Keil S (2009) Short-term and moderate UV-B radiation effects on secondary plant metabolism in different organs of nasturtium (*Tropaeolum majus* L.). Innovative Food Sci Emerg Technol 10(1):93–96

Shih CH, Chu H, Tang LK, Sakamoto W, Maekawa M, Chu IK, Wang M, Lo C (2008) Functional characterization of key structural genes in rice flavonoid biosynthesis. Planta 228(6):1043–1054

Sicard P, Anav A, De Marco A, Paoletti E (2017) Projected global ground-level ozone impacts on vegetation under different emission and climate scenarios. Atmos Chem Phys 17(19):12177–12196

Singh AA, Agrawal SB, Shahi JP, Agrawal M (2014) Assessment of growth and yield losses in two *Zea mays* L. cultivars (quality protein maize and nonquality protein maize) under projected levels of ozone. Environ Sci Pollut Res 21(4):2628–2641

Singh-Sangwan N, Abad Farooqi AH, Singh Sangwan R (1994) Effect of drought stress on growth and essential oil metabolism in lemongrasses. New Phytol 128(1):173–179

Sobuj N, Virjamo V, Zhang Y, Nybakken L, Julkunen-Tiitto R (2018) Impacts of elevated temperature and CO_2 concentration on growth and phenolics in the sexually dimorphic *Populus tremula* (L.). Environ Exp Bot 146:34–44

Solovchenko AE, Merzlyak MN (2008) Screening of visible and UV radiation as a photoprotective mechanism in plants. Russ J Plant Physiol 55(6):719–737

Takshak S, Agrawal SB (2014) Effect of ultraviolet-B radiation on biomass production, lipid peroxidation, reactive oxygen species, and antioxidants in *Withania somnifera*. Biol Plant 58(2):328–334

Takshak S, Agrawal SB (2015) Defence strategies adopted by the medicinal plant *Coleus forskohlii* against supplemental ultraviolet-B radiation: augmentation of secondary metabolites and antioxidants. Plant Physiol Biochem 97:124–138

Tamogami S, Rakwal R, Kodama O (1997) Phytoalexin production elicited by exogenously applied jasmonic acid in rice leaves (*Oryza sativa* L.) is under the control of cytokinins and ascorbic acid. FEBS Lett 412(1):61–64

Tarchoune I, Sgherri C, Baatour O, Izzo R, Lachaal M, Navari-Izzo F (2012a) Phenolic acids and total antioxidant activity in *Ocimum basilicum* L. grown under Na_2SO_4 medium. J Med Plants Res 6(48):5868–5875

Tarchoune I, Sgherri C, Izzo R, Lachaâl M, Navari-Izzo F, Ouerghi Z (2012b) Changes in the antioxidative systems of *Ocimum basilicum* L. (cv. Fine) under different sodium salts. Acta Physiol Plant 10:738–847

Thakur M, Bhattacharya S, Khosla PK, Puri S (2019) Improving production of plant secondary metabolites through biotic and abiotic elicitation. J Appl Res Med Aromat Plants 12:1–12

Thompson B, Cohen MJ (eds) (2012) The impact of climate change and bioenergy on nutrition. Springer

Tomás-Barberán FA, Espín JC (2001) Phenolic compounds and related enzymes as determinants of quality in fruits and vegetables. J Sci Food Agric 81(9):853–876

Tounekti T, Vadel AM, Ennajeh M, Khemira H, Munné-Bosch S (2011) Ionic interactions and salinity affect monoterpene and phenolic diterpene composition in rosemary (*Rosmarinus officinalis*). J Plant Nutr Soil Sci 174(3):504–514

Treutter D (2005) Significance of flavonoids in plant resistance and enhancement of their biosynthesis. Plant Biol 7(06):581–591

Tripathi R, Agrawal SB (2013) Interactive effect of supplemental ultraviolet B and elevated ozone on seed yield and oil quality of two cultivars of linseed (*Linum usitatissimum* L.) carried out in open top chambers. J Sci Food Agric 93(5):1016–1025

Tripathi D, Meena RP, Pandey-Rai S (2021) Short term UV-B radiation mediated modulation of physiological traits and withanolides production in *Withania coagulans* (L.) Dunal under in-vitro condition. Physiol Mol Biol Plants 27(8):1823–1835

Valifard M, Mohsenzadeh S, Kholdebarin B (2017) Salinity effects on phenolic content and antioxidant activity of *Salvia macrosiphon*. Iran J Sci Technol Trans A Sci 41(2):295–300

Vashisth D, Kumar R, Rastogi S, Patel VK, Kalra A, Gupta MM, Gupta AK, Shasany AK (2018) Transcriptome changes induced by abiotic stresses in *Artemisia annua*. Sci Rep 8(1):1–14

Vaughan MM, Christensen S, Schmelz EA, Huffaker A, Mcauslane HJ, Alborn HT, Romero M, Allen LH, Teal PE (2015) Accumulation of terpenoid phytoalexins in maize roots is associated with drought tolerance. Plant Cell Environ 38(11):2195–2207

Verdaguer D, Llorens L, Bernal M, Badosa J (2012) Photomorphogenic effects of UVB and UVA radiation on leaves of six Mediterranean sclerophyllous woody species subjected to two different watering regimes at the seedling stage. Environ Exp Bot 79:66–75

Verma AK, Deepti S (2016) Abiotic stress and crop improvement: current scenario. Adv Plants Agric Res 4(4):00149

Verma N, Shukla S (2015) Impact of various factors responsible for fluctuation in plant secondary metabolites. J Appl Res Med Aromat Plants 2(4):105–113

Vurro E, Bruni R, Bianchi A, di Toppi LS (2009) Elevated atmospheric CO_2 decreases oxidative stress and increases essential oil yield in leaves of *Thymus vulgaris* grown in a mini-FACE system. Environ Exp Bot 65(1):99–106

Wang L, Feng Z, Schjoerring JK (2013) Effects of elevated atmospheric CO_2 on physiology and yield of wheat (*Triticum aestivum* L.): a meta-analytic test of current hypotheses. Agric Ecosyst Environ 178:57–63

Wang P, Marsh EL, Ainsworth EA, Leakey AD, Sheflin AM, Schachtman DP (2017) Shifts in microbial communities in soil, rhizosphere and roots of two major crop systems under elevated CO_2 and O_3. Sci Rep 7(1):1–12

Wang Z, Li G, Sun H, Ma L, Guo Y, Zhao Z, Gao H, Mei L (2018) Effects of drought stress on photosynthesis and photosynthetic electron transport chain in young apple tree leaves. Biol Open 7(11):bio035279

Waraich EA, Ahmad R, Halim A, Aziz T (2012) Alleviation of temperature stress by nutrient management in crop plants: a review. J Soil Sci Plant Nutr 12(2):221–244

Warren JM, Bassman JH, Fellman JK, Mattinson DS, Eigenbrode S (2003) Ultraviolet-B radiation alters phenolic salicylate and flavonoid composition of *Populus trichocarpa* leaves. Tree Physiol 23(8):527–535

Wedow JM, Ainsworth EA, Li S (2021) Plant biochemistry influences tropospheric ozone formation, destruction, deposition, and response. Trends Biochem Sci 46(12):992–1002

WHO (World Health Organization) (2020) The state of food security and nutrition in the world 2020: transforming food systems for affordable healthy diets, vol 2020. Food and Agriculture Organization

WMO (World Meteorological Organization) (2010) Scientific assessment of ozone depletion: 2010. Global Ozone Research and Monitoring Project-Report No 53. 2010

Xie ZM, Zou HF, Lei G, Wei W, Zhou QY, Niu CF, Liao Y, Tian AG, Ma B, Zhang WK, Zhang JS (2009) Soybean Trihelix transcription factors GmGT-2A and GmGT-2B improve plant tolerance to abiotic stresses in transgenic *Arabidopsis*. PLoS One 4(9):e6898

Xu Z, Zhou G, Shimizu H (2010) Plant responses to drought and rewatering. Plant Signal Behav 5(6):649–654

Yadav DS, Rai R, Mishra AK, Chaudhary N, Mukherjee A, Agrawal SB, Agrawal M (2019) ROS production and its detoxification in early and late sown cultivars of wheat under future O_3 concentration. Sci Total Environ 659:200–210

Yadav B, Jogawat A, Rahman MS, Narayan OP (2021) Secondary metabolites in the drought stress tolerance of crop plants: a review. Gene Rep 23:101040

Yang SH, Wang LJ, Li SH, Duan W, Loescher W, Liang ZC (2007) The effects of UV-B radiation on photosynthesis in relation to Photosystem II photochemistry, thermal dissipation and antioxidant defenses in winter wheat (*Triticum aestivum* L.) seedlings at different growth temperatures. Funct Plant Biol 34(10):907–917

Yao X, Liu Q (2006) Changes in morphological, photosynthetic and physiological responses of Mono Maple seedlings to enhanced UV-B and to nitrogen addition. Plant Growth Regul 50(2):165–177

Yin LL, Zhao C, Huang Y, Yang RY, Zeng QP (2008) Abiotic stress-induced expression of artemisinin biosynthesis genes in *Artemisia annua* L. Chin J Appl Environ Biol 14(1):1–5

Yu KW, Murthy HN, Hahn EJ, Paek KY (2005) Ginsenoside production by hairy root cultures of *Panax ginseng*: influence of temperature and light quality. Biochem Eng J 23(1):53–56

Yuan Y, Liu Y, Wu C, Chen S, Wang Z, Yang Z, Qin S, Huang L (2012) Water deficit affected flavonoid accumulation by regulating hormone metabolism in *Scutellaria baicalensis* Georgi roots. PLoS One 7(10):e42946

Zandalinas SI, Balfagón D, Gómez-Cadenas A, Mittler R (2022) Plant responses to climate change: metabolic changes under combined abiotic stresses. J Exp Bot 73(11):3339–3354

Zargar SM, Gupta N, Nazir M, Mahajan R, Malik FA, Sofi NR, Shikari AB, Salgotra RK (2017) Impact of drought on photosynthesis: Molecular perspective. Plant Gene 11:154–159

Zhang L, Wang Q, Guo Q, Chang Q, Zhu Z, Liu L, Xu H (2012) Growth, physiological characteristics and total flavonoid content of *Glechoma longituba* in response to water stress. J Med Plants Res 6(6):1015–1024

Zhao B, Wang L, Pang S, Jia Z, Wang L, Li W, Jin B (2020) UV-B promotes flavonoid synthesis in *Ginkgo biloba* leaves. Ind Crop Prod 151:112483

Złotek U (2017) Effect of jasmonic acid and yeast extract elicitation on low-molecular antioxidants and antioxidant activity of marjoram (*Origanum majorana* L.). *Acta Scientiarum Polonorum Technologia*. Alimentaria 16(4):371–377

Zobayed SMA, Afreen F, Kozai T (2007) Phytochemical and physiological changes in the leaves of St. John's wort plants under a water stress condition. Environ Exp Bot 59(2):109–116

Zvereva EL, Kozlov MV (2006) Consequences of simultaneous elevation of carbon dioxide and temperature for plant–herbivore interactions: a meta-analysis. Glob Chang Biol 12(1):27–41

Chapter 7
Photoprotective Therapeutics: Recent Trends and Future Applications

Atifa Haseeb Ansari, Neeharika Srivastava, Sippy Singh, and Durgesh Singh

Abstract Electromagnetic spectrum consists of various rays—infrared, ionizing, x-ray and ultraviolet radiation. Ozone coverings act as effective protective coat which reflects radiation. Ozone layer present in the atmosphere protects from UV rays as they have the ability to absorb radiation, but continuous emission of chlorofluorocarbons (CFCs) and other gases reacts with ozone molecules and leads to depletion of ozone layer. Light emitted from the sun is the major source of ultraviolet radiation on Earth; besides these, there are also a few artificial sources such as lamps, black light lamps, bulbs and tanning beds. UV radiation can be further subdivided into three types based on their intensity and wavelengths: UV-A, UV-B and UV-C, which are biological threats. UV-A radiation have high penetration property, they can enter deeper into dermis of skin than other. UV radiation are mutagenic and carcinogenic in nature. Although a small amount of these rays is important for synthesis of vitamin D3, long-time exposures cause harmful impact on living organisms. The intensity of rays depends upon time, distance, season and altitude. Exposures are more threatening during summer season. They produce terrifying impact on humans, the skin being most visible outer region has direct exposure to their effects. Many health crises are linked with UV radiation. It causes molecular, cellular changes, DNA damages. There are so many skin abnormalities caused by ultraradiation including wrinkles, dark pigmentation, dryness, premature ageing, fragile skin, tanning, sunburn, erythema, and skin blisters. Skin cancer arises after long-time exposure; melanoma, xeroderma pigmentosum also cause damage to eyes including retina cancer. UV rays impose harmful impacts in the form of creating oxidative stress, generating free radicals and reactive oxygen species. Photoprotection is a biological process which is used for protection from radiation. To reduce their intensity and for prevention purposes, some photoprotective products

A. H. Ansari · S. Singh · D. Singh (✉)
Department of Zoology, S.S. Khanna Girls' Degree College, Prayagraj (A Constituent College of University of Allahabad, Prayagraj), Prayagraj, Uttar Pradesh, India

N. Srivastava
School of Engineering and Sciences, GD Goenka University, Gurugram, Haryana, India

V. K. Kannaujiya et al. (eds.), *Photoprotective Green Pharmacology: Challenges, Sources and Future Applications*,
https://doi.org/10.1007/978-981-99-0749-6_7

are used, especially sunscreen (organic and inorganic filters), sunglasses. Wearing proper clothes, hats are also effective form of protection. They are the most reliable forms of photoprotection. Natural products obtained from plants are used in pharmaceutical preparations. Fruits like red oranges, strawberries, pomegranate and dates have photoprotection property. Secondary metabolites are also used in defence mechanism. Antioxidants prevent cellular damages. Vitamin C and vitamin E inhibit melanin formation and reduce pigmentation. Green tea consists of polyphenols and acts as scavenger for free radicals. Beside these potent inhibitors, extracts of algae and lichen also possess antibacterial, anti-inflammatory and antioxidant properties. Thus, to overcome the impact of harmful UV radiation, one should follow the recommended steps and add supplements to their diets.

Keywords UV radiation · Filters · Radicals · Antioxidants · Oxidative stress · Photoprotections

7.1 Introduction

Electromagnetic radiation have the capability to produce biologic effects in organisms. Ultraviolet (UV) radiation is electromagnetic radiation, with wavelength from 100 nm–400 nm emitted by sun. Ambient sunlight encompasses UV-A, UV-B and UV-C radiation. Depending on the wavelengths, UV spectrum can be divided into three categories: UV-A (315–400 nm), UV-B (280–315 nm) and UV-C (200–280 nm). UV-C has greater effect in comparison to other divisions. Radiation intensity and wavelength depends upon time, season, altitude and latitude. UV rays are much effective during summer and spring season. Sunlight directly falls on the equator, thus the strength of UV exposures is stronger than other regions. UV-C causes sun tanning, premature ageing of skin while UV-A penetrates skin and tends to suppress immune system. Over the years, ozone layer depletion has increased the amount of UV radiation reaching the Earth. As a result, abnormalities in living organisms have increased. UV radiation is the most effective mutagen responsible for causing skin abnormalities including erythema, ageing, wrinkles, blistering and skin cancer (Elwood and Jopson 1997). UV exposures are strongest between 10–11 am and 16–17 pm (Toffetti and De Oliveira 2006). The drastic impact of UV radiation includes photo-ageing, sunburn and carcinogenesis; it is considered to be the main cause of skin diseases. Aerosols and clouds chiefly cause variation in UV rays. UV radiation are culprit to cause skin cancer. The disastrous damage of radiation can be seen on aquatic life in a variety of organisms such as virus, bacteria, fish, amphibians and other zooplankton.

7.2 Sources of UV Exposures

Ambient sunlight is the prime source of ultraviolet radiation on Earth; it constitutes only small proportion of sun rays. Short-period exposure of UV is beneficial especially in synthesis of vitamin D3 (Sampaio and Rivitti 2001). Too much UV exposure to sunlight is very dangerous. The spectrum intensity varies through passage. The stratosphere stops almost all UV radiation as these are absorbed by ozone layer. The continuous reduction in ozone layer leads to increase in intensity of radiation. There are some other artificial or man-made sources of UV rays such as sunlamps and sunbeds depend on specific bulbs, mostly producing UV-A, black-light lamps giving off UV rays, have purple glow and uses fluorescent material. Artificial sources emit spectrum of different wavelengths and energy, including different types of lamps having importance in cosmetics and medicine purposes. Tanning beds emit UV-A and UV-B energies with more advanced features.

7.3 Target Sites

UV rays cause molecular and cellular damages ranging from growth, survival, motility, replication, transcription etc. UV damage mechanism involves two ways either by direct absorption or by generation of highly reactive molecules known as reactive oxygen species (ROS). Biomolecules which tend to absorb radiation with low wavelength like DNA and RNA are prone to direct damage. For UV damage, DNA is the main target because double helixes have capability to absorb shorter wavelength, and the initially produced photoproduct in DNA exposure is cyclobutane pyrimidine dimer (CPDs) and pyrimidine dimer (PD). DNA directly absorbs UV radiation in cyclic structure formation CPDs and 6, 4 pyrimidine-pyrimidone (Kulms and Schwarz 2002; Matsumura and Ananthaswamy 2004). Modifying DNA includes 7, 8-dihydro-8-oxoguanine (8-OH-Dg) that promotes mutation in terms of GC-TA transversion (Schulz et al. 2000). PD photoproducts are more cytotoxic than CPDs; these are responsible to block DNA polymerases activities. Longer wavelengths of ultraviolet radiation such as UV-A is responsible for the production of ROS, especially superoxides and hydrogen peroxides (H_2O_2) leading to oxidative damage and are corrected via base-excision repair mechanism. UV-A radiation have high penetration property into the dermis and increase ROS level. UV direct absorption by amino acids such as tryptophan, tyrosine and histidine leads to damage. Due to UV exposures, lipids are damaged in cell membrane both directly and indirectly.

7.4 Catastrophic Effect of UV Radiation

UV radiation causes physical changes, immunosuppression, oxidative stress, DNA damages, skin abnormalities which induce senescence, mutagenesis and cancers. Skin, being the outermost visible organ of the body in direct contact of environment, is affected severely by UV radiation. UV-induced membrane receptor activates death receptors CD95 for induction of apoptosis. This radiation triggers cellular stress pathways, UV-B promotes sunburn pathway whereas long-time exposures lead to xeroderma pigmentosum (XP) and disease is associated with UV-induced DNA lesion (Setlow et al. 1969). UV radiation are responsible for wrinkles, pigmentation and elastic changes (Leyden 1990). Loss of functions of melanocortin-1 receptors (MC1R) lead to cancer-prone abnormalities (Abdel-Malek et al. 1999, 2008).

Long-time exposure of UV doubles the chances of melanoma (Pfahlberg et al. 2001; Chang et al. 2009; Cust et al. 2011). Skin has been reported to show signs of ageing due to UV exposure (Jenkins 2002) involving both extrinsic and intrinsic pathway of ageing. Reduction in collagen levels makes skin thinner, and it is linked with intrinsic ageing. UV rays cause deleterious impact on cells (Cleaver and Crowley 2002; Wei et al. 2003; Krutmann et al. 2012). Overexposure to UV rays cause cataract affecting cornea, lens and macular degeneration of eyes (Tomany et al. 2004). UV-A is carcinogenic (Paunel et al. 2005) while UV-B rays are mutagenic in nature and induce skin cancer (Grant 2008). Both UV-A, UV-B radiation have potential to affect immune responses. They are responsible for causing damages on both cellular and molecular levels including ageing, photocarcinogenesis and immunosuppression. UV causes photo-ageing (Gilchrest and Yaar 2007; Grant 2008) and skin cells more likely to get wrinkled, fragile, dry with irregular pigmentation. Ultraviolet radiation cause toxic effect on skin and eyes. UV radiation stimulates cell proliferation in skin and induces skin cancer. Corneal sunburn and retinal tissue damage are correlated with overexposure to radiation. It causes damage to eye cells including blurred vision, blindness, pre-orbital cancers (Table 7.1).

Malignant melanoma is rare but most dangerous type of skin cancer. DNA damage is a critical condition for enhancing skin cancers. Besides skin cancer, other skin abnormalities arise by UV exposures such as lupus, eczema. Melanoma in correlated with DNA damages. Overexposure of ultraviolet radiation is key factor in skin abnormalities including tanning, pigmentation, burning sensation, roughness and other severe conditions such as skin and eye cancers. Indoor tanning is also linked with sunburn. Sunburn is the painful erythema condition after exposure to UV radiation. UV-B is more potent than UV-A for sunburn, tanning and blistering. Solar radiation damages cellular metabolism and promotes melanocytes to produce melanin and deposition in keratinocytes-B which is key factor responsible for immunosuppression. UV photons cause skin damages by absorption and sensitization.

Table 7.1 Classification of UV radiation based on their wavelengths, energies and absorption level by ozone layer

Types of UV radiation	Wavelength	Absorption level	Energy	Effects
1. UV-A	315–400 nm	Not absorbed by ozone layer	Least energy	Premature ageing, wrinkles, sunburn, oxidative stress
2. UV-B	280–315 nm	Mostly absorbed by ozone layer	Slightly more than UV-A	Skin cancer, blistering, xeroderma pigmentosum, eye -damage, erythema and hyperpigmentation
3. UV-C	100–280 nm	Completely absorbed by the ozone layer	Highest energy	Severe acute damage to skin, photokeratitis, skin redness

7.5 Photoprotection and Therapeutics

Photoprotection is a biochemical process, used for avoidance of UV exposures caused by sunlight. This effective process helps to mitigate molecular damage. Photoautotrophs have already developed this mechanism to prevent oxidative stress. Exposure of radiation leads to deleterious effect on living organisms including human beings, animals and plants.

7.5.1 Sunscreen with Photoprotective Action

The effectiveness of sunscreens depends upon absorption interval, wavelengths, and the capacity of energy absorption which is in direct proportionality relation to their concentration (Johncock 2000). Sunscreen reduces the intensity of UV radiation reaching skin by means of absorption (Perugini et al. 2002). It reduces the cancer chances and protects from erythema and other skin disorders (Toyoshima et al. 2004). Physical filters are generally mineral-based. They act as physical barrier for skin responsible to reflect light forming protective film (Toffetti and De Oliveira 2006; Flor et al. 2007). Inorganic compounds reflect radiation, while organic compounds absorb radiation in order to protect from radiation exposures. Most commonly used physical filters are zinc oxide (ZnO), titanium dioxide (TiO_2) and magnesium dioxides (Toffetti and De Oliveira 2006; Flor et al. 2007; Violante et al. 2009). Skin problems including wrinkles, tanning, pigmentation and signs of ageing can be minimized by using sunscreens. Sunscreens are classified into two categories based upon their mechanisms, that is, organic and inorganic (Flor et al. 2007). They show low rate of allergeic and sensitization responses. Nanosized zinc oxide and titanium oxide provide superior protection from radiation. Chemical filters are aromatic compounds and electron-releasing substitutes in ortho and para position of benzene rings (Violante et al. 2009). For sunscreen formulation, tonnes of nanoparticles are produced. It has been suggested that nanomaterials provide good safety from ultraviolet rays. Chemical filters are derivatives of salicylic acid, camphor (Jallad 2016). Besides skin cancer prevention, they are also used for medical and cosmetic purposes.

7.5.2 *Proper Clothing and Sunglasses*

Proper clothing is also one of the best ways to protect from exposure and prevent damages of ultraviolet radiation. Fabric plays crucial role in absorption; polyester and wool have more absorption capacity than cotton, linen and rayon. Thicker fabrics provide more protection from exposures (Davis et al. 1997; Crews et al. 1999). UV radiation causes acute and chronic damages to the eye; it causes eyelid and peri-orbital skin cancer (Diepgen and Mahler 2002). It also causes damage to cornea, conjunctiva, lens and retina. Sunglasses can protect eye from radiation and are capable to block UV rays. They protect the eye from harmful impact of rays and provides shade. It would be a better option to reflect rays and prevent eye damages; therefore, it should be used especially during peak hours of exposures. Proper clothes, hats and sunglasses provide protection from both UV-A and UV-B radiation. Gambichler et al. (2001) reported that only 75% of fabrics deliver proper protection. Colours like black and blue enhance the coverage and safety (Gambichler et al. 2001; Wang et al. 2001). Wearing proper clothes and hats are meant for photoprotection (Hatch and Osterwalder 2006).

7.5.3 *Antioxidants*

The major destroying factors for skin are free radicals and to reduce their effect antioxidants are used as they are potent inhibitors against free radicals. Vitamin E is an important antioxidant; it consists of eight major forms, but α-tocopherol is most used by humans. It helps in quenching of peroxyl radicals and provides protection from erythema (Roshchupkin et al. 1979), photo-ageing (Jurkiewicz et al. 1995), immune suppression (Yuen and Halliday 1997) and photo-carcinogenesis (Burke et al. 2000). Vitamin-E also inhibits melanin formation. Silymarin derived from *Silybum marianum* inhibits sunburn (Katiyar et al. 1997). Vitamin-C, a water-soluble vitamin with low molecular weight has antioxidative property. It is an important cofactor for collagen formation (Nusgens et al. 2001) and reduces pigmentation through inhibition of tyrosinase (Maeda and Fukuda 1996). Olive trees have their own protection and provide defence mechanism against oxidative damages (Browden 2009). Borage oil stimulates skin repairment, evening primrose oil has high GLA content for skin protection from radiation. Antioxidants have capacity to prevent and protect cellular damage. UV radiation from reactive oxygen species damages skin cell; these can be treated by enzymatic mechanisms such as glutathione reductase and peroxidase. Antioxidant can be orally supplemented in body. Selenium is another essential element which contains glutathione peroxidase along with thioredoxin reductase enzyme that plays crucial activity in protection against oxidative stress. Green tea contains polyphenols that scavenges hydrogen peroxide and free radicals because they have very high antioxidant property. Thus, antioxidants are potent inhibitor of UV radiation.

7.5.4 Secondary Metabolites

Wavelengths ranging from 240–280 nm and 300–550 nm are absorption peaks of flavonoids while commercial-based UV filters absorb radiation between 290–320 nm and 320–400 nm (Bobin et al. 1995). Genistein is a well-known photoprotective compound, its activity against oxidative stress, molecular damage and carcinogenesis in rats have been observed (Wei et al. 2002). Anthocyanins, colouring pigments found in flowers, vegetables and fruits (Yoshida et al. 2006), possess antioxidative properties and thereby prevent stress (Gould 2004). Catechins are antioxidants that reduce ROS, DNA damage, immune suppression and inflammatory response (Guaratini et al. 2007). Other secondary metabolites are tocotrienols, carotenoids (Guaratini et al. 2009). Flavonoids such as anthocyanins, cinnamic acid can absorb ultraviolet radiation (Ramos et al. 2010). Flavonoids found in green vegetables and medicinal plants, canferol and mircetin have greatest antioxidant activity (Survay et al. 2011).

Isoflavones are present in leguminous plants such as soya beans, chickpeas, green beans (Saewan and Jimtaisong 2013). Extract of *Lippia sericea* contains enough amount of total phenols and have stronger photoprotective property (Polonini et al. 2014). Flavones act as natural chemical protector that absorbs radiation and protect against UV exposures (Bosch et al. 2015). The two main flavones are epigenin and chrysin (Bosch et al. 2015). Apigenin possesses high antioxidant property that shields from UV-A and UV-B radiation (Filho et al. 2016). *Psidium guajava* containing ethyl hexylmethoxycinate is used in sunscreen and increases sun protection factor (Lívia et al. 2018). The most studied secondary metabolites are epicatechin (EC) and its derivatives, for example, epicatechin-3-gallate (ECG). Aloe vera (Liliaceae) has strong potential power against UV radiation (Skarupova et al. 2020).

7.5.5 Plant Products with Photoprotection Activity

Almonds, one of the popular nuts (Menninger 1977), have been reported to have photoprotection property (Sachdeva and Katyal 2011). Caper dates have hydrating property and nourish skin (Lemmi and Rovesti 1979). Propolis is used as traditional medicine as an antiseptic, antioxidant, antimycotic, anaesthetic, anti-inflammatory and anticancer agent (Marcucci 1995; Burdock 1998; Banskota et al. 2001).

Blueberry (*Vaccinium myrtillus*) is rich in anthocyanins and has been reported to be responsible for scavenging free radicals and reducing hydrogen peroxides (Milbury et al. 2002). Grapes (*Vitis vinifera*) seeds are enriched with polyphenols (Zhao et al. 1999) with antioxidant, anti-inflammatory activity. Sunscreens based on the extracts of dates inhibit erythema and are more effective than tocopherol acetate (Bonina et al. 2002). Red oranges *(Citrus sinensis)* found in Italy, commonly called as blood oranges provide an effective protection against photo-oxidative skin damage caused by free radicals (Russo et al. 2002). Red clover is an essential perennial legume distributed throughout the world (Yolcu et al. 2009). Red clover is used in menopause by women and has abundant estrogenic isoflavones, daidzein, genistein and biochannin (Booth et al. 2006) (Table 7.2).

Soya bean has high nutritional values enriched with vitamin A, B, C and minerals (Jing and Zhang 2006). It helps in preventing keratinocyte death caused due to UV exposure and production of hydrogen peroxide (Chiang et al. 2007). Caper dates are used for medicinal purposes (Moghaddasian et al. 2012). Pomegranate contains anthocyanin and tannin which act as potent antioxidant and is used in protection from ultraviolet radiation and helps in prevention of molecular and cellular damages including oxidative stress, damages and skin disorders. Propolis is a natural product

Table 7.2 Extracts obtained from natural plant sources (including secondary metabolites, polyphenolic acid and antioxidant) with effective properties against UV exposures

Natural product	Scientific name	Derivatives	Photoprotection	References
Soya bean	*Glycine max*	Isoflavones, acetylglycitin, genistein	Inhibits keratinocyte death and reduces hydrogen peroxide formation	Kao and Chen (2002), Kao et al. (2005), Chiang et al. (2007)
Caper dates	*Capparis spinosa*	Caffeic, ferulic, cinnamic acid, coumaric	Inhibits erythema, nourishes skin	Lemmi and Rovesti (1979), Bonina et al. (2002)
Almond	*Prunus amygdalus*	Phenolic acid and flavonoids	Stong antioxidant property and decrease peroxidation	Kim et al. (2002), Wijeratne et al. (2006), Sachdeva and Katyal (2011)
Milk thistle	*Silybum marianum*	Silymarin, diastereoisomers	Reduces level of immune suppressor cytokine IL-10	Katiyar et al. (1997), Katiyar (2002)
Foti	*Polygonum multiflorum*	Anthroquinones, stilbenoids, proanthocyanidins	Radical scavenging and antioxidant activity	Yim et al. (1998), Chen et al. (1999)
Turmeric	*Curcuma longa*	Methoxycurcumin, demethoxycurcumin, curcumin	Anticancer, anti-wrinkles, inhibits UV-induced TNFα expression	Reuter et al. (2008), O'Sullivan-Coyne et al. (2009), Jang et al. (2012)

obtained from plant resin. It has been recently reported that Romanian propolis extract has antiapoptotic and antioxidant property (Bolfa et al. 2013).

7.5.6 Algae and Lichens

Thallophytes are found in sea water, river, lakes and sometimes show symbiotic relationship. Harvey was the first botanist who classified algae as red algae (Rhodophyta), green algae (Chlorophyta), brown algae (Heterokontophyta) and diatomaceae (Dixon 1973). Pholorotannins have inhibitory activity against melanogenesis and reduce cellular damage caused by UV exposures. Plastoquinones and sargaquinoic acid are the major component of brown algae. Lichens composed of algal and fungal partner live in symbiotic relationship, and are widely distributed throughout the world. Lichens consist of about 800 secondary metabolites which are aliphatic, aromatic in nature and have medicinal value (Muller 2001; Huneck 2001). Depsidones have antifungal, anticancer and antibacterial activity (Muller 2001; Piovano et al. 2002; Russo et al. 2006). Usnic acid is prominent secondary metabolite derived from lichens and is used in pharmaceutical products against infection, eczema and other skin diseases (Cocchietto et al. 2002; Bezivin et al. 2004). Usnic acid can be considered to be potent UV-B filter (Rancan et al. 2002). Sphaerophorin extracted from *Sphaerophorus globosus* and pannarin from *Psoroma reticulatam* inhibit melanoma cancer and protect from DNA damages (Russo et al. 2008).

7.5.7 Environmental Photoprotection

The ozone layer found in stratosphere is an invisible covering that protects organisms from UV radiation of the sun. Ozone layer is present in the stratosphere through which the rays pass and thereby screened from reaching the earth. Ozone is a gas composed of three oxygen atoms (O_3) with capability to protect and prevent terrifying impacts of harmful radiation. Millions of ozone molecules form layer, acting as natural sunscreen protecting living things from damaging radiation. Other than ozone layer, there are other factors that also play vital role in protection against UV exposures. During the peak hours, especially summer and spring season, intensity of UV radiation increases. Climatic factors chiefly smog, cloud and pollutant reduce UV transmissions (Kromann et al. 1986). Shades are capable of reducing radiation exposures (Moise and Aynsley 1999). On every 300 m elevation, approximately 4% UV exposure rate increases (Rigel et al. 1999).

7.6 Conclusion

UV exposure is an issue of high concern; due to depletion of ozone layer, the intensity of ultraviolet radiation becomes more effective and poses severe threats for organisms on earth. These harmful radiation show disastrous impact on humans including skin abnormalities. In addition to this, UV rays are mutagenic and carcinogenic that leads to skin and eye cancer (retinal and pre-orbital cancer). Cellular damages, oxidative stress, DNA damages, immune suppression, senescence are also correlated. Application of sunscreens, wearing proper clothes and sunglasses are additional steps that help an organism to protect themselves against UV rays. To reduce their impact, nowadays photoprotection is the trend. Plants extract contains enough amount of secondary metabolites, antioxidants, polyphenol, flavonoids that helps in recovery of skin disorders and cellular damage. Many natural photoprotective products are available obtained from plants which protect/shield against radiation. It includes derivatives from strawberry, grapes, dates, blueberry, red oranges, turmeric, red clover, soya bean, almond and nuts which have anticancer, anti-ageing, and antibacterial properties. Antioxidants such as vitamins are potent inhibitor of free radicals, hydrogen peroxide and ROS. This chapter explains harmful impacts of UV radiation and major therapeutics which are reliable and potentially active to reduce disastrous impact of UV exposures. It is expected that by using these easily available and reliable natural products extract like antioxidants, secondary metabolites and poly-phenolic acids from plants including vegetables, fruits, nuts, spices and tea will help in reducing intensity of ultraviolet radiation.

References

Abdel-Malek Z, Suzuki I, Tada A, Im S, Akcali C (1999) The melanocortin-1 receptor and human pigmentation. Ann N Y Acad Sci 885:117–133

Abdel-Malek ZA, Knittel J, Kadekaro AL, Swope VB, Starner R (2008) The melanocortin 1 receptor and the UV response of human melanocytes—a shift in paradigm. Photochem Photobiol 84:501–508

Banskota AH, Tezuka Y, Kadota S (2001) Recent progress in pharmacological research of propolis. Phytother Res 15:561–571

Bezivin C, Tomasi S, Rouaud I (2004) Cytotoxic activity of compounds from the lichen: *Cladonia convolute*. Planta Med 70:874–877

Bobin MF, Raymond M, Martini MC (1995) Propriedades de absorção UVA/UVB de produtos naturais. Cosmet Toil 7:44–50

Bolfa P, Vidrighinescu R, Petruta A et al (2013) Photoprotective effects of Romanian propolis on skin of mice exposed to UVB irradiation. Food Chem Toxicol 62:329–342

Bonina F, Puglia C, Ventura D et al (2002) In vitro antioxidant and in vivo photoprotective effects of a lyophilized extract of *Capparis spinosa* L. buds. J Cosmet Sci 53:321–335

Booth NL, Overk CR, Yao P, Burdette JE, Nikolic D, Chen SN, Bolton JL, van Breemen RB, Pauli GF, Farnsworth NR (2006) The chemical and biologic profile of a red clover (*Trifolium pratense* L.) phase II clinical extract. J Integr Complement Med 12(2):133–139

Bosch R, Philips N, Suárez-Pérez JA, Juarranz A, Devmurari A, Chalensouk-Khaosaat J, González S (2015) Mechanisms of photoaging and cutaneous photocarcinogenesis, and photoprotective strategies with phytochemicals. Antioxidants 4(2):44–50

Browden J (2009) Unleash the amazingly potent anti-aging, antioxidant pro- immune system health benefits of the olive leaf. Freedom Press, Topanga

Burdock GA (1998) Review of the biological properties and toxicity of bee propolis. Food Chem Toxicol 36:347–363

Burke KE, Clive J, Combs GF Jr, Commisso J, Keen CL, Nakamura RM (2000) Effects of topical and oral vitamin E on pigmentation and skin cancer induced by ultraviolet irradiation in Skh :2 hairless mice. Nutr Cancer 38:87–97

Chang YM, Barrett JH, Bishop DT, Armstrong BK, Bataille V, Bergman W, Berwick M, Bracci PM, Elwood JM, Ernstoff MS, Gallagher RP, Green AC (2009) Sun exposure and melanoma risk at different latitudes: a pooled analysis of 5700 cases and 7216 controls. Int J Epidemiol 38:814–830

Chen Y, Wang M, Rosen RT et al (1999) 2, 2-Diphenyl-1-picrylhydrazyl radical-scavenging active components from *Polygonum multiflorum* Thunb. J Agric Food Chem 47:2226–2228

Chiang HS, Wu WB, Fang JY et al (2007) UVB-protective effects of isoflavone extracts from soybean cake in human keratinocytes. Int J Mol Sci 8:651–661

Cleaver JE, Crowley E (2002) UV damage, DNA repair and skin carcinogenesis. Front Biosci 7:d1024–d1043

Cocchietto M, Skert N, Nimis PL (2002) A review on usnic acid, an interesting natural compound. Sci Nat 89:137–146

Crews PC, Kachman S, Beyer AG (1999) Influences on UVR transmission of undyed woven fabrics. Text Chem Color 31:17–26

Cust AE, Jenkins MA, Goumas C, Armstrong BK, Schmid H, Aitken JF, Giles GG, Kefford RF, Hopper JL, Mann GJ (2011) Early-life sun exposure and risk of melanoma before age 40 years. Cancer Causes Control 22:885–897

Davis S, Capjack L, Kerr N, Fedosejevs R (1997) Clothing as protection from ultraviolet radiation: which fabric is most effective? Int J Dermatol 36:374–379

Diepgen TL, Mahler V (2002) The epidemiology of skin cancer. Br J Dermatol 146(61):1–6

Dixon PS (1973) Biology of the rhodophyta. Oliver & Boyd, Edinburgh, pp 285–302

Elwood JM, Jopson J (1997) Melanoma and sun exposure: an overview of published studies. Int J Cancer 73:198–203

Filho JMTA, Sampaio PA, Pereira ECV, de Oliveria Júnior RG, Silva FS, Almeida JRGS, Rolim LA, Nunes XP, Araújo ECC (2016) Flavonoids as photoprotective agents: a systematic review. J Med Plants Res 10(47):848–864

Flor J, Davolos MR, Correa MA (2007) Protetores solares. Quím Nova 30(1):153–158

Gambichler T, Rotterdam S, Altmeyer P, Hoffmann K (2001) Protection against ultraviolet radiation by commercial summer clothing: need for standardised testing and labelling. BMC Dermatol 1:6

Gilchrest BA, Yaar M (2007) Photoagening: mechanism, preventation and therapy. Br J Dermatol 157(5):874–887

Gould KS (2004) Nature's Swiss Army Knife: the diverse protective roles of anthocyanins in leaves. J Biomed Biotechnol 2004(5):314–320

Grant WB (2008) The effect of solar UVB doses and vitamin D production, skin cancer action spectra, and smoking in explaining links between skin cancers and solid tumours. Eur J Cancer 44:12–15

Guaratini T, Medeiros MHG, Colepicolo P (2007) Antioxidante na manutenção do equilíbrio redox cutâneo: uso e avaliação de sua eficácia. Quím Nova 30(1):206–213

Guaratini T, Callejon DR, Pires DC, Lopes JNC, Lima LM, Giannella ND, Sustovich C, Lopes NP (2009) Fotoprotetores derivados de produtos naturais: perspectivas de mercado e interações entre o setor produtivo e centros de pesquisa. Quím Nova 32(3):717–721

Hatch KL, Osterwalder U (2006) Garments as solar ultraviolet radiation screening materials. Dermatol Clin 24:85–100
Huneck S (2001) New results on the chemistry of lichen substances. Fortschr Chem Org Naturst 81:1–276
Jallad KN (2016) Chemical characterization of sunscreens composition and its related potential adverse health effects. J Cosmet Dermatol 16(3):353–357
Jang S, Chun J, Shin EM (2012) Inhibitory effects of curcuminoids from Curcuma longa on matrix metalloproteinase-1 expression in keratinocytes and fibroblasts. J Pharm Invest 42:33–39
Jenkins G (2002) Molecular mechanisms of skin ageing. Mech Ageing Dev 123:801–810
Jing LG, Zhang YZ (2006) Determination of soybean isoflavones extracted from soybean by HPLC. J US-China Med Sci 26:629–632
Johncock W (2000) Interação de formulações com filtro solar. Cosmet Toil 12:40–50
Jurkiewicz BA, Bissett DL, Buettner GR (1995) Effect of topically applied tocopherol on ultraviolet radiation-mediated free radical damage in skin. J Investig Dermatol 104:484–488
Kao TH, Chen BH (2002) An improved method for determination of isoflavones in soybean powder by liquid chromatography. Chromatographia 56:423–430
Kao TH, Lu YF, Chen BH (2005) Preparative column chromatography of four groups of isoflavones from soybean cake. Eur Food Res Technol 221:459–465
Katiyar SK (2002) Treatment of silymarin, a plant flavonoid, prevents ultraviolet light-induced immune suppression and oxidative stress in mouse skin. Int J Oncol 21:1213–1222
Katiyar SK, Korman NJ, Mukhtar H (1997) Protective effects of silymarin against photo carcinogenesis in a mouse skin model. J Natl Cancer Inst 89:556–566
Kim YH, Yang HE, Park BK (2002) The extract of the flowers of *Prunus persica*, a new cosmetic ingredient protects against solar ultraviolet—induced skin damage in vivo. J Cosmet Sci 53:27–34
Kromann N, Wulf HC, Eriksen P, Brodthagen H (1986) Relative ultraviolet spectral intensity of direct solar radiation, sky radiation and surface reflections. Relative contribution of natural sources to the outdoor UV irradiation of man. Photo-Dermatology 3:73–82
Krutmann J, Morita A, Chung JH (2012) Sun exposure: what molecular photodermatology tells us about its good and bad sides. J Investig Dermatol 132:976–984
Kulms D, Schwarz T (2002) Molecular mechanisms involved in UV induced apoptotic cell death. Skin Pharmacol Appl Ski Physiol 15(5):342–347
Lemmi CT, Rovesti P (1979) Experimental studies on the cosmetological uses of *Capparis*. Riv Ital Essenze Profumi Piante off Aromat Syndets Saponi Cosmet Aerosols 61:2–9
Leyden JJ (1990) Clinical features of ageing skin. Br J Dermatol 122(Suppl. 35):1–3
Lívia PGM, Nathália OSG, Mariana CM, Andressa LSD, Nayara LO, Edemilson CC (2018) Extract from byproduct *Psidium guajava* standardized in ellagic acid: additivation of the in vitro photoprotective efficacy of a cosmetic formulation. Rev Bras 28(6):692–696
Maeda K, Fukuda M (1996) Arbutin: mechanism of its depigmenting action in human melanocyte culture. J Pharmacol Exp Ther 276:765–769
Marcucci MC (1995) *Propolis*: chemical composition, biological properties and therapeutic activity. Apidologie 26:83–99
Matsumura Y, Ananthaswamy HN (2004) Toxic effects of ultraviolet radiation on skin. Toxicol Appl Pharmacol 195:298–308
Menninger EA (1977) Edible nuts of the world. Horticultural Books Inc., Stuart, Florida
Milbury BE, Gao G, Prior RL (2002) Bio availablility of elderberry anthocyanins. Mech Ageing Dev 123:997–1006
Moghaddasian B, Eradatmand A, Davood A (2012) Caper the mystique of the recent century. Int J Agric Crop Sci 4:604–608
Moise AF, Aynsley R (1999) Ambient ultraviolet radiation levels in public shade settings. Int J Biometeorol 43:128–138
Muller K (2001) Pharmaceutically relevant metabolites from lichens. Appl Microbiol Biotechnol 56:9–16

Nusgens BV, Humbert P, Rougier A (2001) Topically applied vitamin C enhances the mRNA level of collagens I and III, their processing enzymes and tissue inhibitor of matrix metalloproteinase 1 in the human dermis. J Investig Dermatol 116:853–859
O'Sullivan-Coyne G, O'Sullivan GC, O'Donovan TR (2009) Curcumin induces apoptosis-independent death in oesophageal cancer cells. Br J Cancer 101:1585–1595
Paunel AN, Dejam A, Thelen S, Kirsch M, Horstjann M, Gharini P, Murtz M, Kelm M, de Groot H, Kolb-Bachofen V (2005) Enzyme-independent nitric oxide formation during UVA challenge of human skin: characterization, molecular sources, and mechanisms. Free Radic Biol Med 38:606–615
Perugini P, Simeoni S, Scalia S, Genta I, Modena T, Conti B, Pavanetto F (2002) Effect of nanoparticle encapsulation on the photostability of the sunscreen agent, 2-ethylhexyl-p- methoxycinnamate. Int J Pharm 246(1–2):37–45
Pfahlberg A, Kolmel KF, Gefeller O, Febim Study Group (2001) Timing of excessive ultraviolet radiation and melanoma: epidemiology does not support the existence of a critical period of high susceptibility to solar ultraviolet radiation-induced melanoma. Br J Dermatol 144:471–475
Piovano M, Garbarino JA, Giannini FA (2002) Evaluation of antifungal and antibacterial activities of aromatic metabolites from lichens. Boletin de la Sociedad Chilena de Quimica 47:235–240
Polonini H, Brandao H, Raposo N, Mouton L, Yepremian C, Couté A, Brayner R (2014) Ecotoxicological studies of micro- and nanosized barium titanate on aquatic photosynthetic microorganisms. Aquat Toxicol 154C:58–70
Ramos MFS, Santos EP, Dellamora-Ortiz GM (2010) Avaliação da atividade antissolar e estudos preliminares de fotodegradação da própolis. Revista Fitos, São Paulo 5(3)
Rancan F, Rosan S, Boehm K (2002) Protection against UVB irradiation by natural filters extracted from lichens. J Photochem Photobiol B 68:133–139
Reuter S, Eifes S, Dicato M (2008) Modulation of antiapoptotic and survival pathways by curcumin as a strategy to induce apoptosis in cancer cells. Biochem Pharmacol 76:1340–1351
Rigel DS, Rigel EG, Rigel AC (1999) Effects of altitude and latitude on ambient UVB radiation. J Am Acad Dermatol 40:114–116
Roshchupkin DI, Pistsov MY, Potapenko AY (1979) Inhibition of ultraviolet light-induced erythema by antioxidants. Arch Dermatol Res 266:91–94
Russo A, Bonina F, Acquaviva R (2002) Red orange extract: effect on DNA cleavage. J Food Sci 67:2814–2818
Russo A, Piovano M, Lombardo L (2006) Pannarin inhibits cell growth and induces cell death in human prostate carcinoma DU-145 cells. Anti-Cancer Drugs 17:1163–1169
Russo A, Piovan M, Lombardo L (2008) Lichen metabolites prevent UV light and nitric oxide-mediated plasmid DNA damage and induce apoptosis in human melanoma cells. Life Sci 83:468–474
Sachdeva MK, Katyal T (2011) Abatement of detrimentaleffects of photoaging by *Prunus amygdalus* skin extract. Int J Curr Pharm Res 3:57–59
Saewan N, Jimtaisong A (2013) Photoprotection of natural flavonoids. J Appl Pharm Sci 3(9):129–141
Sampaio SAP, Rivitti EA (2001) Dermatologia, 2nd edn. Artes Médicas. pp 629–642, 839–846, 878–886
Schulz I, Mahler HC, Boiteux S, Epe B (2000) Oxidative DNA base damage induced by singlet oxygen and photosensitization: recognition by repair endonucleases and mutagenicity. Mutat Res 461:145–156
Setlow RB, Regan JD, German J, Carrier WL (1969) Evidence that xeroderma pigmentosum cells do not perform the first step in repair of ultraviolet damages to their DNA. Proc Natl Acad Sci U S A 64:1035–1041
Skarupova D, Vostalova J, Rajnochova Svobodova A (2020) Ultraviolet A protective potential of plant extracts and phytochemicals. Biomed Pap Med Fac Univ Palacký Olomouc Czech Repub 164(1):1–22

Survay NS, Upadhyaya CP, Kumar B, Young KE, Yoon DY, Park SW (2011) New genera of flavonols and flavonol derivatives as therapeutic molecules. J Korean Soc Appl Biol Chem 54:1–18
Toffetti MHFC, De Oliveira VR (2006) A importância do uso do filtro solar na prevenção do fotoenvelhecimento e do câncer de pele. Investigação 6(1)
Tomany SC, Cruickshanks KJ, Klein R, Klein BE, Knudtson MD (2004) Sunlight and the 10-year incidence of age-related maculopathy: the Beaver Dam Eye Study. Arch Ophthalmol 122(5):750–757
Toyoshima M, Hosoda K, Hanamura M, Okamoto K, Kobayashi H, Negishi T (2004) Alternative methods to evaluate the protective ability of sunscreen against photogenotoxicity. J Photochem Photobiol B Biol 73(1–2):59–66
Violante IMP, Souza IM, Venturini CL, Ramalho AFS, Santos RAN, Ferrari M (2009) Avaliação in vitro da atividade fotoprotetora de extratos vegetais do cerrado de Mato Grosso. Rev Bras 19(2):452–457
Wang SQ, Kopf AW, Marx J, Bogdan A, Polsky D, Bart RS (2001) Reduction of ultraviolet transmission through cotton T-shirt fabrics with low ultraviolet protection by various laundering methods and dyeing: clinical implications. J Am Acad Dermatol 44:767–774
Wei H, Zhang X, Wang Y, Lebwohl M (2002) Inhibition of ultraviolet light-induced oxidative events in the skin and internal organs of hairless mice by isoflavone genistein. Cancer Lett 185(1):21–29
Wei Q, Lee JE, Gershenwald JE, Ross MI, Mansfield PF, Strom SS, Wang LE, Guo Z, Qiao Y, Amos CI (2003) Repair of UV light-induced DNA damage and risk of cutaneous malignant melanoma. J Natl Cancer Inst 95:308–315
Wijeratne SS, Abou-zaid MM, Shahidi F (2006) Antioxidant polyphenols in almonds and its coproducts. J Agric Food Chem 54:312–318
Yim TK, Wu WK, Mak DH (1998) Myocardial protective effect of an anthraquinone-containing extract of *Polygonum multiflorum* ex vivo. Planta Med 64:607–611
Yolcu H, Tavlas A, Tan M (2009) Chemical contents of red clover genotypes on crop improvement systems as livestock feed. J Anim Vet Adv 8:506–510
Yoshida K, Kitahara S, Ito D, Kondo T (2006) Ferric ions involved in the flower color development of the Himalayan blue poppy, Meconopsis grandis. Phytochemistry 67(10):992–998
Yuen KS, Halliday GM (1997) Alpha-tocopherol, an inhibitor of epidermal lipid peroxidation, prevents ultraviolet radiation from suppressing the skin immune system. Photochem Photobiol 65(3):587–592
Zhao J, Wang J, Chen Y (1999) Anti-tumor-promoting activity of a polyphenolic fraction isolated from grape seeds in the mouse skin two-stage initiation-promotion protocol and identification of procyanidin B5-30-gallate as the most effective antioxidant constituent. Carcinogenesis 20:1737–1745

Chapter 8
Cancer Therapeutics: Mechanism of Action, Radiation Toxicity, and Drug Formulation

Durgesh Singh, Sippy Singh, and Atifa Haseeb Ansari

Abstract Cancer is a well-known disease that involves complex changes in the genome due to variation in host and environment, and it is also a prime cause of death across the globe. Cancer is characterized by continuous growth that does not respond to signals to cease cell division and shows uncontrolled replication, escape apoptosis, angiogenesis, and metastasis. It has been observed that with an increase in human life expectancy, the cases of cancer have also notched to higher limits. With the increase in cases of cancer, the efforts to combat the situation have led to advent of pharmaceutical industry, making more investment in therapeutic area in search of cancer drugs and better treatment. Various therapies for treatment of cancer have been introduced, including chemotherapy, radiation therapy, thermotherapy, immunotherapy, proton therapy, cryotherapy, differentiation therapy, and thermotherapy. Among the several introduced advancements, the advent of targeted therapies is of high consideration. Two major concerns in cancer drug discovery, drug selectivity and adverse effects, have been modified using monoclonal antibodies. Researchers have been continuously focusing on new approaches for cancer treatment. Radiation therapy is frequently used in treating patients; however, the studies report that this therapy is highly toxic in many cases as it leads to changes in normal cells, converting them into tumorous nature. Drug delivery system is a way to target the drug via drug carriers to desired sites within cells, tissues, or organs. It aims at improving the potential of drugs and to overcome the issues such as low solubility and bioavailability, aggregation of drug, poor bio distribution, lack of selectivity, or to minimize the side effects of drugs. Thus, in course of drug formulation, the above points are taken into consideration. There has been a continuous advent in research field of drug delivery system. The details on mechanism of action suggest for combinational therapies. An important concern in cancer treatment is the development of resistance toward drug in patients with the progression of disease. The recent advancement in drug delivery system has included the use of smart

D. Singh · S. Singh (✉) · A. H. Ansari
Department of Zoology, S.S. Khanna Girls' Degree College, Prayagraj (A Constituent College of University of Allahabad, Prayagraj), Prayagraj, Uttar Pradesh, India

V. K. Kannaujiya et al. (eds.), *Photoprotective Green Pharmacology: Challenges, Sources and Future Applications*,
https://doi.org/10.1007/978-981-99-0749-6_8

nanocarrier-based delivery, also referred to as "smart drug delivery system," which can target drugs to specific sites with reduced dosage and side effects. The smart nanocarriers used for smart drug delivery system include gold nanoparticles, liposomes, carbon nanotubes, micelles, quantum dots, super paramagnetic iron oxide nanoparticles, and mesoporous silica nanoparticles. The prominence of cancer around the globe has resulted in an increase in option for treatment/therapies focused on mechanism of action; however, the toxicity of therapies remains the biggest matter of concern that needs to be addressed. This chapter focuses on different therapies used for treating cancer, their toxicity, drug formulation, and their delivery system.

Keywords Cancer · Therapy · Drug delivery · Toxicity

8.1 Introduction

Cancer is a medical condition where there is an uncontrolled and continuous abnormal growth in cells. It can be better defined as uncontrolled division and growth of cells as they are unresponsive toward the normal cell cycle signals and shows repeated replication, escapes apoptosis, and angiogenesis (Hanahan and Weinberg 2000). A normal cell grows through mitosis in an interdependent way, the division of these cells depend upon external growth factor and are limited to specific number of times (50 times), which is referred to as *Heyflick's limit* of that cell. On the contrary, tumorous cell grow indefinitely and independently of any growth factor (Lum 2005). The normal cells respond to the presence of surrounding cells and thereby exhibit contact inhibition, whereas cancer cells lack phenomenon of contact inhibition, which leads to the formation of an unwanted mass of cells (Hahn 1999).

Cancer involves complex changes in genome and is one of the prime health issue and reason of deaths across globe. The major issue related to cancer is the property of metastasis that refers to the capacity of these abnormal cells to migrate and spread within body, leading to secondary infection. Origin and spread of cancer depend on both internal as well as external factors such as mutation, hormones, immunity, chemicals, infectious agents, and radiations. The most accepted model suggests that major cause of cancer is mutation in tumor suppressor gene and oncogene (Ralph 2010). Such mutations in cell can affect production of specific proteins that plays pivotal role in cell cycle and result in cancer. Both biological and non-biological agents (physical and chemical), which have potential to cause cancer, are known as carcinogens such as viruses, ionizing radiation, chemicals, and so on (Table 8.1).

Oncogenes are the cancer-causing genes that are formed by mutation in proto-oncogenes by either of the five mechanisms: gene amplification, point mutation, enhancer insertion, promoter insertion, and chromosomal translocation. Tumor suppressor genes are the other genes that play crucial role in causation or development of cancer. These are sometimes also referred as recessive oncogene or anti-oncogene. These genes operate in different ways from oncogenes in that their inactivation

Table 8.1 Cancer associated infectious agents

Infectious agents	Type of organism	Associated cancers
Hepatitis B virus (HBV)	Virus	Hepatocellular carcinoma (type of liver cancer)
Human papilloma virus (HPV) types 16, 18, and other HPV	Virus	Cervical cancer, vaginal cancer, vulval cancer, anal cancer, penile cancer, oropharyngeal cancer, squamous cell carcinoma of the skin
Epstein-Barr virus	Virus	Burkitt lymphoma, non-Hodgkin lymphoma, Hodgkin lymphoma, nasopharyngeal carcinoma
Kaposi-sarcoma-associated herpes virus (KSHV)	Virus	Kaposi sarcoma
Helicobacter pylori	Bacterium	Stomach cancer, mucous-associated lymphoid tissue (MALT) lymphoma
Schistosoma hematobium	Parasite	Bladder cancer
Liver flukes	Parasite	Cholangiocarcinoma (type of liver cancer)

Source: National Cancer Institute. Fact sheet: Cancer Vaccine. Bethesda (MD): NCI (updated November 15, 2011; cited November 27, 2011)

removes the barrier/constraint on control of growth. The p53 gene is one of the most important tumor suppressor genes that is well known as "Guardian of the Genome" as it regulates transcription, replication, DNA repair, and apoptosis (Rivlin 2011). The p53 mutation results in the formation of primary tumors, and nearly 60 cases of cancer are associated with p53 abnormalities. Similarly, there are several other tumor suppressor genes that, if mutated, may lead to occurrence of cancer.

8.2 Categories of Cancer

Cancers are named according to the tissue from which they originate. They are broadly categorized into six main types:

(a) *Carcinoma*: Cancer of ectodermal and endodermal origin. They account for nearly 86% of total cancer, e.g., breast cancer, lung cancer, and prostate cancer.
(b) *Sarcoma*: Cancer of mesodermal origin. It is a tumor made up of principally connective tissue, e.g., cancer of bone and cartilage.
(c) *Lymphoma*: Lymphocytes are produced in excessive amount by lymph nodes and spleen, e.g., Hodgkin's lymphoma.
(d) *Leukemia*: In such type of cancer, aplastic division of WBCs occurs, leading to excessive production of abnormal WBCs. Leukemia is also known as blood cancer.
(e) *Lipoma*: Cancer of adipose/fatty tissues. This is also mesodermal in origin.
(f) *Mixed type*: When two or more different tissues of different embryonic origin are involved in formation of cancer, it is called mixed type. Generally, in the later stages of cancer, patients are diagnosed with mixed type of cancer.

8.3 Molecular Mechanism of Cancer

Genetic alterations that can give rise to oncogene and genetic aberrations are translocation of chromosomes, point mutation, deletion, amplification, and insertion. The p53 plays a crucial role in division of cell, apoptosis, formation of new blood vessels, differentiation, and DNA metabolism. The p53 mutation results in formation of unusual protein, which disturbs the normal molecular process. Whenever there is DNA damage, the p53 docks with DNA, thereby stimulating the WAF1 genes (Fogh 2010; Lee and Sokolsky 2010). Further, the p53 connects with CDK2 and ultimately inhibits the effect of p21 protein for next stage of cell cycle. Three routes followed by p53 are stimulation of proteins involved in DNA repair, inducing programmed cell death, and cell cycle arrests in G1/S phase (Bag 2014; Bhosle and Hall 2009). Mutation in p53 promotes its interaction with CDK1-P2 and CDC2 and keeps the cells in G1 and G2 phase of cell cycle (Kano 2017; Hassen-Khodja 2004).

Hypomethylation at specific promoters have potential to induce ectopic expression of oncogenes. Unlike hypomethylation, hypermethylation occurs only specifically in CpG region. Cessation of transcription due to hypermethylation of promoter stimulates the gene participating in apoptosis (WIF-1, etc.), repair (BBRCA1, etc.), and response to vitamin (CRBP1, etc.), which in turn have important role in induction of cancer (EBCTC Group 2005).

8.4 Therapies for Cancer Treatment

Treatment modalities of cancer include surgery and various other therapies that have been discussed as follows (Fig. 8.1):

(a) *Surgery*: It is used for tumors of localized areas such as breast cancer, cancer of liver, pancreas, etc. The infective part of the organ is removed surgically.
(b) *Radiation therapy*: Electromagnetic and particulate forms of radiation are used to treat cancerous cells.
(c) *Chemotherapy*: It involves the use of toxic compounds/chemicals that exert their anti-tumorous effect when used at maximum tolerance.

Fig. 8.1 Therapies for cancer treatment

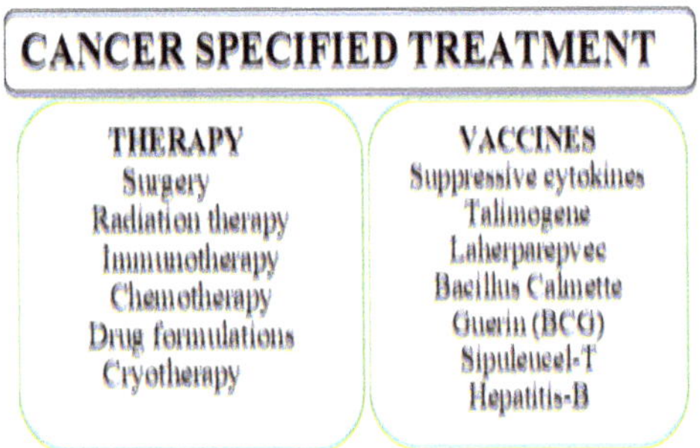

(d) *Immunotherapy*: Cancer treatment can be done by improving the ability of host to reject the tumor immunologically, the stimulant being tumor-associated antigen. The immunotherapy can be active, passive, adaptive, or restorative. In immunotherapy, the immune system of patient is manipulated in order to identify as well as target tumorous tissues. Like other therapies, immunotherapy also leads to toxicity but here the toxicity is slightly unique depending on the mechanism of action as toxicities often require specific management, which can include the use of steroids and immunomodulators. The different approaches of immunotherapy include the use of cytokines or vaccines to elevate the amount of tumor-specific T cells, transfer of specific immune effector cells, and application of a variety of immune checkpoint inhibitors. As our understanding of cancer is burgeoning day by day, there is progression in better and targeted methods of treating this disease.

8.4.1 Radiation Therapy

Radiation oncology deals with treating cancer patients with the use of only ionizing radiation or in combination with other techniques. Radiation therapy can be highly effective if the following points are taken into consideration: tumor type, local and regional extent of infection, geometric accuracy, calculated radiation dose targeted at defined target. Nearly 66% of all cancer patients undergo this therapy as the option of treatment either alone or in conjugation with surgery or chemotherapy. Apart from curing cancer, this therapy also helps in managing symptoms of disease such as pain reduction, preservation of skeletal integrity, reestablishment of organ function, and so on. Radiation causes DNA damage of cancer cells, halting their further growth and division and finally destroys them. The radiation dose recommended to patients differs for different cancers. This dose depends on tumor size, type of surgery, and quality of cancer (Dayes et al. 2006). The biology behind treating cancerous cells with radiation believes in breaking double-stranded nuclear DNA of abnormal cells (Leunens et al. 1992). However, the exact mechanism of cell death through radiation is still under investigation. The break in DNA caused due to radiation treatment is irreversible, eventually leading to death of cells. The mechanism of therapy through radiation depends on the internal ability of normal cells to correct the damage and the capacity/ability of radio-oncologist to take advantage of any structural separation between the transformed and normal cell (Mizer et al. 1986; D'souza et al. 2000). Photoelectric effect, pair production, and Compton effect are of prime interest and consideration in radiation therapy (Benten et al. 1988; Day and Harrison 1995).

8.4.1.1 Types of Radiation Therapy

Teletherapy and brachytherapy are two commonly used clinical radiation techniques. Teletherapy uses distantly located radiation source. The radiation source utilizes radioisotopes (e.g., Co-60) or linear accelerators for generation of radiation. The linear accelerator produces both X-rays and electrons that are targeted on the patient. Brachytherapy involves placement of radiation device within or in proximity to patient. In this technique, radiation source is placed either on surface, within body, into an organ, or tissue (Mudur 2005). During the initial stages of the use of brachytherapy technique, the radioactive substance is kept in vacant tubes or body cavities manually during which the staff member undertaking the technique also receives radiation. Later, the techniques were modified, and in high dose rate brachytherapy, the high dose source is carried along catheters toward the end of a wire by machine to patient. This lowered the exposure of staff to the radiations, as well as the treatment reduced (Fullerton et al. 1978). There has been continuous modification or improvisation in radiation techniques used for cancer patients.

Electron beam therapy utilizes either electron beam or in combination with beam of photons. Beams of electrons are used in treating superficial tissues, including tumors of ear, oral cavity, and ear. The lesions on eyelids, ears, cheeks, and nose are not very invasive and therefore treated with 6–9 MeV electron beams. With increase in thickness of lesion, the voltage is also increased. The energy of electron beams is directly proportional to surface build-up dose and respective skin reactions (Perez et al. 1998; Purdy 1998). For the patients with autoimmune diseases, low-dose total body irradiation (TBI) (less than 2 Gy in single fraction or multiple 0.05–0.15 Gy in multiple fractions) is used. Some side effects have been reported in high-dose TBI such as thyroid dysfunction, whereas thrombocytopenia is the major side effect of low-dose TBI (Bentel 1996; Purdy 1998). Hemi-body irradiation (HBI) helps in treating patients separating the targeted body parts and multiple locations. HBI, specifically single-dose, has been reported to be effective in controlling pain within cancer patients suffering from multiple metastases. Hematological toxicities have been found to vanish within 42–56 days (Purdy 1998). Gamma knife surgery, also known as radiosurgery, is used for precise noninvasive cerebral surgery. It accurately irradiates deeply located targets with the help of collimated beams of ionizing radiation that resemble the effect of a scalpel. This technique obsoletes the surgical and esthetical risks, as well as the lengthy hospitalization. Beams of gamma radiation act as blades that are targeted on the lesion point (Mathieu et al. 2007). The lesion on operation slowly decreases in size and finally dissolves. This technique is advantageous as only the target tissue receives radiation dose while the surrounding cells remain unharmed, and the patients with advanced age can also be conveniently treated (Dogan et al. 2003).

In three-dimensional conformal radiotherapy (3DCRT), a three-dimensional digitized data of tumor of patient and normal anatomy of that individual are created (Bombford et al. 1993; Symonds 2001). Further, the created data sets are applied to develop three-dimensional computer images and complex plan for delivering highly focused radiation, specifically on target cell. The technique may be used to increase control of tumor with decreased side effects. The virtual simulation created in this technique enables the planning group to develop multiple and customized course of therapy. The treatment area can be visualized in three dimensions. The radiation technologist can view and operate from a nearby control room while being protected from radiation (Rath et al. 2006).

Intensity-modulated radiation therapy (IMRT) shapes the radiation beams approximately with the structure of tumor. It enables the physician to manage the amount of radiation within a specific area (Freedman et al. 2006). This therapy utilizes advanced computer programming to decide dosage of 3D radiation based on size, shape, and location of tumor (Oliver et al. 2007). The patients who have undergone radiation treatment at maximum dosage by conventional radiation therapy can also be treated through IMRT. Tumors of breast, lung, kidney, liver, pancreas, spine, tongue, and so on are treated by applying the technique of IMRT. This technique is often used along with or after another primary treatment (Redpath and Muren 2006). Image-guided radio Therapy (IGRT) is another advanced field where there is upgradation of linear accelerator, allowing doctors to target tumors with more precision. An on-board imager automated system is an addition to this technique that utilizes X-rays with high resolution to produce clear images of cancerous cells and surrounding normal cells with contrast (Xing et al. 2006). It also helps in adjusting the position of patients with cancer of central nervous system, neck, and prostate immediately before treatment (Enmark et al. 2006). Radiation therapy had witnessed various advancements and modalities, with tomotherapy being one of them. It is also known as "slice therapy," which is a combination of computer-controlled radiation beam collimation with an on-board computed tomography scanner to view the treatment site. It is very accurate in beam delivery which enables increment in tumor dosage which in turn increases the chances of cancer cure and decrease in treatment complication for healthy cells. Tomotherapy rotates the beam source around the patient, covering 360° and allowing delivery of accurate radiation dose (Storme et al. 2006).

Highly integrated adaptive radiotherapy (HI-ART), an advanced radiation therapy technique that combines advanced IMRT with accurate CT scanning technology, is available in European cancer centers (Lefkopoulos and Mazeron 2006). It can be even used when highest tolerance dose of other radiation has been used or even if it is difficult to reach the deeply suited tumor.

8.4.2 Radiation Toxicity

Radiation therapy has been considered as the main treatment for cancer and nearly 60% of cases require this therapy. Although radiation therapy has provided better treatment option, the radiations used in therapy that target the cancerous cells may also affect the neighboring normal cells. The damage to normal cells surrounding malignant tumor is the major limitation of radiotherapy (Barber et al. 2000; UNSCEAR 2000). Radiation therapy in many cases shows toxicity in the form of secondary malignancies (Lin 2016). Secondary malignancies are tumors that occur after treatment of primary tumors. The ionizing radiations used in technique are carcinogenic in nature, and the cases of second malignant neoplasm (SMN) arise in patients who have been suffering for a long time, which is a major concern (Xi et al. 2013; De Gonzalez et al. 2011). The surrounding normal cells are affected by radiotherapy and tumor microenvironment because they induce the release of inflammatory substances (cytokines, free radicals). In addition, the immune system activates and it causes oxidative DNA damage within the environment. Hence, detecting these DNA damage at primary stage helps in understanding the mechanism of non-targeted effects of radiation (Sprung et al. 2015).

Other major toxicity faced after radiation therapy is infertility. Females who undergo radiation therapy within the abdominal or pelvic region have been reported to be prone to infertility (Cruz and Bellver 2014). High dose of radiation can destroy the eggs within the ovary and even can cause early menopause. Females undergoing treatment of endometrial cancer have been reported to show toxic effects on high dose of radiation such as miscarriage, premature birth, and placental abnormality (Balcerek et al. 2012; Wo and Viswanathan 2009). Radiation within the brain region may affect the pituitary gland, which can alter hormonal balance and affect ovulation in females (Ash 1980). In case of males, radiation therapy in the brain region can lead to testicular failure as radiation can cause damages to the pituitary gland, resulting in an imbalance of follicular stimulating hormone and luteinizing hormone, which in turn decreases the level of testosterone and finally affects the spermatogonia and Leydig cells. The damage caused by radiation impedes the ejaculation process, hence the sperms are not available for fertilization outside the body (Orth et al. 2014).

With the advancement in radiation therapy, there has been decrease in toxicity effects (Nguyen et al. 2012; Yorke and Goodman 2012). Further, the toxicity can be screened down by applying radiosensitizers and radioprotectors (Lu et al. 2016). The former sensitizes the target tumor cells and removes free radicals produced during the process of cell damage, while the latter are mostly antioxidants that are used at the time of radiation therapy to protect normal cells. Some of the examples of radiosensitizers are mitomycin c, taxanes, and tirapazamine, whereas amifostine and nitroxides are some of the commonly used radioprotectors (Prasanna et al. 2015). Ayurvedic formulations such as Triphala, Ashwagandha, and Brahma Rasayana can be used as effective radioprotectors as they can inhibit DNA damage and regenerate bone marrow progenitors and immunomodulatory mechanism by scavenging free radicals and decreasing oxidative stress (Baliga et al. 2013).

8.4.3 Drug Formulation and Delivery System

The expanded research requires characterization of mechanism of action for anti-cancerous agents in both preclinical models as well as in clinical trials wherein the former supported the selection of therapeutic agents, appropriate models of efficacy, and experimental design, as well as rational characterization, whereas the latter supported the screening of markers and the selection of patient subpopulations. Currently, different types of anti-cancerous drugs conjugated with delivery carriers are approved and used for clinical treatment (Wicki et al. 2015). Some of the commonly recommended drugs are DOX, PTX, CUR, DTX, and siRNA. DOX, an anthracycline antibiotic isolated from *Streptomyces peucetius* (Chen et al. 2018), either induces apoptosis or arrests the growth of cells through different molecular mechanisms such as intercalation of DNA, inhibition of topoisomerase II, and production of ROS. DOX is readily used for treating different types of cancers because of its effective antitumorous activity effect and low price (Chen et al. 2018). In order to reduce cardiotoxicity of DOX (Rivankar 2014) and other side effects, it is used with some carriers, which also increases the tolerance to different types of tumors such as Kaposi's sarcoma, ovarian carcinoma, and breast cancer (Slingerland et al. 2012; Rivankar 2014). PTX, a tricyclic diterpenoid obtained from bark and needles of *Taxus brevifolia*, is a mitotic inhibitor that induces polymerization of tubulin and arrests cells in G2 and M phase, causing death of cells (Zhu and Chen 2019; Yardley 2013; Wei et al. 2017). It is widely used for treating breast, cervical, ovarian, brain, lung, liver, and prostate cancers (Wei et al. 2017).

CUR, a polyphenolic compound extracted from *Curcuma longa* rhizome, is well known for its safe, nontoxic, antioxidant, anti-inflammatory, anti-cancerous, and antibacterial properties. It is widely and globally used in preventing different cancers (Wei et al. 2018). CUR poses low bioavailability with poor water solubility, which has been solved by using nanoparticles as delivery carriers, which helps in better internalization, dosage and toxicity reduction, and bioavailability enhancement (Huang et al. 2018). Anitha et al. (2012, 2014) in their studies reported enhanced antitumor activity of CUR-loaded N, O-CMC nanoparticles and combination of N, O-CMC nanoparticles based on 5-FU and CUR. CUR along with liposomes as drug carrier also enhances the stability, bioavailability, antitumor efficacy, and targeting property of CUR (Feng et al. 2017). Esterification of 10-deacetylbaccatin III isolated from needles of European *Taxus baccata* is used for production of DTX (Pazdur et al. 1993; De Weger et al. 2014). It promotes better polymerization of tubulin than PTX and hence is more effective in causing cytotoxicity to tumor cells. This drug has been tested on lung, breast, and ovarian cancers (Nicoletti et al. 1994). O-CMC nanoparticles are readily used as delivery carriers for both PTX and DTX, which enhances the efficacy and bioavailability of drugs. siRNA is used for cancer treatment as they can mediate RNA interference by subduing the carcinogenic genes by precisely targeting the mRNA. This system has been developed to acquire effective anticancer effect (Lee et al. 2016; Singh et al. 2018). Some other drugs such as MET and 5-FU are also used in cancer treatment where the former is a biguanide drug that helps in improvising prognosis of cancer patients and prevents the development of tumors. It can minimize the effect of growth factors such as insulin and IGF-1 and thus inhibit insulin-dependent mechanisms (Morales and Morris 2015). 5-FU, a fluoropyrimidine analog, has anti-cancer properties because it inhibits thymidine synthase and DNA synthesis in cancer cells. This drug has low price and high efficacy, hence used widely either alone or in combination with other anticancer drugs (Wei et al. 2018).

All the drugs used for treating cancer need to be specific and targeted toward the cancer cells, hence the application of nanoparticles as carriers for delivery of drug is a boon in cancer therapy as the nanoparticles elevate the potential on one hand and even protects the normal cells from cytotoxicity on the other. The targeting mechanism of these nanocarriers can be divided into passive and active targeting wherein the former drugs are targeted to specific site for therapeutic action. The readily dividing cancer cells cause neovascularization and formation of large holes in vascular wall (Carmeliet and Jain 2000). The defective blood vessel causes the nanoparticles to leak out and accumulate in tumor tissues and the retention of nanoparticles lead to release of their content in tumor cells. Glycolysis occurs at high rate in cancer cells, which helps in enhanced growth (Pelicano et al. 2006). It also induces an acidic environment, and subsequently pH sensitive nanoparticles release drugs in proximity to cancer cells (Lim et al. 2018). Passive targeting has a few limitations such as nonspecific drug distribution and difference in permeability of blood vessels across different tumors (Jain 1994). Active targeting, on the other hand, specifically targets tumor cells directly through interaction between receptors and ligands. The ligands present on the surface of nanoparticles are selected to target overexpressed molecules present on the surface of cancer cells (Shi et al. 2011; Kamaly et al. 2012). This interaction induces receptor-mediated endocytosis, which in turn helps in internalization of nanoparticles and release of drug (Farokhzad and Langer 2009). This technique is also suitable for drug delivery of macromolecule such as proteins and siRNAs (Danhier et al. 2010). A nanoparticle has been widely used in clinical therapeutics because of their small size, large surface area, ability to bind and carry anticancer agents, drugs and use of nanomedicine has revolutionized cancer diagnosis and therapy. Nanotechnology has the potential and is being used worldwide using nanodrug carriers, which has improved the efficiency and reduced the toxic effects of drugs on normal cells. Nanocarriers have many advantages such as increased bioavailability of drugs and reduced dose and toxicity of normal cells. Both organic and inorganic nanocarriers are used. Organic carriers such as carbon nanotubes, liposomes, dendrimers, lipids, emulsions, and synthetic polymers, and inorganic nanocarriers such as quantum dots have comparatively low side effects and result in controlled drug release. Mesoporous silica, quantum dots, magnetic nanoparticles, and carbon nanotubes are commonly used for cancer treatment (Bharali et al. 2005; Zrazhevskiy et al. 2010; Kairdolf 2013). Some of the metallic and nonmetallic nanoparticles are discussed as follows:

Nonmetallic nanoparticles: Carbon, silicon nanoparticles, grapheme, and grapheme oxide have emerged as useful part of drug delivery system for cancer treatment. Silicon nanoparticles are applied in photodynamic therapy and radiofrequency hyperthermia are used for cancer therapy, while porous silicon nanoparticles are used in chemotherapy, photodynamic, immunotherapy, and gene therapy (Landgraf et al. 2020).

Metallic nanoparticles: Silver, gold, zinc, copper, and aluminum nanoparticles are commonly used in cancer treatment. These nanoparticles are synthesized by bacteria, plants, photoreduction, and chemical electrolysis (Ahmad et al. 2010; Hanan et al. 2018).

Natural polymer nanoparticles: Albumin, chitosan, and ferritin are used as natural material to produce delivery carriers (Wicki et al. 2015). Chitosan is nontoxic, biodegradable, antibacterial, with anti-cancerous and wound-healing properties. Nanoparticles produced from chitosan possess positive surface charge and mucoadhesive property along with tumor-targeting ligand (folic acid, antibodies, biotin, avidin), and they attach to receptor of target cancer cell and release drug (Prabaharan 2015). Albumin is an acidic and hydrophilic protein that can carry anticancer drugs and enhance tumoricidal property (Karimi et al. 2016; Wang and Zhang 2018). Bovine serum albumin (BSA) is used for preparing nanomedicine because of its low cost and stability (Lamichhane and Lee 2020). BSA nanoparticles, a protein carrier for delivery of drug, are nontoxic, non-immunogenic, easily metabolized, and water soluble (Huang et al. 2018). The antitumor drug is either covalently conjugated with BSA or noncovalently conjugated by encapsulation and hydrophobic interaction (Wang and Zhang 2018).

Liposomes, i.e., vesicles carrying both hydrophobic and hydrophilic drugs, are used as carriers (Feng et al. 2017; Riaz et al. 2018). These molecules are biologically inert and biocompatible, therefore preferred for delivering anticancer drugs without causing any toxicity (Wang et al. 2005; Schwendener 2007; Bingham et al. 2010; Feng et al. 2017).

Exosomes, an extracellular vesicle secreted by mammalian cells, contain various labeled proteins and ligand proteins on their surface that can attach and deliver payload such as drugs to target cells (van den Boorn et al. 2013; Batrakova and Kim 2015). They are highly stable, biocompatible, low immunogenic, and low toxic (Batrakova and Kim 2015). Exosome-biomimetic nanoparticles have been recently developed for drug delivery systems. Dendrimers are synthetic dendritic polymers with 3D branching macromolecule (Singh et al. 2008; Zhang et al. 2018). Its structure helps in solubilizing water-insoluble drugs, with three main sites for entrapping the drug. This provides special nanocarrier property utilized for drug delivery system (Chauhan 2015, 2018). Dendrimers increase transdermal permeation and specific drug targeting, and hence are used for cancer treatment (Chauhan 2018). Recently, nanogels have been also used in the field of diagnostic, tissue engineering, chemical and biochemical sensing, and cancer imaging in drug delivery (Oishi et al. 2007; Peng et al. 2010; Wu 2010).

8.4.4 *Future Challenges in Cancer Treatment*

With the advent of new drug delivery system, the cancer treatment has witnessed success in terms of cure. This novel technique promises a bright future, ensuring efficient and proper drug localization with least toxicity to surrounding normal cells. The use of nanocarriers has helped in targeted radiotherapy, imaging-guided radiotherapy, and precision medicine (Mi 2015, 2017). Biodistribution, accumulation at the target site, and the utilization of nanomedicine are all very beneficial and convenient for personalized treatment of tumors (Lammers et al. 2012; Theek et al. 2014). The novel drug delivery system is a promising tool in cancer treatment and provides array of opportunities for complex multifunctional and targeted strategies.

8.5 Conclusion

With the advent of new techniques, there has been burgeoning scope for cancer treatment. Among all the therapies introduced in the field of oncology, radiation therapy remains indispensable in control and management of most cancers. Although some toxicity issues persist, it is one of the best organ-preserving methods because it reduces the chances of surgery. Radiation therapy in conjugation with ayurvedic drugs is obviously a better way to manage cancer.

References

Ahmad N, Sharma S, Alam MK, Singh VN, Shamsi SF, Mehta BR (2010) Rapid synthesis of silver nanoparticles using dried medicinal plant of *Basil*. Colloids Surf B: Biointerfaces 81:81–86

Anitha A, Maya S, Deepa N, Chennazhi KP, Nair SV, Jayakumar R (2012) Curcumin-loaded N, O-carboxymethyl chitosan nanoparticles for cancer drug delivery. J Biomater Sci Polym Ed 23:1381–1400

Anitha A, Sreeranganathan M, Chennazhi KP, Lakshmanan VK, Jayakumar R (2014) In vitro combinatorial anticancer effects of 5-fluorouracil and curcumin loaded N, O-carboxymethyl chitosan nanoparticles toward colon cancer and *in vivo* pharmacokinetic studies. Eur J Pharm Biopharm 88:238–251

Ash P (1980) The influence of radiation on fertility in man. Br J Radiol 53(628):271–278

Bag S (2014) Bio-organic chemistry of natural enediyne anticancer antibiotics. National Programme on Technology Enhanced Learning (NPTEL)

Balcerek M, Reinmuth S, Hohmann C, Keil T, Borgmann-Staudt A (2012) Suspected infertility after treatment for leukemia and solid tumors in childhood and adolescence. Dtsch Arztebl Int 109(7):126–131

Baliga MS, Meera S, Vaishnav LK, Rao S, Palatty PL (2013) Rasayana drugs from the Ayurvedic system of medicine as possible radioprotective agents in cancer treatment. Integr Cancer Ther 12(6):455–463

Barber JB, Burrill W, Spreadborough AR, Levine E, Warren C, Kiltie AE, Scott D (2000) Relationship between in vitro chromosomal radiosensitivity of peripheral blood lymphocytes

and the expression of normal tissue damage following radiotherapy for breast cancer. Radiother Oncol 55(2):179–186
Batrakova EV, Kim MS (2015) Using exosomes, naturally-equipped nanocarriers, for drug delivery. J Control Release 219:396–405
Bentel GC (1996) External beam radiation therapy equipment. In: Bentel GC (ed) Radiation therapy planning. McGraw Hill, New York, pp 21–35
Benten GC, Nelson CE, Noell KT (1988) Pretreatment procedures. In: Bentel GC, Nelson CE, Noell KT (eds) Treatment planning and dose calculation in radiation oncology. Pergamon Press, New York, pp 40–68
Bharali DJ, Lucey DW, Jayakumar H, Pudavar HE, Prasad PN (2005) Folate receptor-mediated delivery of InP quantum dots for bioimaging using confocal and two-photon microscopy. J Am Chem Soc 127:11364–11371
Bhosle J, Hall G (2009) Principles of cancer treatment by chemotherapy. Surgery (Oxford) 27:173
Bingham RJ, Olmsted PD, Smye SW (2010) Undulation instability in a bilayer lipid membrane due to electric field interaction with lipid dipoles. Phys Rev E 81:051909
Bombford CK, Sherriff SB, Kunkler IH (1993) Principles of radiation treatment planning. In: Bomford CK, Sherriff SB, Kunkler IH (eds) Walter and Miller's text book of radiotherapy radiation physics, therapy and oncology. Churchill Livingstone, Edinburgh, pp 97–117
Carmeliet P, Jain RK (2000) Angiogenesis in cancer and other diseases. Nature 407:249–257
Chauhan AS (2015) Dendrimer nanotechnology for enhanced formulation and controlled delivery of resveratrol. Ann N Y Acad Sci 1348:134–140
Chauhan AS (2018) Dendrimers for drug delivery. Molecules 23:938
Chen C, Lu L, Yan S, Yi H, Yao H, Wu D (2018) Autophagy and doxorubicin resistance in cancer. Anti-Cancer Drugs 29:1–9
Cruz F, Bellver J (2014) Live birth after embryo transfer in an unresponsive thin endometrium. Gynecol Endocrinol 30(7):481–484
D'souza H, Ganesh T, Joshi R, Kumar R (2000) Teletherapy concepts and equipment. In: Rath KG, Mohanti BK (eds) Textbook of radiation oncology- principles and practice. New Delhi, B I Churchill Livingstone, pp 113–129
Day MJ, Harrison RM (1995) Cross-sectional information and treatment simulation. In: Bleehan NM, Glatstein E (eds) Radiation therapy planning. Markel-Wekker, New York, pp 95–105
Dayes IS, Whelan TJ, Julian JA, Kuettel MR, Regmi D, Okawara GS, Dubois S (2006) Cross-border referral for early breast cancer: an analysis of radiation fractionation patterns. Curr Oncol 13(4):124
De Gonzalez AB, Curtis RE, Kry SF, Gilbert E, Lamart S, Berg CD, Ron E (2011) Proportion of second cancers attributable to radiotherapy treatment in adults: a cohort study in the US SEER cancer registries. Lancet Oncol 12(4):353–360
De Weger VA, Beijnen JH, Schellens JH (2014) Cellular and clinical pharmacology of the taxanes docetaxel and paclitaxel—a review. Anti-Cancer Drugs 25:488–494
Dogan N, King S, Emami B (2003) Assessment of different IMRT boost delivery methods on target coverage and normal-tissue sparing. Int J Radiat Oncol Biol Phys 57(5):1480–1491
EBCTC Group (2005) Effects of chemotherapy and hormonal therapy for early breast cancer on recurrence and 15-year survival: an overview of the randomised trials. Lancet 365(9472):1687–1717
Enmark M, Korreman S, Nystrom H (2006) IGRT of prostate cancer: is the margin reduction gained from daily IG time-dependent? Acta Oncol 45(7):907–914
Farokhzad OC, Langer R (2009) Impact of nanotechnology on drug delivery. ACS Nano 3:16–20
Feng T, Wei Y, Lee RJ, Zhao L (2017) Liposomal curcumin and its application in cancer. Int J Nanomedicine 12:6027–6044
Fogh SE (2010) Hypofractionated stereotactic radiation therapy: an effective therapy for recurrent high-grade gliomas. J Clin Oncol 28(18):3048–3053
Freedman GM, Anderson PR, Li J (2006) IMRT decreases acute skin toxicity for women receiving radiation for breast cancer. Am J Clin Oncol 29(1):66–70

Fullerton GD, Sewchand W, Payne JT, Levitt SH (1978) CT determination of parameters for inhomogeneity corrections in radiation therapy of the esophagus. Radiology 126(1):167–171
Hahn WC (1999) Inhibition of telomerase limits the growth of human cancer cells. Nat Med 5(10):1164–1170
Hanahan D, Weinberg RA (2000) The hallmarks of cancer. Cell 100(1):57–70
Hanan NA, Chiu HI, Ramachandran MR, Tung WH, Mohamad Zain NN, Yahaya N (2018) Cytotoxicity of plant-mediated synthesis of metallic nanoparticles: a systematic review. Int J Mol Sci 19:1725
Hassen-Khodja R (2004) Gamma knife and linear accelerator stereotactic radiosurgery. Agence d'évaluation des technologies et des modes d'intervention en santé
Huang Y, Hu L, Huang S, Xu W, Wan J, Wang D (2018) Curcumin loaded galactosylated BSA nanoparticles as targeted drug delivery carriers inhibit hepatocellular carcinoma cell proliferation and migration. Int J Nanomedicine 13:8309–8323
Jain RK (1994) Barriers to drug delivery in solid tumors. Sci Am 271:58–65
Kairdolf BA (2013) Semiconductor quantum dots for bioimaging and biodiagnostic applications. Annu Rev Anal Chem 6:143–162
Kamaly N, Xiao Z, Valencia PM, Radovic-Moreno AF, Farokhzad OC (2012) Targeted polymeric therapeutic nanoparticles: design, development and clinical translation. Chem Soc Rev 41:2971–3010
Kano H (2017) Estimating the risks of adverse radiation effects after gamma knife radiosurgery for arteriovenous malformations. Stroke 48(1):84–90
Karimi M, Bahrami S, Ravari SB, Zangabad PS, Mirshekari H, Bozorgomid M (2016) Albumin nanostructures as advanced drug delivery systems. Expert Opin Drug Deliv 13:1609–1623
Lamichhane S, Lee S (2020) Albumin nanoscience: homing nanotechnology enabling targeted drug delivery and therapy. Arch Pharm Res 43:118–133
Lammers T, Rizzo LY, Storm G, Kiessling F (2012) Personalized nanomedicine. Clin Cancer Res 18:4889–4894
Landgraf M, Lahr CA, Kaur I, Shafiee A, Sanchez-Herrero A, Janowicz PW (2020) Targeted camptothecin delivery via silicon nanoparticles reduces breast cancer metastasis. Biomaterials 240:119791
Lee I, Sokolsky O (2010) Medical cyber physical systems. In: 47th ACM/IEEE design
Lee SJ, Kim MJ, Kwon IC, Roberts TM (2016) Delivery strategies and potential targets for siRNA in major cancer types. Adv Drug Deliv Rev 104:2–15
Lefkopoulos D, Mazeron JJ (2006) Presentation of a special cancer radiation therapy volume on the latest developments and clinical applications of IGRT and adaptive radiotherapy (ART). Cancer Radiother 10(5):219–221
Leunens G, Verstraete J, Vanden BW (1992) Human errors in data transfer during preparation and delivery of radiation treatment affecting the final result: "garbage in, garbage out". Radiother Oncol 23:217–228
Lim EK, Chung BH, Chung SJ (2018) Recent advances in pH-sensitive polymeric nanoparticles for smart drug delivery in cancer therapy. Curr Drug Targets 19:300–317
Lin JH (2016) Review structure-and dynamics-based computational design of anticancer drugs. Biopolymers 105(1):2–9
Lu L, Shan F, Li W, Lu H (2016) Short-term side effects after radioiodine treatment in patients with differentiated thyroid cancer. Biomed Res Int 2016:4376720. 5 pages
Lum JJ (2005) Growth factor regulation of autophagy and cell survival in the absence of apoptosis. Cell 120(2):237–248
Mathieu D, Kondziolka D, Cooper PB (2007) Gamma knife radiosurgery in the management of malignant melanoma brain metastases. Neurosurgery 60(3):471–481
Mi P (2015) Hybrid calcium phosphate-polymeric micelles incorporating gadolinium chelates for imaging-guided gadolinium neutron capture tumor therapy. ACS Nano 9:5913–5921
Mi P (2017) Block copolymer-boron cluster conjugate for effective boron neutron capture therapy of solid tumors. J Control Release 254:1–9

Mizer S, Scheller RR, Deye JA (1986) Simulation procedures: general principles. In: Mizer S, Scheller RR (eds) Radiation therapy simulation workbook. Pergamon Press, New York, pp 9–15

Morales DR, Morris AD (2015) Metformin in cancer treatment and prevention. Annu Rev Med 66:17–29

Mudur G (2005) India has some of the highest cancer rates in the world. Br Med J 330(7885):215

Nguyen NP, Chi A, Betz M, Almeida F, Vos P, Davis R, Stevie M (2012) Feasibility of intensity-modulated and image-guided radiotherapy for functional organ preservation in locally advanced laryngeal cancer. PLoS One 7(8):e42729

Nicoletti MI, Lucchini V, D'Incalci M, Giavazzi R (1994) Comparison of paclitaxel and docetaxel activity on human ovarian carcinoma xenografts. *Eur J Cancer* 30a:691–696

Oishi M, Sumitani S, Nagasaki Y (2007) On−off regulation of 19F magnetic resonance signals based on pH-sensitive PEGylated nanogels for potential tumor specific smart 19F MRI probes. Bioconjug Chem 18:1379–1382

Oliver M, Chen J, Wong E (2007) A treatment planning study comparing whole breast radiation therapy against conformal, IMRT and tomotherapy for accelerated partial breast irradiation. Radiother Oncol 82(3):317–323

Orth M, Lauber K, Niyazi M, Friedl AA, Li M, Maihöfer C, Belka C (2014) Current concepts in clinical radiation oncology. Radiat Environ Biophys 53(1):1–29

Pazdur R, Kudelka AP, Kavanagh JJ, Cohen PR, Raber MN (1993) The taxoids: paclitaxel (Taxol) and docetaxel (Taxotere). Cancer Treat Rev 19:351–386

Pelicano H, Martin DS, Xu RH, Huang P (2006) Glycolysis inhibition for anticancer treatment. Oncogene 25:4633–4646

Peng H-S, Stolwijk JA, Sun L-N, Wegener J, Wolfbeis OS (2010) A nanogel for ratiometric fluorescent sensing of intracellular pH values. Angew Chem 122:4342–4345

Perez CA, Brady LW, Roti JLR (1998) Overview. In: Perez CA, Brady LW (eds) Principles and practice of radiation oncology. Lippincott-Raven, Philadelphia, pp 1–78

Prabaharan M (2015) Chitosan-based nanoparticles for tumor-targeted drug delivery. Int J Biol Macromol 72:1313–1322

Prasanna PG, Narayanan D, Hallett K, Bernhard EJ, Ahmed MM, Evans G, Coleman CN (2015) Radioprotectors and radiomitigators for improving radiation therapy: the small business innovation research (SBIR) gateway for accelerating clinical translation. Radiat Res 184(3):235–248

Purdy JA (1998) Three-dimensional physics and treatment planning. In: Perez CA, Brady LW (eds) Principles and practice of radiation oncology. Lippincott-Raven, Philadelphia, pp 343–369

Ralph SJ (2010) The causes of cancer revisited: "mitochondrial malignancy" and ROS induced oncogenic transformation—why mitochondria are targets for cancer therapy. Mol Asp Med 31(2):145–170

Rath GK, Sharma DN, Bahl A (2006) Advances in radiation oncology: time to move ahead. J Cancer Res Ther 2:159–160

Redpath AT, Muren LP (2006) CT guided IMRT for bladder cancer: isocentre shifts, margin and their impact on target dose. Radiother Oncol 87(3):276–283

Riaz MK, Riaz MA, Zhang X, Lin C, Wong KH, Chen X (2018) Surface functionalization and targeting strategies of liposomes in solid tumor therapy: a review. Int J Mol Sci 19:195

Rivankar S (2014) An overview of doxorubicin formulations in cancer therapy. J Cancer Res Ther 10:853–858

Rivlin N (2011) Mutations in the p53 tumor suppressor gene: important milestones at the various steps of tumorigenesis. Genes Cancer 2(4):466–474

Schwendener RA (2007) Liposomes in biology and medicine. Adv Exp Med Biol 620:117–128

Shi J, Xiao Z, Kamaly N, Farokhzad OC (2011) Self-assembled targeted nanoparticles: evolution of technologies and bench to bedside translation. Acc Chem Res 44:1123–1134

Singh P, Gupta U, Asthana A, Jain NK (2008) Folate and folate-PEGPAMAM dendrimers: synthesis, characterization, and targeted anticancer drug delivery potential in tumor bearing mice. Bioconjug Chem 19:2239–2252

Singh A, Trivedi P, Jain NK (2018) Advances in siRNA delivery in cancer therapy. Artif Cells Blood Substitutes Biotechnol 46:274–283
Slingerland M, Guchelaar HJ, Gelderblom H (2012) Liposomal drug formulations in cancer therapy: 15 years along the road. Drug Discov Today 17:160–166
Sprung CN, Ivashkevich A, Forrester HB, Redon CE, Georgakilas A, Martin OA (2015) Oxidative DNA damage caused by inflammation may link to stress-induced non-targeted effects. Cancer Lett 356(1):7281
Storme G, Verellen D, Soete G (2006) From Linac to tomotherapy: new possibilities for cure. Adv Exp Med Biol 587:303–308
Symonds RP (2001) Recent advances in radiotherapy. BMJ 323:1107–1110
Theek B, Rizzo LY, Ehling J, Kiessling F, Lammers T (2014) The theranostic path to personalized nanomedicine. Clin Transl Imaging 2:66–76
UNSCEAR A (2000) United Nations Scientific Committee on the effects of atomic radiation. Sources and effects of ionizing radiation, p 2
van den Boorn JG, Dassler J, Coch C, Schlee M, Hartmann G (2013) Exosomes as nucleic acid nanocarriers. Adv Drug Deliv Rev 65:331–335
Wang J, Zhang B (2018) Bovine serum albumin as a versatile platform for cancer imaging and therapy. Curr Med Chem 25:2938–2953
Wang XY, Ishida T, Ichihara M, Kiwada H (2005) Influence of the physicochemical properties of liposomes on the accelerated blood clearance phenomenon in rats. J Control Release 104:91–102
Wei Y, Pu X, Zhao L (2017) Preclinical studies for the combination of paclitaxel and curcumin in cancer therapy (review). Oncol Rep 37:3159–3166
Wei Y, Yang P, Cao S, Zhao L (2018) The combination of curcumin and 5-fluorouracil in cancer therapy. Arch Pharm Res 41:1–13
Wicki A, Witzigmann D, Balasubramanian V, Huwyler J (2015) Nanomedicine in cancer therapy: challenges, opportunities, and clinical applications. J Control Release 200:138–157
Wo JY, Viswanathan AN (2009) Impact of radiotherapy on fertility, pregnancy, and neonatal outcomes in female cancer patients. Int J Radiat Oncol Biol Phys 73(5):1304–1312
Wu W (2010) In-situ immobilization of quantum dots in polysaccharide-based nanogels for integration of optical pH-sensing, tumor cell imaging, and drug delivery. Biomaterials 31:3023–3031
Xi M, Liu SL, Zhao L, Shen JX, Zhang L, Zhang P, Liu MZ (2013) Prognostic factors and survival in patients with radiation-related second malignant neoplasms following radiotherapy for nasopharyngeal carcinoma. PLoS One 8(12):e84586
Xing G, Thorndyke B, Schreibmann E (2006) Overview of image guided radiation therapy. Med Dosim 31(2):91–112
Yardley DA (2013) Nab-paclitaxel mechanisms of action and delivery. J Control Release 170:365–372
Zhang M, Zhu J, Zheng Y, Guo R, Wang S, Mignani S (2018) Doxorubicin-conjugated PAMAM dendrimers for pH-responsive drug release and folic acid-targeted cancer therapy. Pharmaceutics 10:162
Zhu L, Chen L (2019) Progress in research on paclitaxel and tumor immunotherapy. Cell Mol Biol Lett 24:40
Zrazhevskiy P, Sena M, Gao X (2010) Designing multifunctional quantum dots for bioimaging, detection, and drug delivery. Chem Soc Rev 39:4326–4354

Chapter 9
Role of Nanotechnology in the Development of Photoprotective Formulations

Sonam Dwivedi and Iffat Zareen Ahmad

Abstract Nanotechnology derives from the technology that entailed its designing, production, and application in the nanometer range. Incorporation of nanotechnology in the cosmetic formulation commences the thrust area of research. Nanosized cosmetic formulations offer increased UV protection, penetrate deep into the skin layer, and provide effective release of ingredients, with good solubility and stability. Many of them also exhibit UV protective, antioxidant, and antimicrobial activities. The magnificence of micellar nanoparticles has now become the latest fascinating nanotechnology in the international and local cosmetic market. The micellar nanoparticles effectively enhance the surface area and actively transport the bioactive compounds into the skin. Vesicular nanosystems such as liposome and niosomes are versatile in nature and are able to encapsulate bioactive compounds of different solubilities. Natural compounds with photoprotective activity have created interest in the area of cosmetic formulation since they reduce the oxidative stress, toxicity, and damage caused by radiation. Nanocosmetics can be found in a variety of products ranging from hair care to sunscreen to oral care. The information provided in this chapter about various photoprotection formulations serves as a guide for future research to meet the necessary standards in the cosmeceuticals and cosmetics industries.

Keywords Nanoformulations · Photoprotection · Free radicals · Ultraviolet radiation · Nanocosmeceuticals · Nanopharmaceuticals

---

S. Dwivedi · I. Z. Ahmad (✉)
Natural Products Laboratory, Department of Bioengineering and Biosciences, Integral University, Dasauli, Lucknow, Uttar Pradesh, India
e-mail: iffat@iul.ac.in

V. K. Kannaujiya et al. (eds.), *Photoprotective Green Pharmacology: Challenges, Sources and Future Applications*,
https://doi.org/10.1007/978-981-99-0749-6_9

9.1 Introduction

We are exposed to radiation every day, and its effects on the biological system have the potential to be fatal since they can harm DNA, degrade proteins, and cause oxidative degradation of lipids in humans. Unexpected changes or damage in biological systems can occur as a result of both purposeful and unintentional causes of radiation exposure. There have also been cases of babies with genetic abnormalities whose parents were exposed to substantial amounts of radiation. Damage to the DNA might cause changes in the encoded proteins, which can lead to problems later. It may cause the encoded proteins to become completely inactive or malfunction (Bala et al. 2014). Many types of radiation exist, the majority of which goes undetectable by humans. Both ionizing and nonionizing radiation can harm living things, but high doses of ionizing radiation can harm vital organs and suppress the immune system, leading to cell damage (Omer 2021). Radiation causes cell damages (Tanabe et al. 2022). At low degrees of damage, our system has evolved numerous repair processes to repair damaged biomolecules and preserve cells, while nuclear DNA damage cannot be repaired in most cases. Nanotechnology is the most popular cosmetic technology right now. For cosmetic manufacturing, nanosized materials with improved UV protection activity, great penetration power, stability, and sustained drug release are required to minimize harmful effects such as skin cancer, UV-induced damage, and other skin-related diseases (Kaul et al. 2018; Mohanty et al. 2022). Cosmetic formulations that incorporate nanomaterials for distribution and to improve the efficacy of the given compounds are known as nanocosmeceuticals (Gupta et al. 2022). Cosmetic formulations benefit from nanotechnology in a number of ways (Singh et al. 2013). All nanoformulations have improved UV protection, as well as antiaging and long-lasting effects. Bioactive compounds offer their own set of benefits, and when paired with nanocarriers, they have even more therapeutic potential (Lohani et al. 2014). Micelles are commonly used in nanocosmetics because of their outstanding adsorption characteristics, which allow them to pull pollutants and oils away from the skin. This chemical is commonly found in skin cleansers (Salvioni et al. 2021). When compared to other nanocarriers, micellar nanoparticles are the most suitable and versatile nanoparticles for cosmetic compositions. This also offers excellent encapsulation qualities and a cheaper manufacturing cost (Aziz et al. 2019).

Micellar nanocarriers were employed by several multinational and national businesses to carry active substances. These brands made use of micellar nanotechnology to ensure that they operated well. This technique has the potential to be used to a broader spectrum of cosmetics. The oil-in-water nanoemulsion has been extensively used in cosmetic formulations. By mixing two immiscible liquids and stabilizing the emulsion with a surfactant, micelles are produced. Using a surfactant, which attracts the oil phase with its hydrophobic component and the overall aqueous solution with its hydrophilic component, reduces the space between the phases. Traditional emulsions, microemulsions, and nanoemulsions are all part of this nanosystem. The micellar particle size and thermodynamic stability of the emulsions were used to classify them. Due to their ability to generate smaller micelle particle sizes at lower surfactant concentrations, nanoemulsions are becoming the ideal delivery strategy due to their increased characteristics (Choi and McClements 2020).

A renowned cosmetics company, L'Oreal S.A., has obtained a patent on micellar-based cosmetic formulations. These products use a nanoemulsion system and can be used on a variety of skin and hair care. The cosmeceutical products range in the personal care industry is expected to develop at the quickest rate (Brandt et al. 2011). Safety concerns have been raised with nanoparticles because of their claimed toxicity. Nanocosmeceuticals offer several benefits by obtaining sustained drug release depending upon the manufacturing process, interactions, drug amount, polymer, and additive ratio. They can be found in many hair care products to combat graying and hair loss, thus preventing hair damage. Nanocosmeceuticals, including sprays and perfumes, help smells last longer. These boost the effectiveness of skin care products and increase the UV protection of sunscreens. They also increased the surface area, facilitating easy drug delivery into the skin. Skin occlusion improves penetration while also increasing skin hydration. Cosmeceuticals products having high entrapment efficiency and good sensory properties make them highly stable as compared to other cosmetic products. Drugs can be delivered in both lipophilic and hydrophilic forms by the majority of nanoparticles (Mu and Sprando 2010; Nohynek et al. 2007). Nanocosmecuticals products have some drawbacks. They generated reactive oxygen species (ROS), which damage cells' membranes, DNA, and proteins.

Carbon-based nanomaterials, titanium oxide, copper and silver nanoparticles have the potential to damage cells. Regulatory bodies accepted and regulated nanocosmeceuticals with little scrutiny. According to the Federal Food, Drug, and Cosmetic Act, cosmetics are defined as "articles intended to be applied on the skin for promoting attraction, beauty, smoothness and cleansing." Cosmetics improve the texture and appearance of the skin (Hoang et al. 2021). Cosmetics have a plethora of applications in makeup and personal care (Alves et al. 2020). The beginnings of cosmetics were seen in Egyptian culture and later other countries became involved in using cosmetics (Szalaty and Derda 2020). Cosmetics only became widely available in Western countries after the twentieth century, and they were initially used discreetly (Kaul et al. 2018). Cosmetic market size reaches to 5.5% on yearly basis in the twenty-first century (Szalaty and Derda 2020). The Asia Pacific cosmetics industry is the fastest growing, with a market value of nearly $70 billion (Mohd-Setapar et al. 2022). Cosmetics enriched with pharmacologically significant compounds that have therapeutic benefits on human appearance are known as cosmeceuticals in the cosmetics business (Aguilar-Toalá et al. 2019). Cosmeceuticals are placed between cosmetics and pharmaceuticals. They include sun care, skin care, and hair care (Che Marzuki et al. 2019). Different parameters are described in Fig. 9.1.

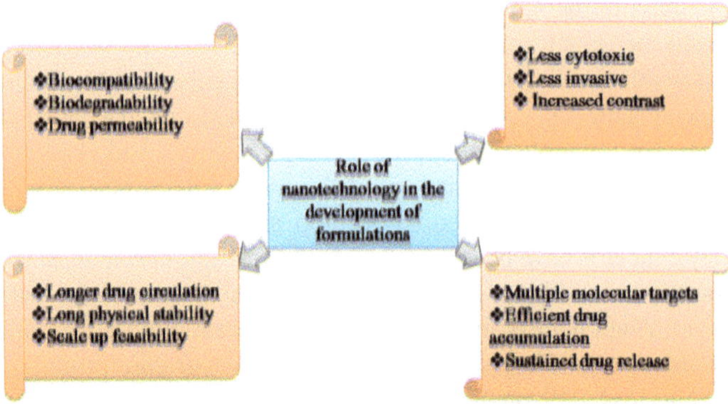

Fig. 9.1 Various roles of nanotechnology in the formulation development

9.2 Types of Nanocarriers for the Development of Photoprotective Formulations Tissue

Drug administration through the skin is depicted in Table 9.1 and Fig. 9.2. Various types of nanocarriers used in cosmetics products are liposomes, emulsions, solid lipid nanocarriers, lipid polymers, and liquid crystalline.

Table 9.1 Different types of nanoformulations and their applications in cosmeceuticals

S. no.	Types of nanoformulations	Properties	Application
1.	Liposomes	Spherical, self-closed vesicles, bioabsorbable, environmental friendly, safe	Skin care and hair care products
2.	Nanocapsule	Encapsulated	Skin care products
3.	Solid lipid nanoparticles	Low toxicity, physical UV blockers	Sunscreen
4.	Nanocrystals	Photoprotective	Skin care product
5.	Dendrimers	Water and sebum resistance, glossiness, tangible sense and bonding agent properties to the skin and hair	Dendrimers in hair, skin, and nail care products
6.	Niosomes	Nonionic surfactants, enhanced chemical strength, and penetration	Drugs and cosmetics
7.	Fullerene	Antioxidative properties	Carbon fullerene preparation of skin rejuvenation cosmeceutical formulations

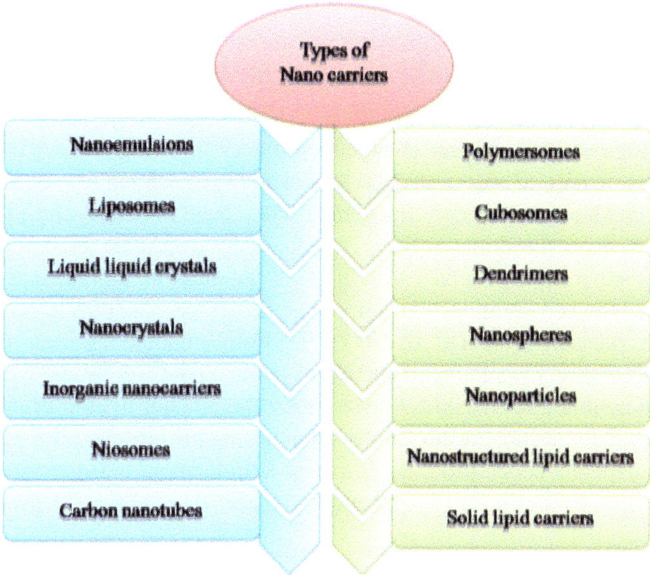

Fig. 9.2 Different types of nanocarriers

9.2.1 Microemulsions/Nanoemulsions

Microemulsions are isotropic dispersive systems that spontaneously form from oil, water, and surfactants and are in a stable state. Nanoemulsions are colloidal dispersion systems stabilized by an emulsifier that resemble an emulsion but contain very small droplets (Azeem et al. 2009). In microemulsions and nanoemulsions, the emulsifier improves drug entrapment efficiency (Sarheed et al. 2020). Resveratrol's solubility in microemulsions, for example, was 23 times that of crystalline resveratrol (Ha et al. 2019). The drug administered through microemulsions travels through the stratum corneum, allowing it to penetrate the tissue. Microemulsions achieve a greater rate of transdermal absorption, resulting in a supersaturated solution that enhances drug diffusion pressure and hence dramatically boosts the transdermal absorption rate (Alkilani et al. 2015). Microemulsion and nanoemulsion have the ability to fluidize the stratum corneum's double-layered structure and enhance drug absorption (Lane 2013). On the other hand, they have the capability to create a "diffuse-distribution-dissolution" effect that may improve epidermal hair follicles, which serve as both a storage location for medications and a route for particles to flow through (Lademann et al. 2008).

9.2.2 *Liposomes*

Liposomes are lipid vesicles, when they come into touch with the skin, release their contents. During fusion, liposomes produce lipid flakes in the stratum corneum's intercellular gap. Liposomes can change membranes, allowing drugs to flow through the skin more easily (Zhou et al. 2021). Niosomes, transfersomes, and ethosomes have all been shown to improve liposomes. They have the capability to delivering drugs deep beneath the skin's surface. Transfersomes overcome the skin penetration problem by adhering themselves with lipids of the outer membrane and expanding the distance between keratinocytes, allowing efficacy components to reach the skin via osmotic pressure (Rajan et al. 2011). Ethosomes enable the drug to reach the systemic circulation by creating disturbances in the outer layer promote penetration, and remain confined to the skin. Furthermore, the ethanol in ethosomes increases the permeability, altering the ethosomes' penetrating physicochemical properties to boost percutaneous penetration. By lowering skin tension, Noisome sustained the ingredients in the outer layer (Verma and Pathak 2010).

9.2.3 *Lipid-Based Liquid Crystal*

Liquid crystalline lipid carrier improved the drug loading capacity and controlled long-term drug release (Rajabalaya et al. 2017). Added lipid liquid crystal compounds (such as surfactants and moisturizers) can also improve skin hydration and lower the cuticle barrier. Liquid lipid crystalline carriers possess the same charge on the skin, so the drug can pass through the repulsive force (Zhou et al. 2021). Due to the presence of an intersecting channel in the liquid lipid crystalline nanocarrier, hydrophilic drugs can penetrate the skin easily. Liquid lipid crystalline nanocarrier permeates through hair follicle into the skin. Liquid crystalline lipid nanocarriers are too viscous, have a high affinity with the skin, and deliver drugs to the target desired, which can improve skin permeability and drug retention while also preserving the skin's wounded and susceptible areas (Rajabalaya et al. 2017). Liquid crystals structurally benefit the skin, provide a soothing effect, and protect against radiation (Kim et al. 2009).

9.2.4 *Nanocrystals*

Drug nanocrystals are sun micron carrier free colloidal suspension system. Reduction in particle size enhances solubility rate. Using nanocrystals, 100% drugs can be considerably improved in solubility and rate of dissolution, making them suitable for high-dose transdermal delivery (Shankar et al. 2020). The increased concentration gradient between the drug and skin nanocrystals can aid in the more rapid distribution of the drug. Nanocrystals remain confined to the skin, permitting a prolonged retention time that improves absorption. Pure drug nanosuspension accumulates in the hair follicle. The drugs are released from the nanocrystals.

9.2.5 Polymer Nanocarriers

Polymer-based nanocarriers are made up of different polymers that are biocompatible and nontoxic. Polymeric nanocarriers remain confined deeper to the skin and are continuously dispersing the drug (Souto et al. 2020). According to various findings, small-diameter nanomaterials accumulate better than larger nanoparticles. Permeability can be achieved by modifying polymeric nanocarrier polymer material. For insoluble drugs, improve the concentration gradient and diffuse the drug directly to the outer layer of the skin. Polymer-based micelles adhere to the skin's outer surface. The maximum amount of drug has been carried out with polymer nanocarriers for transdermal drug delivery. The polymer nanocarriers improve the solubility rate of their design and manufacture to penetrate into the skin for easy delivery. It is necessary while making polymer nanocarriers that their solubility and design ratio be balanced. Nanohydrogels, unlike polymeric nanoparticles and polymeric micelles, have the potential to significantly increase skin moisture. When applying nanohydrogels, keratinocyte swelling was higher, and the distance between keratinocyte cells was wider than in skin treated with water. Furthermore, the cuticle's lipid bilayer has been discovered to be more mobile (Zhou et al. 2021).

9.2.6 Niosomes

These are double-layer nonionic nano-vesicles that can include or exclude cholesterol and its lipids, and solute is completely sealed with in membrane. Its size varies between 100 nm and 2 m. Nonionic surfactants are utilized to make niosomes, as well as cholesterol (Chen et al. 2019). Both hydrophobic and hydrophilic and poorly soluble substances can be transported by niosomes as a unique pharmaceutical delivery strategy. It encapsulates the medication, allowing it to stay in the systemic circulation for longer and more efficiently enter target site. They are more stable, low cost, and oxidation susceptibility as compare to liposomes. Niosomes are used in skin care products because they can reversibly lower the horny layer's barrier resistance, allowing the component to penetrate living tissues faster. Chemicals that have been poorly adsorbed have better bioavailability, while entrapped molecules have improved stability (Kaul et al. 2018). Niosome formation is influenced by surfactant type and structure, active compounds nature, membrane, and temperature, collectively formed the small nanovesicles. Pro-niosomes are non-hydrated nonionic surfactant vesicles; they must be hydrated in order to transform into niosomes (Kazi et al. 2010). In 1970, L'Oreal pioneered the manufacture of niosomes by studying and developing synthetic liposomes. In 1987, L'Oreal invented niosomes, which were then sold under the Lancome brand.

9.2.7 Carbon-Based Nanomaterials

These are allotropic modifications of carbon formed into cylindrical tubes, with nanometer-sized diameters and millimeter-long lengths. They have strong powers to flow heat and electricity, as well as good mechanical strength. They have a high absorption rate, allowing them to load more drugs. They directly reach the drug at the target site. They are widely used in sensing, imaging, tissue repair, extraction of bioactive molecules, and measurement of pollution (He et al. 2013).

9.2.8 Lipid-Based Polymer Nanocarrier

These are lipid-based polymeric vesicles with an amphiphilic inner layer structure, making it a better vehicle for drug delivery with a highly stable and good drug release rate (Poschenrieder et al. 2017). It could be as small as 50 nm or as large as 5 m or more. It can encapsulate biomolecules. Polymersomes have traditionally been made up of repeating units of monomers. Diverse properties have been created in polymer vesicles by varying the molecular weight of the polymer. Polymersomes can target and manage drug release due to the flexibility of their membrane. They are more stable than liposomes due to their thick, double-layered structure (Lee and Feijen 2012).

9.2.9 Cubosomes

These are liquid-crystalline nanocarriers made up mostly of certain amphiphilic lipids in a precise proportion. Hydrating the liquid phase, followed by dispersion of the solid phase, forms the cubic structure. They have a solid rheology with certain characteristics that are practical in practice (Rao et al. 2018). They have a thermodynamically stable, three-dimensional structure and are compactly arranged in a double layer, which carries more drugs into it. Cubosomes can encapsulate any type of particle, whether it has a positive, negative, or neutral charge. Cubosomes help in the solubilization of all types of compounds, including hydrophobic, hydrophilic, and amphipathic compounds, and, therefore, help in drug design, development, and targeted drug delivery. Because of their properties, cubosomes are versatile systems that can be given orally, subcutaneously, or parenterally. Even though electron microscopy, light scattering, X-rays, and NMR may all be used to study cubosome structure, just a few researchers have investigated their potential as delivery vehicles (Rao et al. 2018).

9.2.10 Dendrimers

Dendrimers are excellent solubility enhancers; drugs can be incorporated into them to improve solubility and stability. It offers a variety of biomaterial delivery methods, ranging from oral to transdermal. Dendrimers have the ability to control drug release, entrap many drugs, good stability, a faster dissolution rate, site-specific delivery, and good penetration power (Chauhan 2015). Dendrimers loaded the anticancer drug doxorubicin and released it at different pH ranges under radiation; this suggests that dendrimer is also suitable for drug delivery under radiation (Wu et al. 2018).

9.2.11 Polymeric Matrix Nanoparticle

Small vesicle used to carry drugs. The sizes of the particles range from 10 to 200 nm. The drug is stored in nanospheres and entrapped, dissolved, linked, or encapsulated in a polymer matrix, which protects it from chemical and enzymatic destruction (Singh and Lillard Jr 2009). In the polymer matrix system, the drug is physically and uniformly disseminated, with a good absorption rate and high efficacy. Nanospheres contain enzymes, genes, and drugs in their cores for drug delivery (Modicade et al. 2019). Biodegradable and nonbiodegradable nanospheres are the two types of nanospheres. Polymer matrix modified with biodegradable and nonbiodegradable materials (Rao and Kumari 2020). Nanospheres used in skin care to carry biomolecules deeper into the skin's layers, allowing them to deliver their therapeutic benefits to the affected area with better precision and efficiency. These tiny bits are crucial in the prevention of actinic ageing. Nanospheres are increasingly being used in cosmetics (Kaul et al. 2018).

9.2.12 Nanoparticles

Nanogold is available in a variety of sizes ranging from 5 to 400 nm. Nanospheres, nanoshells, nanoclusters, nanorods, nanostars, nanocubes, branching nanotriangles, and other forms are among them. Gold nanoparticles have good dielectric constant and resonance. Nanogold is nonbleaching following membrane staining and is exceptionally durable in liquid or dry form; conjugated and unconjugated variants are available. They have antimicrobial properties that can be used as a suitable delivery vehicle for skin care. The key properties of nanogold in beauty care include increased blood circulation, anti-inflammatory and antibacterial properties, lift augmentation, and tissue regeneration stimulation.

9.2.13 *Liquid Lipid Carriers*

Solid lipid carriers were the first to be developed, and liquid lipid carriers were the next step. Liquid lipids, which are similar to liposomes, were the first cosmetics to hit the market. There are used as excipients. Liquid lipids are found in a wide range of products (Müller et al. 2007).

9.2.14 *Solid Lipid Carriers*

These are the advance form of traditional lipoidal carriers. They have dimension between 50 and 1000 nm. They contained lipids inside, and the matrix drug is made up of solid lipids or lipid mixes that are disseminated (Naseri et al. 2015). The hydrophobic chains of phospholipids are absorbed lipid core. Active compounds that are lipophilic, hydrophilic, or mildly water soluble can be included in Solid lipid nanoparticles (SLNs) made up of physiological and biocompatible lipids (Puri et al. 2009). Toxicity issues are minimized by preparing SLN with biocompatible compounds. High-pressure homogenization and precipitation are the two most popular procedures for making SLNs. Solid lipids have low dissolution rate and are nontoxic. Because of their small size, it allows active chemicals to penetrate deeper into the skin (Mishra et al. 2018). Solid lipids are UV resistant and can be combined with sunscreen to provide sun protection with fewer side effects. To improve UV protection, they can be mixed with vitamins and their occlusive property used as a moisturizing agent. Solid lipids are also used in perfume formulations since they extend the fragrance's release time and are perfect for usage in day care. They are stable, safe, and suitable for drug carriers (Kaul et al. 2018).

9.3 Mechanism of Photoprotection

9.3.1 *Scavenging of Free Radicals*

The biological system's defense against free radicals results in the formation of thiols and vitamins. Ionizing radiation produces free radicals, which are scavenged by radioprotective substances prevalent in biological systems (Mishra et al. 2018). Low molecular weight intracellular thiol mostly found in all cells. It is worth noting that the majority of medicinal herbs include antioxidants. Natural antioxidants, on the other hand, can help protect against oxidative stress by scavenging free radicals produced by ionizing radiation. Phytochemicals have been shown to help patients with a variety of complicated conditions in a number of studies, including heavy metal toxicity. A number of radioprotective drugs, including cellular antioxidants, have this property. As a result, finding new radioprotective compounds based on this property should be beneficial.

9.3.2 DNA Repair

DNA susceptibility to radiation is the primary target for cell death and transformation. Various repair mechanisms are used by cells to restore genomic integrity and maintain cell survival (Rastogi et al. 2010; Smith et al. 2017). Radioprotective chemicals are designed to help with both DNA repair and DNA damage reduction. It promotes homologous recombination repair by increasing BRCA1 stability and interaction (Volcic et al. 2012; Huang and Zhou 2020).

9.3.3 Synchronization of Cells

Radiosensitivity affects cell division rate directly and cell differentiation indirectly. Radiosensitivity depends on cell cycle. According to cell synchronization research, the G2 and M cells are the most risk to radiation, followed by the G1 cells, while the S cells are resistant (Pawlik and Keyomarsi 2004). Despite the fact that G2 and M have closely packed DNA and limited restoration, cells in the S are insensitive to radiation and have a greater potential for healing because repair enzymes are more easily accessible in the uncompacted DNA. As a result, natural compounds may reversibly inhibit the cells in the S phase (Chen and Deng 2018).

9.3.4 Modification of Antioxidants and Redox Responsive Genes

The production of various antioxidant enzyme genes lowers the oxygen species. Antioxidant enzymes regulate oxygen species and maintain balance. Antioxidants in our diet fight oxygen species and increase enzyme expression (MnSOD) and cellular redox efficiency in many circumstances. Several antioxidant-based radioprotective drugs alter redox state by maintaining the regulation of genes. Different redox sensitive genes exist in the cellular system (Fig. 9.3). Natural antioxidant chemicals change the expression of these genes due to their redox sensitivity. The

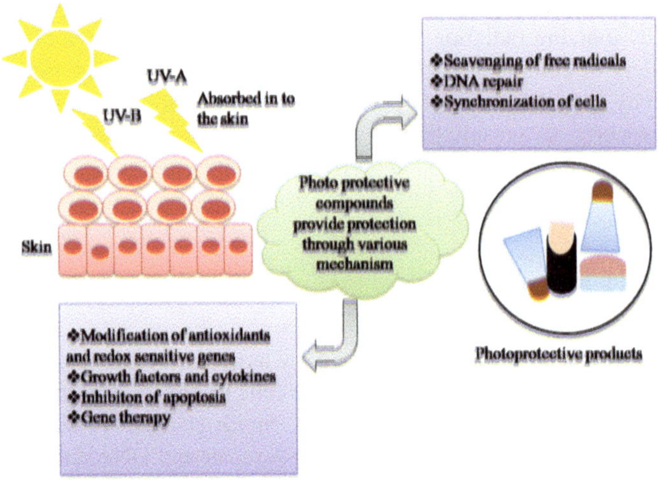

Fig. 9.3 Overview of the mechanism of photoprotection to guard the skin cells from UV radiation

xenobiotic response activates the induction of natural antioxidants and antioxidant enzymes to prevent from oxidative damage.

9.3.5 Role of Cytokines and Growth Factors in Photoprotection

The primary focus of radioprotectors lies on the initiation of events that occur in the cells after radiation. Different cytokines play an important role in inflammation (Fig. 9.3). The tissue response has been linked to IL6 and IL1 immediately after irradiation (Schaue et al. 2012). Many FDA-approved growth factors have been observed in mice where they protect the bone marrow, causing new tissue formation in mice suffering from blood cell-related disease. The mobilization of naive blood cells can aid in the reduction of radiation damage (Mishra et al. 2018).

9.3.6 Inhibition of Apoptosis

ROS generates many oxygen species in cells, and an imbalance in the ions results in an uncharged cell membrane. Complex mitochondria initiate cell death and activate caspase cascades after initiating mitochondria degrade. The apoptotic family proteins unite, creating a hole in the mitochondria that allows the release of cytochrome *c*. The p53-mediated apoptosis is another key apoptosis mechanism. A growing body of data from tumor cell investigations suggests that p53 is required to check cell cycle and apoptosis following radiation exposure (Chen 2016) (Fig. 9.3). Small compounds that decrease apoptosis without jeopardizing DNA repair might be an effective radiomitigator. According to a study, oligomers of *N*-alkyl glycines (peptides) prevent apoptosis by interacting with the apoptosome complex (Lademann et al. 2003). The findings of the preceding investigations imply that apoptosis may be halted, either by mitochondrial or p53-mediated mechanisms, and that these substances can attenuate the effects of radiation, particularly in the GI tract. However, more studies are required to determine if apoptosis inhibitors can be used for radioprotection.

9.3.7 Gene Therapy

Radioprotectors, or substances capable of shielding tissue from radiation by avoiding radiation damage or lowering cell death in the presence of radiation damage, have been sought by researchers to lessen normal tissue damage as shown in Fig. 9.3. While much of the early research focused on small molecule radioprotectors, gene therapy for radioprotection is becoming more popular. Gene therapy is a great technique for radioprotection because of the targeting capabilities of gene therapy vectors and the flexibility of gene therapy to achieve ablation or augmentation of physiologically important genes. Future advancements in vector targeting and distribution should considerably improve gene therapy's radioprotection (Everett and Curiel 2015).

9.4 Signaling Pathways Involved in the Photoprotection

9.4.1 NFkB Signaling

Nuclear factor (NF)-B is a transcription factor that induces genetic networks to regulate immunological function, cellular survival, and apoptosis suppression. Cells that send signals to members of the NF-B-related family produce a variety of structures (Fig. 9.4). These signal-activated complexes (Es) translocate to the promoter region and bind to the 10 bp. NF-B is known to reduce cell death in response to radiation. Continuous involvement of NF-B increases resistance in cells, while deletion promotes hypersensitivity to IR-induced GI injury in mice. Following IR exposure, NF-B also regulated the production of interleukins, which help in the recovery of tissue regeneration in the progenitor cells and organs. Several additional cytokines have been shown to have powerful pro-inflammatory actions, which may contribute to increased tissue damage after ionizing radiation exposure. There have been a number of molecules or groups of chemicals studied for their radioprotective potential, but only a handful has progressed to clinical trials. It is possible that the failure or lack of effectiveness of discovered agents in humans is related to their lower radiation-protective efficacy (Vijay et al. 2015).

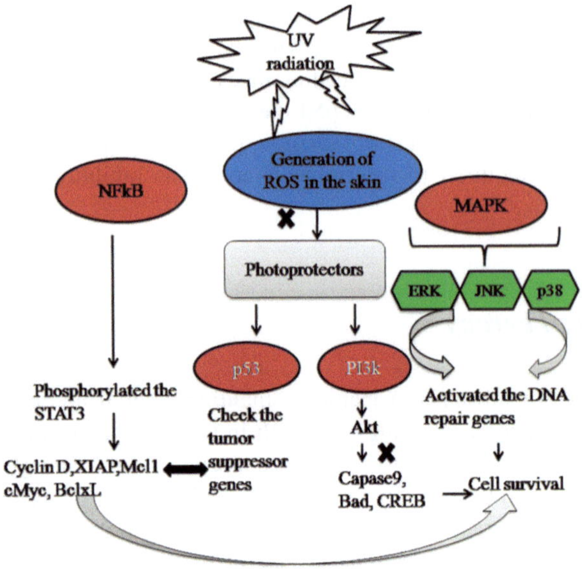

Fig. 9.4 Different signaling pathways of photoprotection to prevent harmful UV radiation

9.4.2 *MAPK Signaling*

This pathway controls many cellular processes. Multiple stressors present in the environment collectively activate the pathway (Fig. 9.4). After irradiation, cells signal to activate the pathways that initiate various downstream signaling processes that help the cell recover from the damage. Growth factors boost MAPK pathway responses in cells. Basal signaling was boosted by proto-oncogenes. This can happen through the PI3K route in many cell types (Dent et al. 2003).

9.4.3 *PI3K/Akt Signaling*

PI3K divided into three groups, with type I PI3K being the most researched. It is a two-component heterodimer with a catalytic and a regulatory component (Jean and Kiger 2014). SH2 and SH3 domains in the regulatory subunits interact with target proteins with suitable locations. p85 is the regulatory component that has many isotypes with different molecular weight. p110 are the catalytic subunit. Others are isolated to leukocytes and are broadly disseminated in various cells. Tyrosine kinases present in the receptors activate Akt signaling, and further dephosphorylation of Akt can inhibit cell growth and enhance apoptosis sensitivity. PTEN inhibits this signaling pathway; otherwise, PI3K/Akt remains activated indefinitely (Fig. 9.4) (Teng et al. 2021).

9.4.4 p53 Signaling

In the DNA replication or cell division process, this signaling process responds to stress. The p53 protein receives a stress signal via posttranslational modifications as shown in Fig. 9.4. Six feedback loops affect p53 activity via the MDM-2 protein. The significance of redundancy is described by the way gene products interact with p53. As we understand more about cancer, the connections between signal transduction pathways will become increasingly crucial (Harris and Levine 2005). The p53 halts the cell cycle at the G1 phase when DNA is damaged. The role of p53 in the response to UV radiation is unraveling. A study was performed under UV radiation, and delays in the cell cycle were found in the normal cells, but cells lacking the repair genes remained arrested. (Geyer et al. 2000).

9.5 Potential Application of Nanotechnology in the Developments of Formulations

9.5.1 Nanopharmaceuticals

Nanotechnology is a relatively new field of therapeutics that is quickly expanding. Several medications may be administered effectively and safely using nanocarriers. Good absorption, a longer release rate, and good penetration power are several of the benefits of nanocarriers. Radiopharmaceuticals, or radioactive substances, are already utilized in a variety of medical treatments. Radiopharmaceuticals are used in the treatment of a variety of diseases, and they have a tendency to accumulate preferentially on certain organ targets. Nanocarriers used in the radiopharmaceuticals improved their release kinetics, rate of absorption, and penetration (Fig. 9.5). Radionuclides, on the other hand, might be used in pharmacological evaluations of nanosystems in experimental research. However, more research has been carried out to explore the radiopharmaceutical applications mediated by nanosystems (Mironidou-Tzouveleki and Tsartsalis 2010).

9.5.2 Nanocosmeceuticals

Cosmetics make extensive use of nanocarriers. In the cosmeceutical business, nanotechnology has had a considerable influence. Product diversity, increased rate of drug release, and long-term benefits are all advantages of nanotechnology-based cosmeceuticals (Fig. 9.5). The increased use of nanocarriers in cosmeceuticals results in the accumulation of nanoparticles in the skin and associated health risks. Different nanocarriers used in cosmeceutical products pose a potential threat to humans, so recent rules have been initiated to reduce them (Lohani et al. 2014).

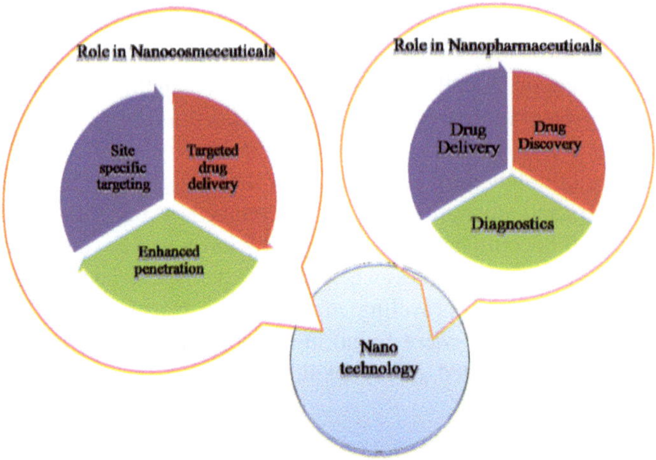

Fig. 9.5 Role of nanotechnology in the development of nanocosmeceuticals and nanopharmaceuticals

9.5.2.1 Moisturizing Agent

The outer layer of the skin keeps it hydrated. Moisturizers that promote skin suppleness can help prevent dehydration when dryness and dehydration occur. Moisturizers provide humidity to the skin, which aids in the retention of moisture and improves the texture. Moisturizing agents are mainly nanoemulsions due to their high retention capacity.

9.5.2.2 Sun Care

Sunscreen shields the skin from harmful UV rays. Sun protection prevents harmful rays from penetrating deeper into the skin. Older sunscreens formed a layer on the skin; this problem has been solved with the development of nanoparticle-incorporated sunscreen. They are nonsticky, have a longer retention time, are soothing, and have a pleasant scent. Octyl-*p*-methoxycinnamate and bemotrizinol were used to create a sunscreen system, and rosmarinic acid was added to boost antioxidant activity. The multifunctional new formulations have been developed to provide safety and better results (Cândido et al. 2022).

9.5.2.3 Antiaging Products

Chemical compounds, pollution, and various types of stress all contribute to skin aging. Collagen helps in recovery and wrinkle reduction. As we become older, the amount of collagen in our skin diminishes. The antiaging products work on lifting the skin and removing wrinkles and spots. Nanotechnology is now used in antiaging cosmetics, which are the most popular cosmeceuticals on the market. Nanocarriers improve skin elasticity and reduce wrinkle appearance. Retinol can improve collagen production while also increasing epidermal water content, hyperplasia, and cell renewal. The therapeutic effects reduce fine lines and wrinkles. There are creams on the market that are nanoencapsulated with triceramide to help skin look younger. A liposome containing coenzyme Q10 aids in the prevention of aging (Souto et al. 2022).

9.5.2.4 Hair Care

Nanotechnology has also shown promise in the field of hair care. Nanoemulsion in hair cosmetics, unlike traditional hair-straightening treatments, does not disrupt the surface structure of hair fibers, known as cuticles, in order to permeate into the hair strands. Hair cosmeceuticals contain sericin (which is made up of cationic sericin nanoparticles). Hair cosmeceuticals containing sericin nanoparticles have shown good results. As a result, crude plant material infusion has become a common approach for raising the quantity of bioactive compounds in seed oils, resulting in greater cell-based anti-oxidative impact, cell proliferation, and so on. Calendula flowers, which contain a range of seed oil chromatography, were used to analyze and assess the change in phytochemical content following the infusion of oils. An in vitro assay was carried out to investigate the biological impact on living cells. Based on the results of in vitro analysis, seed oils infused with calendula have higher bioactivity than their simple equivalents (wound healing and radioprotective action). Phyto-nanoemulsion was created by extracting herbal oils using an effective extraction method and then combining them with nanoemulsion systems; the biological activities of the recommended formulations were investigated. Finally, these preparations could be a potential choice for therapeutic inclusion in cutaneous cosmetics or nutritional supplements, especially following radiotherapy or chemotherapy (Gumus et al. 2015).

9.5.2.5 Cleansing Agent

The skin is naturally protected against harmful organisms while also attracting dirt and impurities. This causes sweat and body odor. As a result, maintaining skin health requires cleaning on a regular basis to remove debris, pollution, and odor. Regular washing of the skin is necessary to remove dirt. Nanoparticles are used as decontaminants and disinfectants. There are numerous skin cleanser products on the market that effectively address the issue of dirt and impurities.

9.5.2.6 Lip Care

Another new cosmeceutical segment is lip care. Lipstick and lip gloss containing nanoparticles help to maintain the moisture of the lips. A patent shows how to manufacture pigments while preserving their color for a long duration. The color dispersion of lipsticks containing silica nanoparticles is more consistent.

9.5.2.7 Nanotechnology in Nail Products

According to a study, nanosized particles in nail polish improve mammalian nails (Kaul et al. 2018). A patent on nail polish incorporating nanocarriers with additional benefits has also been obtained. Incorporating antifungal nanoparticles in nail polish to prevent infections is one of the new strategies added to nail cosmetics.

9.6 Conclusion and Future Prospects

Cosmetics production is now dominated by nanotechnology; the future will be shaped by omics science advances at the cell and tissue level. Cosmeceuticals products are always scientifically proven to have a preventative or therapeutic function. Due to social media transmission, this claim is growing in popularity and affecting customers and the market. Customers demand for the inclusion of new natural products in cosmetics to stimulate cellular rejuvenation. Diversification is a new source of these natural and long-lasting molecules. Assessment of novel compounds and actions is now the focus of substantial scientific efforts. Bioactive chemicals and secondary metabolites are abundant in plants, algae, and many microbes. As a result, natural bioactive chemicals have been intensively studied and incorporated into cosmeceutical products. They provide relaxation from stress, stop aging, and provide good health. The active compounds from the plant source are allergic and do not easily penetrate the skin. Prior to commercialization, it is vital to thoroughly analyze both the good effects and any potential difficulties associated with plant compounds. Traditional extraction methods have drawbacks such as composition conversion risk, extraction time, high productivity, and toxicity. Nanobiotechnology has created sustainable and healthy cosmetics to allow for proper compound transport; their small size influences efficacy across the skin and product without preservatives. More advancement will be required to prepare the biofeasible material, emphasizing the importance of process sustainability.

References

Aguilar-Toalá JE, Hernández-Mendoza A, González-Córdova AF, Vallejo-Cordoba B, Liceaga AM (2019) Potential role of natural bioactive peptides for development of cosmeceutical skin products. Peptides 122:170170

Alkilani AZ, McCrudden MT, Donnelly RF (2015) Transdermal drug delivery: innovative pharmaceutical developments based on disruption of the barrier properties of the stratum corneum. Pharmaceutics 7(4):438–470

Alves TF, Morsink M, Batain F, Chaud MV, Almeida T, Fernandes DA, da Silva CF, Souto EB, Severino P (2020) Applications of natural, semi-synthetic, and synthetic polymers in cosmetic formulations. Cosmetics 7(4):75

Azeem A, Rizwan M, Ahmad FJ, Iqbal Z, Khar RK, Aqil M, Talegaonkar S (2009) Nanoemulsion components screening and selection: a technical note. AAPS PharmSciTech 10(1):69–76

Aziz ZAA, Mohd-Nasir H, Ahmad A, Mohd. Setapar SH, Peng WL, Chuo SC, Khatoon A, Umar K, Yaqoob AA, Mohamad Ibrahim MN (2019) Role of nanotechnology for design and development of cosmeceutical: application in makeup and skin care. Front Chem 7:739

Bala K, Saini P, Katare DP (2014) Current status and future potential of herbal radioprotectants. World J Pharm Pharm Sci 3(8):1341–1366

Brandt FS, Cazzaniga A, Hann M (2011) Cosmeceuticals: current trends and market analysis. Semin Cutan Med Surg 30(3):141–143. WB Saunders

Cândido TM, Ariede MB, Pinto CASDO, Lourenço FR, Rosado C, Velasco MVR, Baby AR (2022) Prospecting in vitro antioxidant and photoprotective properties of rosmarinic acid in a sunscreen system developed by QbD containing Octyl p-methoxycinnamate and bemotrizinol. Cosmetics 9(2):29

Chauhan AS (2015) Dendrimer nanotechnology for enhanced formulation and controlled delivery of resveratrol. Ann N Y Acad Sci 1348(1):134–140

Che Marzuki NH, Wahab RA, Abdul Hamid M (2019) An overview of nanoemulsion: concepts of development and cosmeceutical applications. Biotechnol Biotechnol Equip 33(1):779–797

Chen J (2016) The cell-cycle arrest and apoptotic functions of p53 in tumor initiation and progression. Cold Spring Harb Perspect Med 6(3):a026104

Chen G, Deng X (2018) Cell synchronization by double thymidine block. Bio-protocol 8(17)

Chen S, Hanning S, Falconer J, Locke M, Wen J (2019) Recent advances in non-ionic surfactant vesicles (niosomes): fabrication, characterization, pharmaceutical and cosmetic applications. Eur J Pharm Biopharm 144:18–39

Choi SJ, McClements DJ (2020) Nanoemulsions as delivery systems for lipophilic nutraceuticals: strategies for improving their formulation, stability, functionality and bioavailability. Food Sci Biotechnol 29:149–168

Dent P, Yacoub A, Fisher PB, Hagan MP, Grant S (2003) MAPK pathways in radiation responses. Oncogene 22(37):5885–5896

Everett WH, Curiel DT (2015) Gene therapy for radioprotection. Cancer Gene Ther 22(4):172–180

Geyer RK, Nagasawa H, Little JB, Maki CG (2000) Role and regulation of p53 during an ultraviolet radiation-induced G1 cell cycle arrest. Cell Growth Differ 11(3):149–156

Gumus ZP, Guler E, Demir B, Barlas FB, Yavuz M, Colpankan D, Senisik AM, Teksoz S, Unak P, Coskunol H, Timur S (2015) Herbal infusions of black seed and wheat germ oil: their chemical profiles, in vitro bio-investigations and effective formulations as phyto-nanoemulsions. Colloids Surf B: Biointerfaces 133:73–80

Gupta V, Mohapatra S, Mishra H, Farooq U, Kumar K, Ansari MJ, Aldawsari MF, Alalaiwe AS, Mirza MA, Iqbal Z (2022) Nanotechnology in cosmetics and cosmeceuticals—a review of latest advancements. Gels 8(3):173

Ha ES, Sim WY, Lee SK, Jeong JS, Kim JS, Baek IH, Choi DH, Park H, Hwang SJ, Kim MS (2019) Preparation and evaluation of resveratrol-loaded composite nanoparticles using a supercritical fluid technology for enhanced oral and skin delivery. Antioxidants 8(11):554

Harris SL, Levine AJ (2005) The p53 pathway: positive and negative feedback loops. Oncogene 24(17):2899–2908

He H, Pham-Huy LA, Dramou P, Xiao D, Zuo P, Pham-Huy C (2013) Carbon nanotubes: applications in pharmacy and medicine. Biomed Res Int 2013:578290

Hoang HT, Moon JY, Lee YC (2021) Natural antioxidants from plant extracts in skincare cosmetics: recent applications, challenges and perspectives. Cosmetics 8(4):106

Huang RX, Zhou PK (2020) DNA damage response signaling pathways and targets for radiotherapy sensitization in cancer. Signal Transduct Target Ther 5(1):1–27

Jean S, Kiger AA (2014) Classes of phosphoinositide 3-kinases at a glance. J Cell Sci 127(5):923–928

Kaul S, Gulati N, Verma D, Mukherjee S, Nagaich U (2018) Role of nanotechnology in cosmeceuticals: a review of recent advances. J Pharma 2018:1–19

Kazi KM, Mandal AS, Biswas N, Guha A, Chatterjee S, Behera M, Kuotsu K (2010) Niosome: a future of targeted drug delivery systems. J Adv Pharm Technol Res 1(4):374

Kim IY, Nakagawa S, Ri K, Hashimoto S, Masaki H (2009) Liquid crystal 0/W emulsions to mimic lipids and strengthen skin barrier function. Cosmet Toiletries 124(7)

Lademann U, Cain K, Gyrd-Hansen M, Brown D, Peters D, Jäättelä M (2003) Diarylurea compounds inhibit caspase activation by preventing the formation of the active 700-kilodalton apoptosome complex. Mol Cell Biol 23(21):7829–7837

Lademann J, Knorr F, Richter H, Blume-Peytavi U, Vogt A, Antoniou C, Sterry W, Patzelt A (2008) Hair follicles—an efficient storage and penetration pathway for topically applied substances. Skin Pharmacol Physiol 21(3):150–155

Lane ME (2013) Skin penetration enhancers. Int J Pharm 447(1–2):12–21

Lee JS, Feijen J (2012) Polymersomes for drug delivery: design, formation and characterization. J Control Release 161(2):473–483

Lohani A, Verma A, Joshi H, Yadav N, Karki N (2014) Nanotechnology-based cosmeceuticals Int Sch Res Notices 2014:1–14

Mironidou-Tzouveleki M, Tsartsalis S (2010) Nanotechnology and radiopharmaceuticals: diagnostic and therapeutic approaches. Curr Drug Deliv 7(2):168–174

Mishra V, Bansal KK, Verma A, Yadav N, Thakur S, Sudhakar K, Rosenholm JM (2018) Solid lipid nanoparticles: emerging colloidal nano drug delivery systems. Pharmaceutics 10(4):191

Modicade ML, Keating AV, de Fátima Pina M, Raimi-Abraham BT (2019) Engineering of nanofibrous amorphous and crystalline solid dispersions for oral drug delivery. Pharmaceutics 11(1):7

Mohanty S, Badhei L, Pal A, Panda P (2022) Novel cosmeceutical formulations: a better approach to photoprotection. Int J Appl Pharm 14(4):9–17

Mohd-Setapar SH, John CP, Mohd-Nasir H, Azim MM, Ahmad A, Alshammari MB (2022) Application of nanotechnology incorporated with natural ingredients in natural cosmetics. Cosmetics 9(6):110

Mu L, Sprando RL (2010) Application of nanotechnology in cosmetics. Pharm Res 27(8):1746–1749

Müller RH, Petersen RD, Hommoss A, Pardeike J (2007) Nanostructured lipid carriers (NLC) in cosmetic dermal products. Adv Drug Deliv Rev 59(6):522–530

Naseri N, Valizadeh H, Zakeri-Milani P (2015) Solid lipid nanoparticles and nanostructured lipid carriers: structure, preparation and application. Adv Pharm Bull 5(3):305

Nohynek GJ, Lademann J, Ribaud C, Roberts MS (2007) Grey goo on the skin? Nanotechnology, cosmetic and sunscreen safety. Crit Rev Toxicol 37(3):251–277

Omer H (2021) Radiobiological effects and medical applications of non-ionizing radiation. Saudi J Biol Sci 28(10):5585–5592

Pawlik TM, Keyomarsi K (2004) Role of cell cycle in mediating sensitivity to radiotherapy. Int J Radiat Oncol Biol Phys 59(4):928–942

Poschenrieder ST, Schiebel SK, Castiglione K (2017) Polymersomes for biotechnological applications: large-scale production of nano-scale vesicles. Eng Life Sci 17(1):58–70

Puri A, Loomis K, Smith B, Lee JH, Yavlovich A, Heldman E, Blumenthal R (2009) Lipid-based nanoparticles as pharmaceutical drug carriers: from concepts to clinic. Crit Rev Ther Drug Carrier Syst 26(6):523
Rajabalaya R, Musa MN, Kifli N, David SR (2017) Oral and transdermal drug delivery systems: role of lipid-based lyotropic liquid crystals. Drug Des Devel Ther 11:393
Rajan R, Jose S, Mukund VB, Vasudevan DT (2011) Transferosomes—a vesicular transdermal delivery system for enhanced drug permeation. J Adv Pharm Technol Res 2(3):138
Rao YS, Kumari PK (2020) Biodegradable nanopsheres—current status. Indian Drugs 57(5):7–18
Rao SV, Sravya BN, Padmalatha K (2018) A review on cubosome: the novel drug delivery system. GSC Biol Pharm Sci 5(1):76–81
Rastogi RP, Kumar A, Tyagi MB, Sinha RP (2010) Molecular mechanisms of ultraviolet radiation-induced DNA damage and repair. J Nucleic Acids 2010:592980
Salvioni L, Morelli L, Ochoa E, Labra M, Fiandra L, Palugan L, Prosperi D, Colombo M (2021) The emerging role of nanotechnology in skincare. Adv Colloid Interf Sci 293:102437
Sarheed O, Dibi M, Ramesh KV (2020) Studies on the effect of oil and surfactant on the formation of alginate-based O/W lidocaine nanocarriers using nanoemulsion template. Pharmaceutics 12(12):1223
Schaue D, Kachikwu EL, McBride WH (2012) Cytokines in radiobiological responses: a review. Radiat Res 178(6):505–523
Shankar SJ, Jaswanth Gowda BH, Akshatha RS, Metikurki B, Rehamathulla M (2020) A review on the role of nanocrystals and nanosuspensions in drug delivery systems. Int J Appl Pharm 12(1):10–16
Singh R, Lillard JW Jr (2009) Nanoparticle-based targeted drug delivery. Exp Mol Pathol 86(3):215–223
Singh R, Tiwari S, Tawaniya J (2013) Review on nanotechnology with several aspects. Int J Res Comput Eng Electron 2(3):1–8
Smith TA, Kirkpatrick DR, Smith S, Smith TK, Pearson T, Kailasam A, Herrmann KZ, Schubert J, Agrawal DK (2017) Radioprotective agents to prevent cellular damage due to ionizing radiation. J Transl Med 15(1):1–18
Souto EB, Fernandes AR, Martins-Gomes C, Coutinho TE, Durazzo A, Lucarini M, Souto SB, Silva AM, Santini A (2020) Nanomaterials for skin delivery of cosmeceuticals and pharmaceuticals. Appl Sci 10(5):1594
Souto EB, Jäger E, Jäger A, Štěpánek P, Cano A, Viseras C, de Melo Barbosa R, Chorilli M, Zielińska A, Severino P, Naveros BC (2022) Lipid nanomaterials for targeted delivery of dermocosmetic ingredients: advances in photoprotection and skin anti-aging. Nano 12(3):377
Szalaty PM, Derda I (2020) "Revolutionizing beauty industry": co-creation and customer engagement for brand image development. Case study research of a crowdsource-driven cosmetics company volition beauty. Int J Market Commun New Media 8(14)
Tanabe S, O'Brien J, Tollefsen KE, Kim Y, Chauhan V, Yauk C, Huliganga E, Rudel RA, Kay JE, Helm JS, Beaton D (2022) Reactive oxygen species in the adverse outcome pathway framework: toward creation of harmonized consensus key events. Front Toxicol 4:81
Teng Y, Fan Y, Ma J, Lu W, Liu N, Chen Y, Pan W, Tao X (2021) The PI3K/Akt pathway: emerging roles in skin homeostasis and a group of non-malignant skin disorders. Cells 10(5):1219
Verma P, Pathak K (2010) Therapeutic and cosmeceutical potential of ethosomes: an overview. J Adv Pharm Technol Res 1(3):274
Vijay S, Damodar G, Rajesh A (2015) NF-kB as a key player in regulation of cellular radiation responses and identification of radiation countermeasures. Discover 3(1):1–22
Volcic M, Karl S, Baumann B, Salles D, Daniel P, Fulda S, Wiesmüller L (2012) NF-κB regulates DNA double-strand break repair in conjunction with BRCA1–CtIP complexes. Nucleic Acids Res 40(1):181–195
Wu SY, Chou HY, Yuh CH, Mekuria SL, Kao YC, Tsai HC (2018) Radiation-sensitive dendrimer-based drug delivery system. Adv Sci 5(2):1700339

Zhou H, Luo D, Chen D, Tan X, Bai X, Liu Z, Yang X, Liu W (2021) Current advances of nanocarrier technology-based active cosmetic ingredients for beauty applications. Clin Cosmet Investig Dermatol 14:867

Chapter 10
Photoprotective Effects of Nutraceuticals and Food Products

Urmila Jarouliya and Meenu Jain

Abstract Ionising radiation is the energy produced from natural or artificial sources and is continously affecting the individual. Absorption of ionizing radiation for extended period of time can directly damage DNA by initiating the formation of reactive oxygen species and free radicals, which causes the damage to DNA, protein, and lipid membrane. This cellular damage can lead to cancer, ageing process, necrosis, or apoptosis. Photoprotective agents function through various cellular pathways to reduce the formation of free radicals caused by ionizing radiation. These photoprotective agents may be useful in clinical imaging to reduce patients' morbidity and mortality rate caused due to the exposure of ionizing radiation. A large number of chemical and natural nutraceutical compounds were screened globally to develop and evaluate the photoprotective agents such as artificial and natural tablets like antioxidants, sulfhydryl compounds, cytoprotective agents, immunomodulators, metallo-factors, vitamins, and polysaccharides. Natural remedies, such as plants or a combination of plants, natural substances, and goods, have been used for thousands of years before the development of a new drug regime that is currently popular. Many researches have mentioned that plants have several medicinal properties including antioxidant, anti-inflammatory, anticancer, antimicrobial, analgesic, and antibiotic effects. Apart from this, it suggests that a particular diet and food products can also be useful against harmful impact of ionizing radiations. This chapter describes the biological consequences of ionization radiation at organelle and cellular levels, the mechanisms of photoprotection, and the role of plant-based food products and their components in presenting protection and restoration of cellular damage triggered by radiation.

U. Jarouliya (✉)
School of Studies in Biochemistry, Jiwaji University, Gwalior, Madhya Pradesh, India

M. Jain
Viral Research & Diagnostic Lab (VRDL), GR Medical College,
Gwalior, Madhya Pradesh, India

V. K. Kannaujiya et al. (eds.), *Photoprotective Green Pharmacology: Challenges, Sources and Future Applications*,
https://doi.org/10.1007/978-981-99-0749-6_10

Keywords Ionizing radiation · Photoprotective agents · Antioxidants · Vitamins · Nutraceuticals

10.1 Introduction

Radiation is the energy that travels through space in the form of particles or electromagnetic waves such as radio waves, microwaves, infrared, visible light, ultraviolet, alpha particles, X-rays, and gamma-rays, among others. The radiation that has enough strength to remove tightly bound electrons from the orbit of atoms, causing that atom to become ionize refer as ionizing radiation (IR). The sources of ionizing radiation can be from natural background radiation along with radon and thoron, cosmic and terrestrial radiation, or man-made radiation, including that from X-ray or nuclear medication (NM). Human beings are exposed to natural sources of ionizing radiation such as soil, water, and plant life, as well as man-made sources such as X-rays and medical gadgets. At a low level, living creatures are continuously exposed to ionizing radiation from natural sources. This form of radiation is referred to as natural background radiation, and its essential sources are ultraviolet light, visible light, and infrared light (daylight). This is the form of radiation through which we are encouraged on its benefits to generate electric energy and to kill maximum cancer cells through various methods.

Exposure to effluents and solid waste from military, aviation, nuclear power, commercial, or workplace activities, as well as from diagnostic radiology, nuclear medication, and radiotherapy, will be probably destructive (Von Sonntag 1987) for human beings. Ionizing radiation (IR) is used in lots of diverse functions, including therapeutic, industrial, and different applications, in addition to the generation of nuclear power and the development of new high-yielding varieties of vegetation, and the improvement of storage duration of food substances (Michael and Amer 2010; Lee et al. 2010; Zhou et al. 2014). Even though IR has wide applications, it has the potential for health risks if not used properly (Mosse 2012; Chen 2014). Moreover, it (ionization radiation) is one of the most extreme reasons of oxidative stress mediated via free radicals. These radicals interfere with oxidation/reduction-based physiological mechanisms in the mammalian body system.

Ionizing radiation (IR) can affect the human beings in any one of the ways, either directly or indirectly. Direct effect of ionization radiations is mediated by way of direct interaction of IR with components of DNA, while indirect effects are derived from the types of active oxygen species (ROS) produced in the molecules around the DNA (Wang et al. 2018). Recognized risks related to human exposure to ionizing radiation evolves the induction of cellular death, genetic mutations, and carcinogenesis. Similarly, to direct cellular effects, radiation exposure also can damage cells via producing the reactive oxygen species (i.e., hydroxyl radical (OH^{-1}), superoxide (O^{2-}), hydrogen peroxide (H_2O_2)) (Greenstock 1993).

In the production of free radicals, small molecules (normally water) surrounding cell bio-macromolecules absorb ionizing radiations and generate reactive oxygen species (ROS). These generated ROS react with cellular macromolecules (such as DNA and proteins) and other contents of cell. ROS are continuously produced in mammalian cells together with neurons, as by-products of normal and neurotransmitter metabolic processes, possibly threatening neuronal probity and induces the lipid peroxidation (LPO) process of cellular membrane, protein oxidation, and DNA damage (Girotti 1990). Long-term exposure to radiation elevates the seriousness of free radical formation inside the brain. To prevent the adverse effects of ROS, an endogenous antioxidant (e.g., superoxide dismutase, glutathione, and catalase) is available inside the body to lower or eliminate the damage that free radicals bring to cell structures.

Other than DNA and protein, a few different primary targets of radiation inactivation are the plasma membranes of cytoplasmic organelles. Oxidative damage to membrane is generally mediated via the degradation of phospholipids, which are the important components of the membrane. Membrane lipids are easily peroxidized through ROS produced by ionizing radiation, inflicting structural and functional disfigurement of the membrane (Nair et al. 2001; Barrera 2012).

In 1930, Müller determined the chromosomal aberrations because of ionization radiation and shown that immoderate radiation causes large translocations with very rare point mutations (Giardi et al. 2013). Further, these high-energy ionizing radiation-triggered oxidative stress that results in multiple single point mutations or small nucleotide deletions affects DNA indirectly and have chronic effects.

10.1.1 Radiations and Cell Phones

Nowadays, the use of cellular phones especially influences the human beings to a large extent (Lin 2012). The advancement in the area of wireless communication made human beings to speak while in motion without any hinderance. Based on the observation carried out on college students from distinct engineering, medical, and pharmacy colleges, the impact of the radiations by using cellular phones results in loss of sleep, stress in eyes, headache, migraine, irritability induced while attending phone calls, restlessness, eye stress, non-stop watching and skipping of food, hearing problem and sometimes road accident occurred using cell phones at the same time as riding and so on. Another common adverse effect is digital thumb, which is brought on by the constant use of the thumb to type on the small display screen of cellular phones. The improvement needs to be carried out to reduce its terrible effect (Vijayalakshmi and Nirmala Devi 2020).

10.1.2 Radiations and 5G Technology

The arrival of 5G has significantly progressed the rate of information transmission in wireless mobile technology. However, it has placed society in suspense due to ailments that got here collectively with its deployment. These days the primary reason of maximum cancers is because of the emission of 5G radiations, which is due to the fact 5G makes use of radio waves to switch information at a faster rate than present-day mobile wi-fi technology. From the technological point of view, the 5G mobile wireless network technology is located at the nonionizing region of the electromagnetic spectrum. The higher frequencies of electromagnetic (EM) radiation, collectively with X-rays and gamma rays, are types of ionizing radiation. The lower frequency radiation, which includes ultraviolet (UV), infrared (IR), microwave (MW), radio frequency (RF), and extremely low frequency (ELF), are types of nonionizing radiation. Though various solutions are provided by the 5G technology, it is nevertheless seen as a threat to human life because of the ailments that came in conjunction with the deployment of the wi-fi technology. It is far great an assumption that the electromagnetic radiation emitted from 5G technology is risky as soon as human beings are exposed to it through the use of their smartphones, computer devices, distinctive mobile, and wireless devices. Despite this, it has been claimed that 5G technology causes a few health issues, which has no longer been confirmed. Consequently, the reduced electromagnetic strength is incapable of breaking the chemical bonds of any biological tissue of deoxyribonucleic acid (DNA) or causes changes to cells with a view to give rise either cancer or any other health-related ailment (Aru et al. 2021). The scientific study is going on to confirm whether 5G radiation has any negative impact on human health.

10.2 Role of Photoprotectors

The long-term exposure to ionization radiations produces ROS and free radicals, which causes harm to cells. The generated free radicals cause DNA damage by disturbing bonds of nitrogenous bases, breaking DNA double strands (DSBs), and interrupting DNA–protein cross-links, which in turn can alter gene expression and is the main reason of protein mutations, cell death, genomic instability, and physical appearance (Iqbal et al. 2022). Hall and Giaccia (2012) found in their study that exposure of cells at the normal dose of IR has been shown to result in the breakdown of 1000 single-strand DNAs (SSB), 40 double-strand DNAs, and 3000 damaged bases per gray matter. Given that free radicals generated by IR cause harm to biomolecules by interacting with them, these free radicals are short-lived. Antioxidants are agents that can destroy these free radicals or inhibit their formation, as well as block biomolecule-damaging pathways and act as protective radicals

(Lobo et al. 2010). For this reason, during IR exposure, suitable photoprotective active compounds need to be present within the cellular system at an appropriate proportion. The use of herbal remedies can improve the body's immune system and provide protection against harmful radiations.

10.3 Plant Products/Herbs as Photoprotectors

During the last two decades, modulation or removal of radiation-induced harmful effects has attracted a lot of research, which recommends the use of plant products and herbs against IR-induced damage. More than 80% of the world's population has used complementary and alternative drugs, which includes herbalism and botanical drugs. Herbalism is the preparations that contain completely plant material; botanical drugs consist of enormous active components that may have identical molecular goals such as pharmaceutical drugs (Mun et al. 2018). As a result, herbal and botanical medicines are effective, easily available, and less expensive than other existing medicines.

Photoprotectors are also required to reduce damage to normal tissues during radiotherapy for various forms of cancer. In fact, various photoprotectors are directly effective and risk-free. In the last decade, the use of natural products as possible radioprotectors has gained widespread momentum because of less toxicity, lower cost, and other health-related benefits (Kumar 2016).

Several plant-based products had been applied efficiently for the treatment of autoimmune-mediated (generated by free radical) human's diseases, including rheumatoid arthritis, Alzheimer's disease, Parkinson's disease, aging, and several other conditions such as inflammatory diseases (Fischer et al. 2018). This is because herbs contain active components or compounds that provide protection against radiation-induced ROS-mediated damage in nervous tissue, most likely through their effective antioxidant, immunostimulant, cell proliferation stimulation, antimicrobial, and anti-inflammatory effects.

Protection from radiation is an area of great significance because of its possible applications in planned or unplanned radiation exposure. Many authors have proposed the use of a variety of herbal agents to regulate the cell damage caused by radiation exposure. Researchers from all around the world have set out to screen the variety of bioactive chemical and biological components of photoprotectors.

Numerous drugs from herbal or synthetic origin were evaluated drastically for their radioprotective potentials each in vivo and in vitro models (Smith et al. 2017). However, the fact remains that there are very few photoprotective drugs that meet all of the conditions for a remarkable photoprotector, i.e., produces no harmful or irremediable toxic effect, provides long-term protection, has a constant shelf life, and may be given effortlessly (Arora et al. 2006).

Natural treatments had been implemented since historic period to cure several diseases and other health-related issues. Even nowadays, more than 70% of the world's population still depends upon on plant-based remedies to fulfill their health-care needs. Plants represent an important supply of natural products that vary widely in their chemical structures, biological properties, and mechanism of action. Numerous plant-derived phytochemical additives, particularly polyphenols, flavonoids, phenolic acids, and so forth, are responsible for the scavenging of free radicals and enhancing antioxidant activity. Polyphenols have many biological effects, specifically attributed to their antioxidant activities in scavenging free radicals, inhibiting peroxidation, and chelating transition metals (Farag 2013). It is, therefore, anticipated that plants can protect toward radiation-triggered reactive oxygen species (ROS) and reactive nitrogen species (RNS) damage.

Naturally occurring antioxidants are best photoprotectors. Even though a huge variety of effective photoprotectors, mainly those having sulfhydryl as function group, can be considered as antioxidants, these functional groups work especially via receptor-mediated mechanisms, e.g., bioactive lipids, cytokines, and growth factors. Photoprotectors can cause similar effects via particular signaling pathways whether they act through chemical methods or through receptor-mediated mechanisms. Sulfhydryl compounds may defend cells from cellular DNA damage through a variety of mechanisms, including hydrogen donation, free radical scavenging and repair mechanism modulation (Moding et al. 2013), whereas cytokines might also result in cellular antioxidant activity, collectively with superoxide dismutase (SOD) and metallothionein, via receptor-mediated mechanisms. Few man-made photoprotectors (e.g., *N*-acetylcysteine, cysteamine, and dithiolthiones) or pharmacological compounds are similar to endogenous, or naturally occurring radioprotectors.

However, many herbal compounds, including vitamins and minerals, additionally provide a degree of protection against radioprotection. Vitamins and minerals help to keep the immune system strong and assist to repair healthy tissues. Other natural components such as glutamine, ubiquinone, arginine, and hydroquinone had been investigated to guard the immune system from IR-mediated damage (Gonzalez et al. 2018; Tabeie et al. 2017). Therefore, attention has shifted to the testing of natural compound-derived photoprotectors, given that these natural compounds are having less toxicity, more effective, and low cost. A number of secondary metabolites such as polyphenols and flavonoids are detected from different parts of plants having photoprotective qualities (Citrin et al. 2010; Pal et al. 2013). The primary focus of this chapter is to learn about plant-derived protective compounds against IR-induced damage.

10.3.1 Vitamin E as Photoprotector

Vitamin E or tocopherol is a natural product referred to an antioxidant, which has the ability as photoprotector. It is a group of eight structurally associated fat-soluble nutrients and four tocopherols (α, β, γ, and δ). Antioxidants block the circulation of

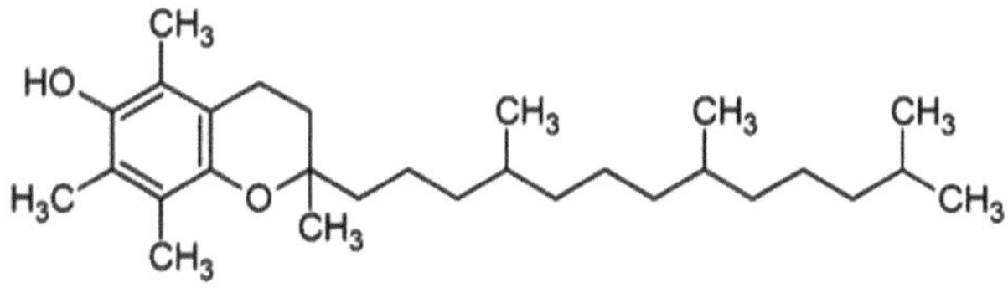

Fig. 10.1 Chemical structure of vitamin E (alpha-tocopherol)

free radicals via donating hydrogen from their phenolic group to stabilize the radicals by destroying the chain of events leading to oxidative damage (Darlina et al. 2017). Vitamin E is found in plant-based oils, nuts, seeds, fruits, and vegetables. The preventive effect of vitamin E (alpha-tocopherol) against IR-triggered oxidative damage was proven in vitro, and the study had proven that the level of lipid peroxidation was high when triggered by radiation. After the supplementation of vitamin E, it lowered the lipid peroxidation level compared to that of higher rate of radiation (Maurya et al. 2007). In a study by Zarei et al. (2021), they found that the presence of vitamin E in IR-BSA samples efficiently neutralizes ROS and stops its binding to BSA via the inherent antioxidant activity. Administration of vitamin E in irradiated cells elevated the expression of antioxidant enzymes and inhibited the expression of oncogenes (Singh et al. 2013). It also protects the structure and function of human cell membranes (Vasilyeval and Bespalov 2015; Niki 2014; Satyamitra et al. 2011). It shows the protective effects of rat bone marrow against chromosomal degradation caused by radiation and micronuclei. In addition to protecting against fatal effects, vitamin E exhibits various radioprotective actions in mice (Fig. 10.1)

10.3.2 *Vitamin A as Photoprotector*

Vitamin A or retinoids is found in dairy products, liver, fish, and fortified cereals; other sources of provitamin A consist of carrots, broccoli, cantaloupe, and squash. The administration of 100 mg/kg vitamin A minimizes the harmful effects of ionizing radiation on DNA, as free radicals are trapped by antioxidants, which in turn reduces the genetic damage to the bone marrow (Changizi et al. 2019). It has been proven that the supplementation of vitamin A prevents in vitro malignant mutations caused by X-ray radiations. The addition of 2.5 mg/kg beta carotene to mice inhibited chromosomal damage caused by radiation measured through micro-nucleated polychromatic erythrocytes. A team of scientists from the University of Chicago Medical Centre completed a number of cancer-based research and found that combining radiotherapy with all-trans retinoic acid (ATRA) significantly inhibits the growth of not only radiation-induced tumors, but also other untreated tumors. Integrated radiation therapy with ATRA alters the tumor microenvironment and improves radiation effects at both local and systemic levels (Fig. 10.2) (Rao et al. 2021).

10.3.3 Vitamin C as Photoprotector

Vitamin C, more commonly known as ascorbic acid, is widely found in nature and used as antioxidant and free radical scavenger. It is particularly abundant in citrus and stone fruits, berries, peppers, and green vegetables, and its supplementation provides protection from radiation-induced chromosomal damage in mice. Radiation-induced apoptosis in bone marrow cells and restored hematopoietic characteristic were found after treated with vitamin C (Sato et al. 2015). Oral administration of vitamin C after 24 h of radiation exposure averted the fatal gastrointestinal syndrome in mice (Brown et al. 2010), making this vitamin interesting for postexposure photoprotection. Furthermore, reduced level of free radical metabolites or inflammatory cytokines has been determined after ascorbic acid treatment (Sato et al. 2015). Moreover, vitamin C along with *N*-acetylcysteine, lipoic acid, and beta carotene significantly reduced the various double-strand breaks in peripheral blood mononuclear cells of sufferers undergoing bone scans in comparison to the control group, which is due to the antioxidants' potential of ascorbic acid to eliminate free radicals that are generated via irradiations (Velauthapillai et al. 2017). As per the recommendation by WHO, vitamin C remarkably decreased the number of double-strand breaks by 25% ($p < 0.0001$) in peripheral blood lymphocytes when incubated 1 h earlier than 10 mGy γ-irradiation (Fig. 10.3) (Brand et al. 2015).

10.3.4 Role of Polyphenols Against Irradiation

It has been determined that polyphenolic compounds found in plants possess antioxidant properties and can neutralize free radicals (as a result of inhibiting radiation-induced chain reactions of oxidation in cells), scavenge and neutralize lipoperoxyl radicals, and chelate metallic ions (which might be important as free radical response initiators). The primary mechanism of radiation protection provided by polyphenols is detoxification of free radicals, modulation of inflammatory responses, DNA repair, activation of hematopoietic restoration, and stimulation of immune functions.

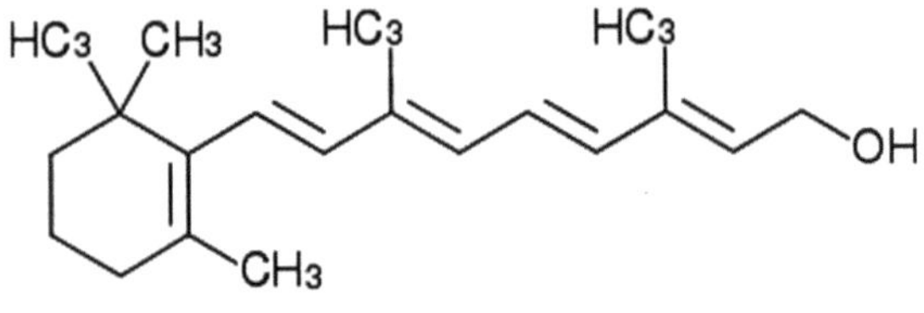

Fig. 10.2 Chemical structure of vitamin A (retinol)

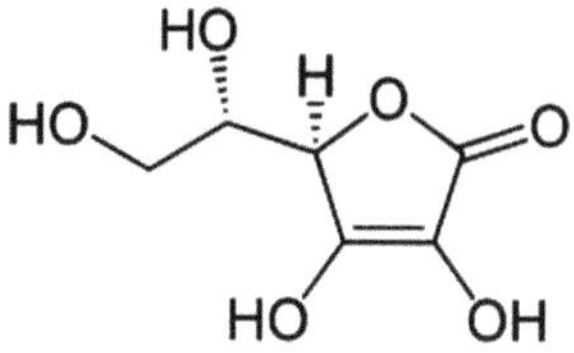

Fig. 10.3 Chemical structure of vitamin C (ascorbic acid)

Beneficial resources of treatments for radiation-triggered damage prevention include aromatic plants. They significantly enhance the flavor of dishes and are extensively used as drugs. In recent years, a number of research has been carried out on the radioprotective properties of some fragrant herbs and spices. This makes it possible to use these natural agents as a potential drug to prevent radiation-triggered damage (Oliinyk et al. 2020).

10.3.5 Caffeine as Photoprotector

Caffeine, a methyl xanthine derivative and a nervous system stimulator, a main component of coffee, and other beverages like tea which also includes some quantity of caffeine. It has a stimulatory action on the central nervous system, elevates mood, and underlies alertness through blocking off the inhibition of cAMP phosphodiesterase of cells within the brain.

Caffeine acquires each anti-inflammatory and antioxidant activity (Hall et al. 2015), which may additionally make a contribution that it functions as immunostimulatory against IR-induced damage. Particularly, caffeine degrades hydroxyl radicals (Brezova et al. 2009) in IR-activated cells. Further, caffeine intake reinstates IR-triggered normal cell cycle in the G2 phase of mouse embryos (Jung and Streffer 1992). In another study, it was demonstrated that caffeine consumption decreased the amount of protein exposed to UV radiation in cancer cells (Ravi et al. 2008). It has been used as an adjuvant analgesic in combination with acetaminophen, aspirin, and ibuprofen in clinics. The caffeine has been proven to exert a wide form of outcomes on DNA damage caused by UV and ionizing radiations, depending on its concentration and pre- or postirradiation administration (Kamat et al. 2000; Hebbar et al. 2002; Kumar et al. 2001) (Fig. 10.4).

10.3.6 Curcuma longa *(Curcumin) as Photoprotector*

Curcumin, a substance in turmeric, is extracted from the roots of *Curcuma longa* and has been proven to have an antioxidant, anti-inflammatory, antimutagenic, antimicrobial, anti-carcinogenic, and radioprotective activity (Perrone et al. 2015). Curcumin blocks the IR from producing ROS and hence acts as a free radical scavenger (Verma 2016). Curcumin also confirmed the dual action of normal cancer-free

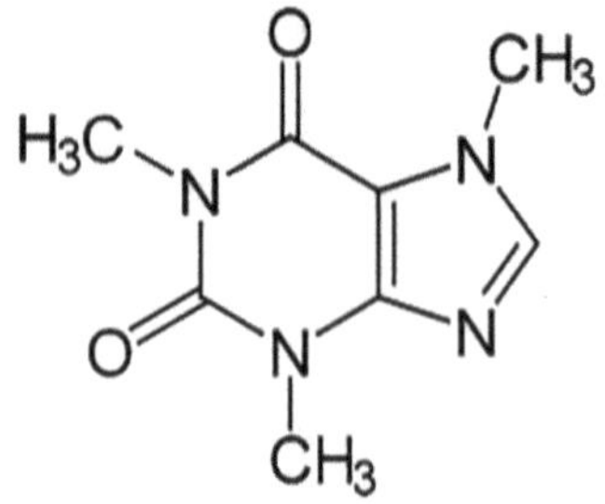

Fig. 10.4 Chemical structure of caffeine

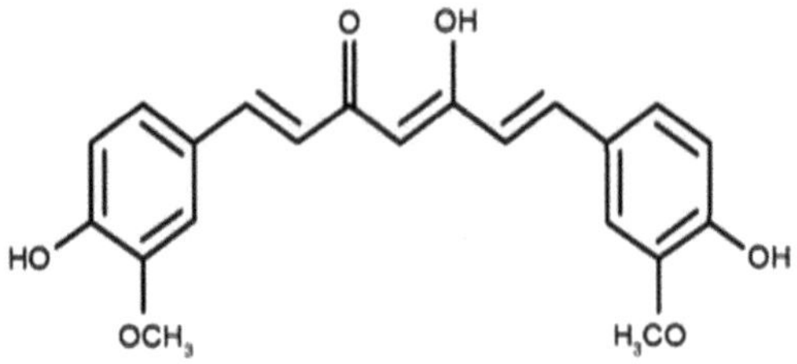

Fig. 10.5 Chemical structure of curcumin

cells that protect the photosensitive tumor cells, which is thought to be mediated by their strong detoxification structures and cell cycle. Recently, some researchers (Sebastià et al. 2011) have recognized and proved that for therapeutic purposes, curcumin-based nanoparticles shield normal cells by releasing free radicals (Xie et al. 2017). The hydrophobic components of curcumin had the capability to defend cells from DNA damage. This compound directly passes through the plasma membrane into cytosol and degrades free radicals such as superoxide anion and hydroxyl radicals, and it acts as a chain-terminating antioxidant due to its lipid soluble properties.

In addition, curcumin consumption enhanced IR-induced dementia, which includes learning and memory impairment, by improving Nrf2 antioxidant expression mechanisms, which regulate genetic expression of their protein products involved in detoxification and elimination of reactive oxidants (Xie et al. 2014). Balasubramanyam et al. (2003) confirmed that in isolated human lymphocytes, curcumin inhibits thapsigargin and protein kinase-induced free radical production. Protection from chemically induced DNA damage with curcumin was determined by Polasa et al. (2004), and they had shown that curcumin remarkably protects against the action of benzo(a)pirene that caused the rupture of a single-strand DNA in human blood lymphocytes (Fig. 10.5).

10.3.7 Mushroom as Photoprotector

Mushrooms have developed an excellent enchantment not only for their dietary and medicinal value but also for their various bioactive compounds. *Morchella esculenta*, generally known as Morel, is safe for human consumption and is a nutritious mushroom (Nitha et al. 2020). The medicinal mushrooms available in India are *Ganoderma lucidum*, *Phellinus rimosus*, *Pleurotus florida*, and *Pleurotus pulmonaris*, which have an excellent antioxidant activity.

An in vitro study carried out on aqueous extract of *Ganoderma lucidum* (reishi mushroom) polysaccharides and triterpenes in mice exposed to γ-rays shows that they may reduce radiation harm as the open circular form increases, in addition to increase the DNA extension. Ex vivo studies performed in human blood cells show that *Ganoderma lucidum* aqueous extract β-glucan has a radioprotective effect (González et al. 2020).

Revskaya et al. (2012) studied *Auricularia auricula-judae* (common name Jelly Ear) as a source of edible melanin, an oral radioprotector based entirely on the natural pigment melanin that could act as an internal defender and protect tissues via Compton scattering followed by free radical scavenging.

10.3.8 Organosulfur and Nitrogen Compounds

In *Allium sativum*, alliin (cysteine sulfoxide) and allicin (cysteine thiosulfinate) are essential organosulfur compounds found in immoderate concentrations (Liang et al. 2012). Allicin is formed by the action of the enzyme alliinase on alliin, and it is actually an amino acid (cysteine) by-product. This action induces the protection of cardiac activity via a ROS-based mechanism associated with more than one intracellular signaling; such properties suggest a probable safety against cardiovascular disease in patients exposed to ionizing radiation. Allicin provides protection from cardiac hypertrophy and fibrosis by attenuating ROS-based signaling pathways, and it improves aging-related cognitive, learning, and memory deficits by enhancing Nrf2 antioxidant signaling pathways (Liu et al. 2010; Touloupakis and Ghanotakis 2010). Evidently, sulfur-based compounds impede tumor cell growth and cell proliferation, affecting DNA repair mechanisms and decreasing chromosomal aberrations.

10.3.9 Delphinidin as Photoprotector

Delphinidin is an anthocyanidin compound with several divergent natural functions including anti-inflammatory and antioxidant effects (Jeong et al. 2016), and it is found in green vegetables and fruits such as tomatoes, carrots, red onions, cranberries, and concord grapes. Among anthocyanins, delphinidin has the most effective antioxidant activity due to the presence of a variety of hydroxyl radicals in its structure (Watson and Schonlau 2015). It prevents normal cell tissues from the direct and indirect effects of radiation (Kim et al. 2018). Collectively, these properties recommend delphinidin as a promising photoprotector compound.

10.3.10 Green Tea (Epigallocatechin-3-Gallate)

Green tea contains epigallocatechin-3-gallate (EGCG) as a major polyphenolic component and an antioxidant against free radicals. Research has proven that EGCG is a potent component in treating many health problems (Zhu et al. 2016). Particularly, it has been tested against aging, arthritic, angiogenic, inflammatory disease, viral infection, and neurol disorder. Researcher reported that (both in vitro and in vivo) intake of EGCG elevated the levels of several antioxidant enzymes such as superoxide dismutase (SOD), heme oxygenase-1 (HO1), and glutamate cysteine ligase (Zhu et al. 2014).

10.3.11 Lycopene as Photoprotector

Lycopene is a carotenoid that has a protective effect against IR damage and is widely present in red vegetables and fruits such as tomatoes, pink guava, pink grape fruit, apricots, papaya, and watermelon (Gupta et al. 2015). The antioxidant activity of lycopene has been broadly tested, and it has the ability to eliminate free radicals (Srinivasan et al. 2009). Various research has also proven that lycopene can reduce IR-induced damage by eliminating singlet oxygen and liberating free radicals (Kelkel et al. 2011). The preventive effect of lycopene was found in irradiated human lymphocytes and mice hepatocytes, and its consumption drastically reduced the frequency of micronuclei, dicentric aberration, and translocation in damaged DNA. The use of lycopene pretreatment reduces IR-triggered lipid peroxidation level and elevates the level of antioxidant enzymes such as glutathione peroxidase, catalase, and superoxide dismutase (SOD) (Gajowik and Dobrzynska 2017; Srinivasan et al. 2009). Therefore, cells treated with lycopene before exposure to an oxidative stress, oxidative molecules, or ionizing radiation may be an effective approach in reducing the unwanted side effects of radiation.

10.3.12 Sesamol as Photoprotector

Sesamol (3,4-methylenedioxyphenol) is a phenolic compound found in sesame seeds and sesame oil and is regarded as a major antioxidant component. Sesamol has strong ROS-eliminating activity (Kanimozhi and Prasad 2009) as well as the ability to prevent against IR-induced damage in human lymphocyte cells. Further, sesamol extensively reduces IR-triggered DNA damage in the rat's hematopoietic system (Kumar et al. 2018) and lowers the genotoxicity in bone marrow cells (Kumar et al. 2015). Similarly, in mice it prevents the gastrointestinal and hematopoietic systems against IR-induced damage (Khan et al. 2015). The radioprotective outcomes of sesamol are mediated through free radicals' eliminating activity.

10.3.13 Zingerone as Photoprotector

Zingerone is an essential oil extracted from ginger. Ginger, rhizome of *Zingiber officinale roscoe*, has effective therapeutic anti-inflammatory, anticancer, antioxidant, and antibacterial properties (Ahmad et al. 2015; Rao et al. 2011). Zingerone was also determined to reduce cellular sensitivity and reduce the formation of pro-inflammatory cytokines generated by UV-B irradiation (Lee et al. 2018). In addition, zingerone shows protection against IR-triggered DNA damage and oxidative stress to Chinese hamster fibroblast cells, demonstrating its capability as a potential radioprotector.

10.3.14 Probiotic as Photoprotector

Probiotics are useful living microorganisms that enhance the diet to promote health resources by stimulating the immune response and acting as a therapeutic potential in various diseases. The probiotic treatment has been clinically evaluated and used as a remedy for human gastrointestinal infections, pouchitis, irritable bowel syndrome, and antibiotic-related diarrhea. In inflammatory bowel disease, many probiotic studies have been very detailed, despite the fact that recent research in mice described apoptotic and anti-inflammatory mechanisms of the probiotic, *Lactobacillus rhamnosus* GG (LGG) (Yan et al. 2007). The research suggests that probiotics could be used as radioprotectors in humans and offers a theory on the route of administration, timing, and mechanism of probiotic-mediated intestinal photoprotection. Probiotics, like photoprotectants, have the advantage of providing greater luminal coverage and acting locally in the mucous membrane to protect the normal epithelium while not having a systematic distribution. Even though probiotics, including LGG46, have a remarkable human protection profile (Matthew et al. 2012). Experimental and scientific studies have confirmed that probiotics can effectively control intestinal infections by altering the structure, metabolic processes, and function of indigenous intestinal flora.

10.4 Conclusion

The use of photoprotectors is a completely challenging task for both preventive and therapeutic exposure, and there is no photoprotectors found till date that can provide complete prevention from IR-induced toxicity. Currently, there is a long list of pharmaceutical products in the market that can protect, reduce, or treat IR-induced toxicity. A complete photoprotector needs to be easily available, without any side effects, safe for oral administration, protective and preventive effect on all body organs, and should exhibit effectiveness towards various forms of radiation (X-ray, gamma, electron, and neutron). It should have a cost-effective reduction factor and capable to work in more than one way (Cheki et al. 2016). Consequently, the usage of naturally occurring compounds can be an excellent approach for the development of entire photoprotectors. A lot of research has been conducted on a variety of nutraceutical and phytochemical compounds that can measure radiation responses and have a protective effect against IR. Some of the antioxidants as discussed above (such as LPO, SOD, CAT, and many more) showed antimutagenic properties when delivered after exposure to radiation. It is generally well known that a diet rich in vegetables and fruits reduces health risks. However, specific randomized controlled trials using antioxidants, including certain vitamins, to prevent certain health problems are still ongoing.

Antioxidants can interfere with the early mediation of apoptosis by scavenging ROS and subsequently inhibiting the peroxidation of membrane lipids, which is a factor in radiation-induced apoptosis. This chapter shows that treatment using plant-based antioxidants and phytochemicals after exposure to radiation is a promising area for a variety of health-related complications. In the future, significant efforts will be required to expand natural photoprotectors.

References

Ahmad B, Rehman MU, Amin I, Arif A, Rasool S, Bhat SA, Afzal I, Hussain I, Bilal S, Mir M (2015) A review on pharmacological properties of zingerone (4-(4-Hydroxy-3-methoxyphenyl)-2-butanone). Sci World J 816364:1

Arora RR, Chawla S, Singh RK, Kumar AK, Sharma SC, Puri AK, Tripathi RP, Sharma RK (2006) Radioprotection by Himalayan high-altitude region plants. In: Sharma RK, Arora R (eds) Herbal drugs: a twenty-first century perspective. Jaypee Brothers Medical Publishing Limited, New Delhi, pp 301–325

Aru EO, Adimora KC, Nwankwo FJ (2021) Investigating the impact of 5G radiation on human health using machine learning. Niger J Technol (NIJOTECH) 40(4):694–702

Balasubramanyam M, Koteswari AA, Sampath Kumar R, Finny Monickaraj S, Uma Maheswari J, Mohan V (2003) Curcumin-induced inhibition of cellular reactive oxygen species generation: novel therapeutic implications. J Biosci 28:715–721

Barrera G (2012) Oxidative stress and lipid peroxidation products in cancer progression and therapy. ISRN Oncol 2012:137289

Brand M, Sommer M, Ellmann S, Wuest W, May MS, Eller A, Vogt S, Lell MM, Kuefner MA, Uder M (2015) Influence of different antioxidants on X-ray induced DNA double-strand breaks (DSBs) using gamma-H2AX immunofluorescence microscopy in a preliminary study. PLoS One 10:e0127142

Brezova V, Slebodova A, Stasko A (2009) Coffee as a source of antioxidants: an EPR study. Food Chem 114:859–868

Brown SL, Kolozsvary A, Liu J, Jenrow KA, Ryu S, Kim JH (2010) Antioxidant diet supplementation starting 24 hours after exposure reduces radiation lethality. Radiat Res 173:462–468

Changizi V, Haeri SA, Abbasi S, Rajabi Z, Mirdoraghi M (2019) Radioprotective effects of vitamin A against gamma radiation in mouse bone marrow cells. MethodsX 3(6):714–717

Cheki M, Mihandoost E, Shirazi A, Mahmoudzadeh A (2016) Prophylactic role of some plants and phytochemicals against radio-genotoxicity in human lymphocytes. J Cancer Res Ther 12:1234–1242

Chen M (2014) Radiation protection and regulations for the nuclear medicine physician. Semin Nucl Med 44:215–228

Citrin D, Cotrim AP, Hyodo F, Baum BJ, Krishna MC, Mitchell JB (2010) Radioprotectors and mitigators of radiation-induced normal tissue injury. Oncologist 15:360–371

Darlina ALD, Alatas Z, Kisnanto T, Syaifudin M (2017) Capability of vitamin E as a radioprotector in suppressing DNA damage determined with comet assay. Biosaintifika 9(2):201–208

Farag FMS (2013) Evaluation of radio protective effects of coriander (*Coriandrum sativum* L.) in male rats. Arab J Nucl Sci Appl 46(1):240–249

Fischer N, Seo E-J, Efferth T (2018) Prevention from radiation damage by natural products. Phytomedicine 47:192–200

Gajowik A, Dobrzynska MM (2017) The evaluation of protective effect of lycopene against genotoxic influence of X-irradiation in human blood lymphocytes. Radiat Environ Biophys 56:413–422

Giardi MT, Touloupakis E, Bertolotto D, Mascetti G (2013) Preventive or potential therapeutic value of nutraceuticals against ionizing radiation-induced oxidative stress in exposed subjects and frequent fliers. Int J Mol Sci 14(8):17168–17192

Girotti AW (1990) Photodynamic lipid peroxidation in biological systems. Photochem Photobiol 51:497

Gonzalez E, Cruces MP, Pimentel E, Sanchez P (2018) Evidence that the radioprotector effect of ascorbic acid depends on the radiation dose rate. Environ Toxicol Pharmacol 62:210–214

González A, Atienza V, Montoro A, Soriano JM (2020) Use of *Ganoderma lucidum* (Ganodermataceae, Basidiomycota) as radioprotector. Nutrients 12(4):1143

Greenstock CL (1993) Radiation and ageing: free radical damage, biological response and possible antioxidant intervention. Med Hypotheses 41(5):473–482

Gupta S, Jawanda MK, Arora V, Mehta N, Yadav V (2015) Role of lycopene in preventing oral diseases as a nonsurgical aid of treatment. Int J Prev Med 6:70

Hall EJ, Giaccia AJ (2012) Radiobiology for the radiologist, 7th edn. Lippincott Williams & Wilkins, Wolters Kluwer Health, Philadelphia, p 576

Hall S, Desbrow B, Anoopkumar-Dukie S, Davey AK, Arora D, Mcdermott C, Schubert MM, Perkins AV, Kiefel MJ, Grant GD (2015) A review of the bioactivity of coffee, caffeine and key coffee constituents on inflammatory responses linked to depression. Food Res Int 76:626–636

Hebbar SA, Mitra AK, George KC, Verma NC (2002) Caffeine ameliorates radiation-induced skin reactions in mice but does not influence tumour radiation response. J Radiol Prot 22(1):63–69

Iqbal S, Shah MA, Rasul A, Saadullah M, Tabassum S, Ali S, Zafar M, Muhammad H, Uddin MS, Batiha GE, Vargas-De-La-Cruz C (2022) Radioprotective potential of nutraceuticals and their underlying mechanism of action. Anticancer Agents Med Chem 22(1):40–52

Jeong MH, Ko H, Jeon H, Sung GJ, Park SY, Jun WJ, Lee YH, Lee J, Lee SW, Yoon HG, Choi KC (2016) Delphinidin induces apoptosis via cleaved HDAC3-mediated p53 acetylation and oligomerization in prostate cancer cells. Oncotarget 7:56767–56780

Jung T, Streffer C (1992) Effects of caffeine on protein phosphorylation and cell cycle progression in X-irradiated two-cell mouse embryos. Int J Radiat Biol 62:161–168

Kamat JP, Boloor KK, Devasagayam TP, Jayashree B, Kesavan PC (2000) Differential modification by caffeine of oxygen-dependent and independent effects of gamma-irradiation on rat liver mitochondria. Int J Radiat Biol 76(9):1281–1288

Kanimozhi P, Prasad NR (2009) Antioxidant potential of sesamol and its role on radiation-induced DNA damage in whole-body irradiated Swiss albino mice. Environ Toxicol Pharmacol 28:192–197

Kelkel M, Schumacher M, Dicato M, Diederich M (2011) Antioxidant and anti-proliferative properties of lycopene. Free Radic Res 45(8):925–940

Khan S, Kumar A, Adhikari JS, Rizvi MA, Chaudhury NK (2015) Protective effect of sesamol against ^{60}Co γ-ray-induced hematopoietic and gastrointestinal injury in C57BL/6 male mice. Free Radic Res 49(11):1344–1361

Kim HM, Kim SH, Kang BS (2018) Radioprotective effects of delphinidin on normal human lung cells against proton beam exposure. Nutr Res Pract 12:41–46

Kumar N (2016) Prospects in the development of natural radioprotective therapeutics with anticancer properties from the plants of Uttarakhand region of India. J Ayurveda Integr Med 7(1):62–68

Kumar SS, Purva JV, Pasupathy K (2001) Radioprotective action of caffeine: use of Saccharomyces cerevisiae as a test system. Indian J Exp Biol 39:1254–1257

Kumar A, Choudhary S, Adhikari JS, Chaudhury NK (2018) Sesamol ameliorates radiation induced DNA damage in hematopoietic system of whole-body gamma-irradiated mice. Environ Mol Mutagen 59:79–90

Kumar A, Selvan TG, Tripathi AM, Choudhary S, Khan S, Adhikari JS, Chaudhury NK (2015) Sesamol attenuates genotoxicity in bone marrow cells of whole-body gamma-irradiated mice. Mutagenesis 30:651–661

Lee TK, O'Brien KF, Wang W, Johnke RM, Sheng C, Benhabib SM, Wang T, Allison RR (2010) Radioprotective effect of American ginseng on human lymphocytes at 90 minutes postirradiation: a study of 40 cases. J Altern Complement Med 16(5):561–567

Lee J, Oh SW, Shin SW, Lee KW, Cho JY, Lee J (2018) Zingerone protects keratinocyte stem cells from UVB-induced damage. Chem Biol Interact 279:27–33

Liang G, Qiao X, Bi Y, Zou B, Zheng Z (2012) Studies on purification of allicin by molecular distillation. J Sci Food Agric 92(7):1475–1478

Lin JC (2012) Are cell phones or radio-frequency electromagnetic fields possibly carcinogenic to humans? [Telecommunications health and safety]. IEEE Antennas Propag Mag 54(1):210–212

Liu C, Cao F, Tang QZ, Yan L, Dong YG, Zhu LH, Wang L, Bian ZY, Li H (2010) Allicin protects against cardiac hypertrophy and fibrosis via attenuating reactive oxygen species-dependent signaling pathways. J Nutr Biochem 21:1238–1250

Lobo V, Patil A, Phatak A, Chandra N (2010) Free radicals, antioxidants and functional foods: impact on human health. Pharmacogn Rev 4:118–126

Matthew AC, Terrence ERM, Suprada RCM, Ee X, Gerardo MN, Monica RW, Jeffrey MM, Thaddeus SS, William FS (2012) *Lactobacillus* probiotic protects intestinal epithelium from radiation injury in a TLR-2/cyclo-oxygenase-2 dependent manner. Gut 61:829–838

Maurya DK, Adhikari S, Nair CK, Devasagayam TP (2007) DNA protective properties of vanillin against γ-radiation under different conditions: possible mechanisms. Mutat Res 634:69–80

Michael MI, Amer MM (2010) Is the use of antioxidants mixture ameliorating the double stress effect of radiation and hyperlipidemia? J Radiat Res Appl Sci 3(3):855–874

Moding EJ, Kastan MB, Kirsch DG (2013) Strategies for optimizing the response of cancer and normal tissues to radiation. Nat Rev Drug Discov 12(7):526–542

Mosse IB (2012) Genetic effects of ionizing radiation—some questions with no answers—low radiation doses. J Environ Radioact 112:70–75

Mun G-I, Kim S, Choi E, Kim CS, Lee Y-S (2018) Pharmacology of natural radioprotectors. Arch Pharm Res 41:1033–1050

Nair CKK, Parida DK, Nomura T (2001) Radioprotectors in radiotherapy. J Radiat Res 42(1):21–37

Niki E (2014) Role of vitamin E as a lipid-soluble peroxyl radical scavenger: in vitro and in vivo evidence. Free Radic Biol Med 66:3–12

Nitha B, Strayo D, Devasagayam TPA, Janardhanan KK (2020) Edible mushroom *Morchella esculenta* (L.) Pers. mycelium protects DNA and mitochondria from radiation induced damages. Indian J Exp Biol 58:842–847

Oliinyk P, Voronenko D, Lysiuk R (2020) Radioprotective properties of polyphenolic compounds of edible and aromatic plants. Food Sci Technol 14(3):62–72

Pal S, Saha C, Dey SK (2013) Studies on black tea (*Camellia sinensis*) extract as a potential antioxidant and a probable radioprotector. Radiat Environ Biophys 52:269–278

Perrone D, Ardito F, Giannatempo G, Dioguardi M, Troiano G, Lo Russo L, Del A, Laino L, Lo Muzio L (2015) Biological and therapeutic activities, and anticancer properties of curcumin. Exp Ther Med 10:1615–1623

Polasa K, Naidu NA, Ravindranath I, Krishanaswamy K (2004) Inhibition of B(a)P induced strand breaks in presence of curcumin. Mutat Res 557:203–213

Rao BN, Archana PR, Aithal BK, Rao BS (2011) Protective effect of zingerone, a dietary compound against radiation induced genetic damage and apoptosis in human lymphocytes. Eur J Pharmacol 657:59–66

Rao E, Hou Y, Huang X, Wang L, Wang J, Zheng W, Yang H, Yu X, Yang K, Bugno J, Ding X, Vokes E, Fu YX, Weichselbaum RR, Liang HL (2021) All-trans retinoic acid overcomes solid tumor radio resistance by inducing inflammatory macrophages. Sci Immunol 6(60):eaba8426

Ravi D, Muniyappa H, Das KC (2008) Caffeine inhibits UV-mediated NF-kappaB activation in A2058 melanoma cells: an ATMPKCdelta-p38 MAPK-dependent mechanism. Mol Cell Biochem 308:193–200

Revskaya E, Chu P, Howell RC, Schweitzer AD, Bryan RA, Harris M, Gerfen G, Jiang Z, Jand T, Kim K, Ting LM, Sellers RS, Dadachova E, Casadevall A (2012) Compton scattering by internal shields based on melanin-containing mushrooms provides protection of gastrointestinal tract from ionizing radiation. Cancer Biother Radiopharm 27(9):570–576

Sato T, Kinoshita M, Yamamoto T, Ito M, Nishida T, Takeuchi M, Saitoh D, Seki S, Mukai Y (2015) Treatment of irradiated mice with high-dose ascorbic acid reduced lethality. PLoS One 10:e0117020

Satyamitra M, Kulkarni S, Ghosh SP, Mullaney CP, Condliffe D, Srinivasan V (2011) Hematopoietic recovery and amelioration of radiation-induced lethality by the vitamin E isoform, delta-tocotrienol. Radiat Res 175(6):736–745

Sebastià N, Montoro A, Almonacid M, Villaescusa JI, Cervera J, Such E, Silla MA, Soriano JM (2011) Assessment in vitro of radioprotective efficacy of curcumin and resveratrol. Radiat Meas 46(9):962–966

Singh VK, Singh PK, Wise SY, Posarac A, Fatanmi OO (2013) Radioprotective properties of tocopherol succinate against ionizing radiation in mice. J Radiat Res 54(2):210–220

Smith TA, Kirkpatrick DR, Smith S, Smith TK, Pearson T, Kailasam A, Herrmann KZ, Schubert J, Agrawal DK (2017) Radioprotective agents to prevent cellular damage due to ionizing radiation. J Transl Med 15(1):232

Srinivasan M, Devipriya N, Kalpana KB, Menon VP (2009) Lycopene: an antioxidant and radioprotector against gamma radiation-induced cellular damages in cultured human lymphocytes. Toxicology 262:43–49

Tabeie F, Tabatabaei SM, Mahmoud-Pashazadeh A, Assadi M (2017) Radioprotective effect of beta D-glucan and vitamin E on gamma irradiated mouse. J Clin Diagn Res 11:08–11

Touloupakis E, Ghanotakis DF (2010) Nutraceutical use of garlic sulfur-containing compounds. Adv Exp Med Biol 698:110–121

Vasilyeval IN, Bespalov VG (2015) Release of extracellular DNA after administration of radioprotective combination of α-tocopherol and ascorbic acid. Radiats Biol Radioecol 55(5):495–500

Velauthapillai N, Barfett J, Jaffer H, Mikulis D, Murphy K (2017) Antioxidants taken orally prior to diagnostic radiation exposure can prevent DNA injury. J Vasc Interv Radiol 28:406–411

Verma V (2016) Relationship and interactions of curcumin with radiation therapy. World J Clin Oncol 7:275283

Vijayalakshmi L, Nirmala Devi P (2020) Impacts of RF radiation from mobile phones on human health and its remedies. J Appl Res Technol 18:269–278

Von Sonntag C (1987) The chemical basis of radiation biology. Taylor and Francis, London

Wang H, Mu X, He H, Zhang XD (2018) Cancer radiosensitizers. Trends Pharmacol Sci 39(1):24–48

Watson RR, Schonlau F (2015) Nutraceutical and antioxidant effects of a delphinidin-rich maqui berry extract Delphinol(R): a review. Minerva Cardioangiol 63(2):1–12

Xie Y, Zhao QY, Li HY, Zhou X, Liu Y, Zhang H (2014) Curcumin ameliorates cognitive deficits heavy ion irradiation-induced learning and memory deficits through enhancing of Nrf2 antioxidant signaling pathways. Pharmacol Biochem Behav 126:181–186

Xie J, Yong Y, Dong X, Du J, Guo Z, Gong L, Zhu S, Tian G, Yu S, Gu Z, Zhao Y (2017) Therapeutic nanoparticles based on curcumin and bamboo charcoal nanoparticles for chemo-photothermal synergistic treatment of cancer and radioprotection of normal cells. ACS Appl Mater Interfaces 9(16):14281–14291

Yan F, Cao H, Cover TL, Whitehead R, Washington MK, Polk DB (2007) Soluble proteins produced by probiotic bacteria regulate intestinal epithelial cell survival and growth. Gastroenterology 132(2):562–575

Zarei H, Bahreinipour M, Sefidbakht Y, Rezaei S, Gheisari R, Ardestani SK, Uskoković V, Watabe H (2021) Radioprotective role of vitamins C and E against the gamma ray-induced damage to the chemical structure of bovine serum albumin. Antioxidants 10(12):1875

Zhou R, Si J, Zhang H, Wang Z, Li J, Zhou X, Gan L, Liu Y (2014) The effects of x-ray radiation on the eye development of zebrafish. Hum Exp Toxicol 33(10):1040–1050

Zhu W, Xu J, Ge Y, Cao H, Ge X, Luo J, Xue J, Yang H, Zhang S, Cao J (2014) Epigallocatechin-3-gallate (EGCG) protects skin cells from ionizing radiation via heme oxygenase-1 (HO-1) over-expression. J Radiat Res 55(6):1056–1065

Zhu W, Jia L, Chen G, Zhao H, Sun X, Meng X, Zhao X, Xing L, Yu J, Zheng M (2016) Epigallocatechin-3-gallate ameliorates radiation-induced acute skin damage in breast cancer patients undergoing adjuvant radiotherapy. Oncotarget 7(30):48607–48613

Chapter 11
Pharmacology of Natural and Synthetic Phytoprotectants: Application and Consequences in Cancer Therapies

Sneha Singh, Pajeb Saha, Nidhi Rai, Sabitri Kumari, and Shashi Pandey-Rai

Abstract Cancer represents one of the most fatal health issues, claiming the lives of millions of people each year. Tumorous growths can develop in almost any portion of the body and migrate to different parts. There are numerous treatment approaches available for cancer such as radiation therapy, photodynamic therapy, and cancer vaccinations. However, in most cases, they have adverse side effects. Thus, anticancer medications with higher efficiency and fewer side effects are desperately needed. Plants are a prospective source of such compounds. Natural plant bioactive substances have been used in traditional medicine since the dawn of humanity. These metabolites have also been implicated in providing protection to plants under various environmental influences, such as the influence of UV-B. Plant-based natural secondary metabolites/phytochemicals and their derivatives have great potential in the suppression of cancer development and metastasis. These biologically active compounds can be isolated from various plant parts, such as leaves, stems, barks, flowers, rhizomes, roots, and seeds. The natural bioactive compounds produced by plants during secondary metabolism have great pharmacological importance, especially as anticancer agents. Therefore, this chapter is an attempt to summarize the importance of various plant-derived compounds and their mechanism of action, which can be used in cancer therapies as anticancer agents.

Keywords Anticancer · Apoptosis · Cell proliferation · Phytoprotectants · Secondary metabolites

---

S. Singh · P. Saha · N. Rai · S. Kumari · S. Pandey-Rai (✉)
Laboratory of Morphogenesis, Centre of Advanced Study in Botany, Institute of Science, Banaras Hindu University, Varanasi, Uttar Pradesh, India

V. K. Kannaujiya et al. (eds.), *Photoprotective Green Pharmacology: Challenges, Sources and Future Applications*,
https://doi.org/10.1007/978-981-99-0749-6_11

11.1 Introduction

Globally, malignant diseases such as cancer have been reported to be the second major reason for human fatalities. Cancer comprises a multitude of diseases that can emerge in nearly any body part and metastasize to other organs. Studies indicate that cancer originates in an individual cell that proliferates, and the affected cells are then transported to various body parts via lymphatic or blood vessels. Expeditious production of aberrant cells that multiply uncontrollably, as well as evading cell death, is the hallmark of cancerous tissues. Other characteristics include resistance to growth suppressors and a number of genetic and epigenetic modifications that have the potential to alter gene expression and facilitate tumor progressive signaling (Dai et al. 2016). This process of invasion of various organs by cancerous cells is the foremost cause of mortality.

There are more than 100 types of cancer, which affected approximately 19.3 million individuals across the globe in 2020. An estimated ten million died due to cancer in the same year. Despite substantial research and technological developments in the field, cancer remains a persistent burden. Moreover, studies estimate a 47% rise in the number of cancer incidences from 2020 to 2040 (Sung et al. 2021).

In the view of the fact that even today cancer continues to be an aggressively fatal disease worldwide, a great deal of advancement in the area of preventative and remedial therapies continues. Surgical removal of tumor is performed if the cancer has not metastasized. However, in cases where the cancer has progressed, surgery must be paired with other treatment plans for successful results. The most prevalent and sought-after approaches to cancer treatment include chemotherapy and radiation therapy. Chemotherapy is a drug-based treatment intended to shrink or retard the growth of cancerous tissue. Radiotherapy utilizes high doses of radiation to achieve similar outcomes. Although both these methods are highly effective, they have severe ramifications on the patient's health and in some cases, might not even ameliorate the prognosis. Therefore, there is a persistent need for developing alternative and affordable approaches to treat cancer.

Over the decades, in indigenous therapies, several plant-derived medicines with varying pharmacological properties have been utilized for the treatment of various ailments and disorders. The extensive curative properties of medicinal plants are largely reliant on their secondary metabolite constitution. A secondary metabolite is defined as "a metabolic intermediate or product, found as a differentiation product in restricted taxonomic groups, not essential to growth and life of the producing organism, and biosynthesized from one or more general metabolites by a wider variety of pathways than is available in general metabolism" (Bennett and Bentley 1989). These phytochemical constituents can be extracted from various plant parts, such as the root (*Panax ginseng*), bulb (*Drimia maritima*), bark (*Cinchona* species), resin (*Resina mastix*), and seed (*Ricinus communis*) (Van Wyk and Wink 2018). In traditional medical practices, several phytomedicines have been used for the treatment of Parkinson's disease (*Mucuna pruriens*), neurogenerative disorders such as

Alzheimer's disease (*Galanthus nivalis*), sexually transmitted diseases (*Senna petersiana*), and so on.

Many medicinal plants, with pharmacological properties, have been used as an alternative for chemoprevention and treatment of cancer. Vinblastine, paclitaxel, irinotecan, and camptothecin are a few examples of compounds isolated from plants that show promising results in the treatment of cancer. Anti-cancerous phytomedicines can be natural or synthetically derived. In this chapter, we discuss about the pharmacology of various plant-based anti-cancerous therapeutics.

11.2 History of Plant-Based Anticancer Medications

The struggle with cancer is an ongoing fight which started many years ago. Cancer is an extremely complex disease that causes local cell damage and inflammation. Before the modern way of treatment of cancer such as chemotherapy, people looked to Ayurveda for a solution. Ayurveda is a natural, holistic system of medicine, which was developed in India thousands of years ago. Those learned in the field of ayurveda identified different forms and causes of cancer in their own ways. They saw cancer as inflammatory or noninflammatory swelling, which they further divided into two types. Minor neoplasm was referred to as "Granthi" and major neoplasm was called "Arbuda." Weak tissue regeneration and cell division, which is called Shukra Dhatu, along with imbalanced Kapha and/or Vata Dosha lead to minor neoplasm, and when all three of these forces are uncoordinated, it leads to the formation of a major tumor. There are three main aspects of a body in ayurveda—Vata, Pitta, and Kapha. They are referred to as Tridoshas. When these are affected due to any reason, such as bad habits or unhealthy lifestyle, it leads to the formation of tumor. Prevention in this growth can be achieved through Sama Dhatu Paramapara, which essentially means practicing good habits and lifestyle. Ayurveda involves using natural products to achieve a better life and fight diseases. The same was used in the treatment of cancer. Plants have bioactive properties, which was the basis of using them as nutraceuticals (Bhandari et al. 2015; Balachandran and Govindarajan 2005).

Another field where plants were actively used to deal with cancer was in Traditional Chinese Medicine, or TCM. Because of their active constituents and specific natural chemicals with anticancer activity, several specific herbs, animals, minerals, and combination formulations were recorded and exploited in TCMs. TCM had specific combinations of formulas and prescriptions for dealing with tumors in a way that involves multiple targets and signals at the same time, instead of just attacking at singular molecular level. Two of these proven prescriptions are Shi Quan Da Bu Tang and Yanshu injection. It has been seen that these show outstanding effects on the cancer. TCM has been used to effectively deal with various types of cancer through different mechanisms. The findings suggest that TCM works against cancer by inducing apoptosis, inhibiting proliferation, suppressing metastasis, reversing multidrug resistance, and regulating immunological function.

Moreover, TCM also improves the overall quality of life of an afflicted individual (Wang et al. 2014; Yan et al. 2017).

People later shifted their attention to natural products due to the toxic effects of chemotherapy, and they found a lot of what they were looking for. Out of the hundreds of thousands of plant species on our planet, nearly a thousand were found to have anticancer properties. Medicinal plants have been an abundant source of anticancer, chemotherapeutic agents, with plant-based products making up to 60% of the final products used in the treatment of cancer. While there are several plant-based compounds still being used only in preclinical and clinical trials, there have been certain compounds that are now being used in clinical practice. These include terpenoid indole alkaloids vinblastine and vincristine extracted from *Catharanthus roseus*, paclitaxel and docetaxel isolated from plants of the genus *Taxus*, and etoposide isolated from family Podophyllaceae (Balachandran and Govindarajan 2005). These compounds have played a significant role in the management of cancer, leading to the conclusion that it is important that we explore and study medicinal plants in the area of cancer treatment (Fig. 11.1). Previous studies have indicated the presence bioactive compounds that frequently target cell cycle and cell death components.

11.3 Examples of Phytochemicals with Anticancer Properties

In impoverished countries, the herbal medicine system has long been and continues to be used as the major source of medical treatment. Antiseptic qualities of different secondary metabolites found in plants have been widely used in medicine since ages. Various anticancer phytocompounds isolated from different plant parts that have been discovered in various plant species and documented by scientists are listed in Table 11.1.

11.3.1 Vinca Alkaloids

Vinca alkaloids are extracted from the evergreen herbaceous plant *Catharanthus roseus* (also known as the Madagascar periwinkle). The plant belongs to the Apocynaceae family and is native to Madagascar, as described in Table 11.1. Vinca alkaloids were initially identified as a versatile group of phytochemicals by Thomas Beer and Noble. They were the first plant-based mitosis inhibiting compounds to be used commercially in the drug industry.

The action of the dimeric indole alkaloids vincristine and vinblastine leads to cell cycle arrest at the metaphase stage by means of depolymerization of microtubules (Sharifi-Rad et al. 2019). This disruption caused by the binding of vinca alkaloids to tubulin heterodimers leads to apoptosis during mitosis. Because of this antitumor activity, these naturally occurring compounds have been used as chemotherapeutic agents for the treatment of several types of cancers (liver, lung, breast, etc.).

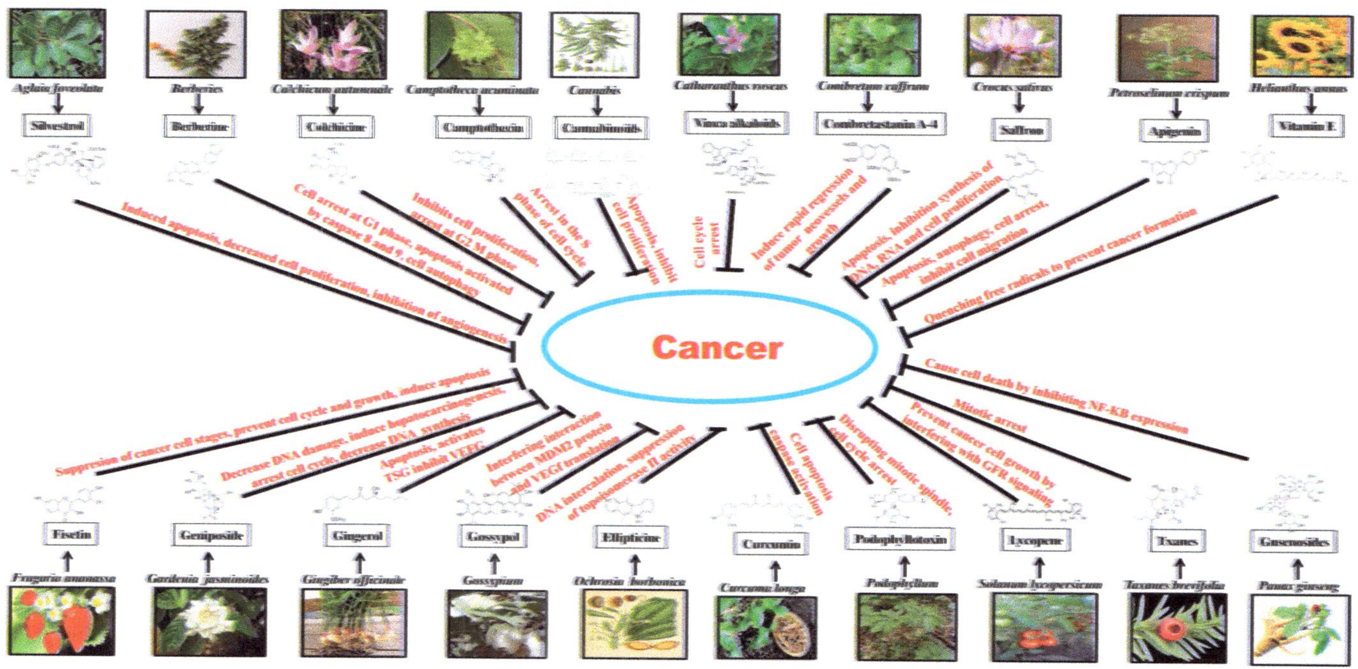

Fig. 11.1 Some major medicinal plants and their bioactive phytochemicals. The cancer chemo-preventive role of various phytochemicals via different modes

Table 11.1 Summary of the phytochemicals discussed in this chapter

Sr. no.	Phytochemicals	Plant	Family	Plant part	Chemical formula	Molecular weight	References
1.	Vinca alkaloids	*Catharanthus roseus*	Apocynaceae	Petiole and leaves			
	Vinblastine				$C_{46}H_{58}N_4O_9$	810.42	Pathania et al. (2015)
	Vincristine				$C_{46}H_{56}N_4O_{10}$	824.39	
2.	Taxanes			Bark of trees			
	Taxol	*Taxus brevifolia*	Taxaceae		$C_{47}H_{51}NO_{14}$	853.33	Harborne (1999)
	Paclitaxel	*Taxus canadensis*	Taxaceae		$C_{47}H_{51}NO_{14}$	853.33	Petzke et al. (2004)
3.	Camptothecin	*Camptotheca acuminata*	Cornaceae or Nyssaceae	Bark and stem of tree	$C_{20}H_{16}N_2O_4$	348.11	Harborne (1999)
4.	Podophyllotoxin	*Podophyllum* spp.	Berberidaceae	Roots and rhizomes	$C_{22}H_{22}O_8$	414.13	Zhao et al. (2006)
5.	Colchicine	*Colchicum autumnale*	Colchicaceae	Bulb like corms	$C_{22}H_{25}NO_6$	399.16	Harborne (1999)
6.	Curcumin	*Curcuma* spp.	Zingiberaceae	Rhizome (roots)	$C_{21}H_{20}O_6$	368.12	Harborne (1999)
7.	Genosides	*Panax ginseng* *Panax notoginseng*	Araliaceae	Rhizomes (roots)	$C_{48}H_{82}O_{18}$	946.55	Harborne (1999)
8.	Berberine	*Berberis aristata* *Berberis aquifolium* *Berberis aemulans*	Berberidaceae	Roots and rhizomes	$C_{20}H_{18}NO_4$	336.12	Kumar et al. (2015)
9.	Ellipticine	*Ochrosia acuminate* *Ochrosia elliptica* *Ochrosia moorei*	Apocynaceae	Leaves	$C_{17}H_{14}N_2$	246.11	Salim et al. (2004)
10.	Gingerol	*Zingiber officinale*	Zingiberaceae	Roots and rhizomes	$C_{17}H_{26}O_4$	294.18	Harborne (1999)
11.	Cannabinoids	*Cannabis sativa*	Cannabaceae	Flower and fruits	$C_{21}H_{30}O_2$	314.22	Harborne (1999)
12.	Lycopene	*Solanum lycopersicum* *Capsicum annuum* *Daucus carota*	Solanaceae Apiaceae	Fruit Fruit Roots	$C_{40}H_{56}$	536.43	Enzell (1969) Harborne (1999)

(continued)

Table 11.1 (continued)

Sr. no.	Phytochemicals	Plant	Family	Plant part	Chemical formula	Molecular weight	References
13.	Crocetin	*Crocus sativus*	Iridaceae	Stigma of flower	$C_{20}H_{24}O_4$	328.16	Pfander and Rychener (1982)
14.	Apigenin	*Matricaria chamomilla* *Artemisia afra* *Artemisia diffusa*	Acanthaceae Asteraceae Asteraceae	Flowers Aerial part	$C_{17}H_{14}O_5$	298.08	Harborne (1999)
15.	Geniposide	*Gardenia jasminoides* *Gardenia* spp.	Rubiaceae	Fruit	$C_{17}H_{24}O_{10}$	388.13	Uesato et al. (1986) Harborne (1999)
16.	Gossypol	*Gossypium* sp.	Malvaceae	Seed	$C_{30}H_{30}O_8$	518.19	Harborne (1999)

Vinorelbine, vinfosiltine, and vindesine are semisynthetic analogs derived from vinca alkaloids that can be used in combination with other compounds to show activity against various types of cancers (Nirmala et al. 2011). A novel formulation of cholesterol and sphingomyelin-encapsulated vincristine, commercially known as vincristine sulfate liposome injection, is used to treating relapsed Philadelphia chromosome-negative acute lymphoblastic leukemia (Sharifi-Rad et al. 2019). The use of vinflunine, which is obtained as a result of the bifluorination of vinorelbine, results in improved efficacy in treating cancers.

11.3.2 Taxanes

Taxanes are anticancer cytotoxics derived from the bark of the Taxus tree (Pacific yew) that have an effective role in regulating cell division. These diterpenes can be derived from various species such as *Taxus brevifolia*, *Taxus baccata*, and *Taxus canadensins* (Nirmala et al. 2011). Their primary mechanism involves the binding of taxanes to polymerized microtubules, leading to mitotic arrest and apoptosis. αβ-Tubulin heterodimers comprise the microtubule polymers. The dynamic property of these polymers is stabilized as a consequence of the conformational changes brought about by taxane association as well as the reduced binding affinity of MAPs (microtubule-associated proteins) (Fridlender et al. 2015). This prevents normal spindle formation during mitosis, triggering cell cycle arrest at metaphase (Sharifi-Rad et al. 2019).

Paclitaxel (Taxol) and docetaxel are first-generation taxanes. Taxol is a complex diterpene alkaloid. Promising results have been observed via its therapy in patients of breast, lung, cervical, and ovarian cancer (Spencer and Faulds 1994). However, paclitaxel shows poor solubility in water, thus its semisynthetic, water-soluble derivative, docetaxel, has a greater cytotoxic activity. It is used in curing breast, prostate, lung, and pancreatic cancers. When used with prednisone, it is administered to castration-resistant prostate cancer patients. A second-generation taxane, Cabazitaxel, can traverse the blood–brain barrier and is thus more potent for treating metastases of the central nervous system (Sharifi-Rad et al. 2019).

11.3.3 Camptothecin

Isolated from *Camptotheca acuminata* (Nyssaceae), the camptothecin class of monoterpene compounds possesses strong anticancer potential. They are modified indole alkaloids that have been extensively utilized in traditional Chinese medicine. Camptothecin is synthesized from tryptophan precursors and produced in higher amounts from cell suspensions of roots and leaves than of shoots (Roja and Rao 2000).

Camptothecin acts by bringing about collisions between topoisomerase I cleavable complexes and progressing replication forks. This is responsible for the construction of a covalent binary complex between the double helical structure of DNA and topoisomerase I, resulting in the arrest of cell cycle in S phase. Hence, camptothecins are often referred to as a family of topoisomerase I poisons (Kim et al. 2015).

However, it shows high toxicity and has low solubility in water. Thus, comptothecin's water-soluble analogs, irinotecan, CPT-11, and topotecan, with intact lactone rings, were developed. Topotecan is utilized in treating ovarian and lung cancer (Creemers et al. 1996). CPT-11 has shown promising results in the treatment of lymphomas and leukemias, and irinotecan is given in combination with fluoropyrimidines to colorectal cancer patients.

11.3.4 Podophyllotoxin

Podophyllotoxin is the most abundant lignan with a wide spectrum of antiviral and antineoplastic characteristics. It is isolated from a resin, podophyllin, which is an ethanolic extract produced from the rhizomes of *Podophyllum* genera members (Gordaliza et al. 2004).

Podophyllotoxin is an aryltetralin lignan that exhibits chemotherapeutic properties by virtue of its ability to inhibit polymerization of tubulin, thereby controlling mitotic spindle formation by its disruption along with cell cycle arrest. Epipodophyllotoxin, a β-configurated biological compound derived from podophyllotoxin, functions by impeding topoisomerase II activity (Xiao et al. 2020).

This antitumor lignan also acts as the starting material for its semisynthetic derivatives etoposide, teniposide, and etopophos. Synergistic and combination interactions with other cytotoxic agents have been implemented as therapeutics in lymphomas, genital tumors, and various types of lung cancer (Gordaliza et al. 2004).

11.3.5 Colchicine

Colchicinc is a secondary metabolite derived from the poisonous *Colchicum autumnale*, commonly known as meadow saffron (Kumar et al. 2017). It is an alkaloid used for its anti-inflammatory effects in patients suffering from gout, arthritis, and so on. Colchicine is a cheap microtubule destabilizer that inhibits cell proliferation. It perturbs the microtubule assembly dynamics by binding strongly to tubulin, bringing about mitotic arrest at the G2/M phase (Kumar et al. 2017). It can also increase cellular free tubulin, which inhibits the mitochondrial membrane anion channels, limiting its metabolism in tumorous cells.

As a consequence of its severe toxicity, colchicine's clinical use has been limited. Colchicine has the potential to be exploited as a lead chemical in the development of anticancer drugs. As a result, various colchicine analogs have been created in the hopes of producing novel medications with better pharmacological characteristics. These include colchicinamide, colchicoside, deacetylcolchicine, and thiocolchicocide. Breast, esophageal, and bronchial cancers are among the cancers for which it has notable effects (Kumar et al. 2017).

11.3.6 Curcumin

Curcumin is isolated from the Indian spice, *Curcuma longa*. It is a polyphenol that has been used in treating numerous inflammatory disorders for generations (Rodrigues et al. 2021). It is a promising turmeric-derived diferuloylmethane that has the potential to be used in the treatment of a range of malignancies.

The turmeric powder is yellow-orange in color and accumulates in the plant's rhizome. Curcuminoids make up roughly 3–5% of turmeric's yellow pigment (Fridlender et al. 2015). Curcumin exerts its anticancer benefits by modulating various molecular pathways involved in the extensive carcinogenesis process, including induction of cell apoptosis (caspase activation, cyclin D1), scavenging reactive oxygen species (ROS), decreasing the inflammatory cancer microenvironment, and blocking proliferative signaling (HER-2, PI3K/AKT pathway).

Curcumin can act as a chemoprotective agent for treating a few cancers, such as leukemia, gastrointestinal cancers, genitourinary cancers, colorectal cancer, lung cancer, lymphoma, melanoma, neurological cancers, and sarcoma, owing to its ability to regulate essential transcription factors, protein kinases, adhesion molecules, cytokines, and so on (Fridlender et al. 2015).

11.3.7 Ginsenosides

Ginseng was first discovered in China and has since been revered as a powerful medicinal herb. The three most widely utilized ginseng herbs are Asian ginseng (*Panax ginseng*), notoginseng (*Panax notoginseng*), and American ginseng (*Panax quinquefolius*) (Sun et al. 2011).

Ginsenosides, a type of triterpenoid saponin, have been identified as the principal bioactive molecules of ginseng root. Hepatoprotection, cardiovascular protection, neuroprotection, immunity boost, antioxidative and antitumor properties have all been demonstrated for ginsenoside (Zhu et al. 2021). Numerous ginsenosides have been isolated of which Rc, Rg1, Re, Rb1, Rb2, and Rd. constitute nearly 80%.

Panaxadiol is a powerful chemoprotective drug derived from ginseng and exhibits anti-invasion and anti-migration properties in tumorous cells and signaling pathways. Ginsenoside Rg3 causes cell death by inhibiting NF-kB expression. Administration of Rg3 with other anticancer drugs ameliorates the efficacy of the treatment.

11.3.8 Berberine

Berberine is a naturally occurring quaternary ammonium salt purified from various species of the traditional Chinese herb *Berberis*. It belongs to the protoberberine class of alkaloids. It is extracted from the stem, bark, roots of the plant (Gaba et al. 2021).

Berberine regulates the cell cycle (G1 phase cell cycle arrest), cell apoptosis (activates caspases-8,9), and cell autophagy in order to inhibit the metastasis of cancerous cells. It is also responsible for the transcriptional regulation of certain oncogenes and the suppression of enzymes such as *N*-acetyltransferase and topoisomerase.

It has been reported that berberine can be used in the treatment of liver, lung, breast, and other types of cancers. However, the intestines are not able to absorb berberine efficiently due to its hydrophilic nature (Lo et al. 2013). Thus, to enhance its inhibitory function, its structure must be modified to develop compounds with higher stability and biological activity. Studies show that various compounds have been produced with structural modifications at the C-8, C-9, C-12, C-13 positions, 9-*O*-alkylberberine derivatives with increased alkyl chain length, and terpenyl substitutes (Lo et al. 2013).

11.3.9 Ellipticine

Ellipticine is a potent antitumor compound first extracted from the leaves of *Apocyanaceae* plants such as *Ochrosia borbonica* and *Excavatia coccinea* (Stiborova et al. 2006). The most prevalent modes of action of ellipticine are based on DNA intercalation and suppression of topoisomerase II activity. It also carries out selective inhibition of p53 phosphorylation.

The excellent efficacy of ellipticine and its more soluble derivatives (such as 9-hydroxyellipticine, 9-chloro-*N*-2-methylellipticinium) against numerous types of cancers, their relatively low toxic side effects, and its absence of hematological toxicity are the key reasons for their therapeutic interest (Stiborova et al. 2006).

11.3.10 Gingerol

The rhizome of ginger (*Zingiber officinale*) has been utilized as a condiment as well as for its medicinal properties since ancient times. It comprises a number of phenolics, the most pungent of which are gingerols (Yahyazadeh et al. 2021).

Gingerol regulates several cell-signaling pathways such as STAT3, MAPK, nuclear factors, and AP-1 (Nafees et al. 2021). It works by causing cell cycle arrest, resulting in apoptotic cell death in cancer cells expressing mutant p53 in pancreatic cells. It causes downregulation of mammary carcinoma cell proliferation in humans and mice via cell cycle arrest in the S phase (Bernard et al. 2017). Anticancer properties of [6]-gingerol, [8]-gingerol, and [10]-gingerol have been studied in colon, pancreatic, ovarian, and breast malignancies.

11.3.11 Cannabinoids

The biologically active chemical compounds of *Cannabis sativa* are cannabinoids. Cannabinoids exert palliative effects on cancer-associated symptoms such as chemotherapy-induced vomiting and nausea (Mangal et al. 2021). Research suggests that cannabinoids and their derivatives can also help in decreasing tumor growth in laboratory animals.

Cannabinoids have been shown to downregulate cancer cell migration, suppress tumor angiogenesis, and activate the endoplasmic reticulum stress-related pathway, leading to cell autophagy. In an individual, the effects of both natural and synthetic cannabinoids are exerted by mimicking the endogenous receptor ligands (endocannabinoids), which activate the G-protein-coupled cannabinoid receptor types 1 and 2 (Kovalchuk and Kovalchuk 2020).

Cannabinoids have been used for mediating their antitumor properties in treating various types of cancers, including neuroblastoma, carcinomas, gliomas, breast cancer, and lymphomas.

11.3.12 Lycopene

Tomatoes and tomato-based products are known to have a variety of health benefits. Epidemiological studies have shown that increasing the intake of tomatoes in one's diet increases the antioxidant capacity, thereby reducing the risk of diseases caused by reactive oxygen species (ROS), such as cardiovascular diseases and cancer (Marzocco et al. 2021). Lycopene, with a singlet oxygen quenching activity, is the active component found in tomatoes, watermelon, guava, pink grapefruit, and other red fruits and vegetables. It is a stable antioxidant carotenoid known for mediating anti-inflammatory and cardioprotective effects along with being a potent chemoprotective agent. Lycopene exerts its effects through a series of strategies, including interference with growth factor signaling, inducing cell cycle arrest, suppression of cell invasion, angiogenesis inhibition, and antioxidant enzyme modifications. Research shows that lycopene lowers oxidative damage by upregulating GST and GSH enzyme activities in models with gastric cancer (Kim and Kim 2015). In humans, it has been shown to lower the risk of lung and prostate cancer.

11.3.13 Saffron

Crocus sativus L. (saffron) is the source of one of the costliest spices. Saffron is extracted from the dried stigma of the plant's flowers. Its principal ingredients, crocetin, crocin, and safranal, have piqued the scientific community's attention for application in the pharmaceutical industry (Bakshi et al. 2010).

Crocetin is a diterpenic carotenoid with a symmetrical structure that is considered to be the most potent compound derived from saffron. It has considerable effects on coronary heart disease, neurogenerative diseases, and maintaining kidney and retinal functions.

Crocetin protects cells from oxidative damage caused by lipid peroxidation, and these antioxidant properties may be important for tumor suppression. It has been demonstrated to bring about inhibition of breast cancer cell proliferation as well as suppress DNA, RNA, protein synthesis in cervical cancer in a dose-dependent manner. It also mediates its therapeutic effects in leukemia, colorectal, liver, and pancreatic cancers (Gutheil et al. 2012).

11.3.14 Combretastatin A-4

The African bush willow tree *Combretum caffrum*, belonging to the Combretaceae family, is the source of a potent antimitotic compound, combretastatin A-4. Tubulin has been regarded as a potential target for anticancer therapeutic approaches. Thus, combretastatin A-4, being a negative regulator of tubulin polymerization, finds applications in chemoprevention and treatment.

Combretastatin A-4 is a class of stilbenes that is active in its *cis* form. Owing to its structural simplicity, various analogs and derivatives have been developed, of which the most examined in clinical studies is combretastatin A-4 phosphate.

The dynamic stability between α and β tubulin heterodimers is disrupted as a result of combretastatin A-4 interacting with tubulin at colchicine binding sites. Consequently, the formation of microtubules is hampered, causing abnormalities in the mitotic spindles. In this manner, the cell cycle is arrested in the M-phase and apoptosis takes place (Chaudhary et al. 2007).

Combretastatin A-4 also mediates an antivascular effect through its antitubulin activity. It causes rapid depolymerization of the cellular microtubule, which disrupts endothelial cells present around blood vessels. As a consequence, blood flow to the tumor cells is affected, leading to depletion of nutrient and oxygen supplies to these cells and eventually necrosis (Chaudhary et al. 2007).

11.3.15 Apigenin

Apigenin (4′,5,7-trihydroxyflavone) is an edible natural flavonoid. It is abundant in a number of fruits and vegetables, including dried flowers of *Matricaria chamomilla*, celery seeds, *Artemisia afra* dried artichokes, and vine spinach (Shankar et al. 2017). Owing to its potential anti-inflammatory and antioxidant activities and relatively low cytotoxicity in comparison to other structurally similar flavonoids, apigenin has sparked a lot of interest as a potential chemotherapeutic agent.

By regulating numerous signaling pathways (such as the WNT/β-catenin and JAK/STAT pathways), apigenin causes cell cycle arrest and death in a variety of malignancies, including breast, liver, blood, skin, lung, colon, pancreatic, cervical, oral, prostate, and stomach tumors (Ozbey et al. 2019). It is responsible for positively regulating the expression of caspases-3, -8 and TNF-α (extrinsic caspase-dependent pathway), thereby inducing apoptosis.

11.3.16 Geniposide

Geniposide is a bioactive substance found in a number of species from various families. Its primary source is the fruit of *Gardenia jasminoides* Ellis (Rubiaceae). It has been applied in the Chinese medicine system since ancient times. Geniposide has shown its cytotoxic activity in numerous cell lines. Genepin, a hydrolysis product of geniposide (aglycone of geniposide), is also found to have anti-cancerous properties (Habtemariam and Lentini 2018).

Apoptosis induction by geniposide involves the activation of caspases that exert changes in the morphology, including fragmentation of DNA, membrane blebbing, and apoptotic body formation. Geniposide shows a cytotoxic effect by various mechanisms, some of them are via poisoning by topoisomerase I, ROS formation by NADPH oxidase, UCP2 pathways, and targeting specific enzymes involved in the process of carcinogenesis.

11.3.17 Silvestrol

Silvestrol is a cyclopenta[b]benzofuran derived from the twigs and bark of *Aglaia foveolate*, a plant indigenous to Borneo in Southeast Asia (Kim et al. 2007). It belongs to the Flavagline family and has been found to show cytotoxic activity against various cancer cell lines.

In human cancerous cells, Silvestrol suppresses cell growth and causes cell death by inducing early autophagy and caspase-mediated apoptosis. It also causes cytochrome *c* release into the cytoplasm, thereby disrupting the mitochondrial transmembrane potential. Silvestrol selectively suppresses the eukaryotic translation initiation factor eIF4A. Silvestrol and episilvestrol show synergistic effects (Cencic et al. 2009).

11.3.18 Fisetin

Fisetin is an active dietary flavonoid (3,3′,4′,7-tetrahydroxyflavone) sourced from apples, strawberries, grapes, persimmons, cucumbers, and onions. As per previous studies, strawberries had the highest concentration of fisetin (160 g/g), followed by apples (26.9 g/g) and persimmons (10.5 g/g) (Kimira et al. 1998). According to recent data, fisetin may be effective in reducing the risk of many types of human cancers.

According to the data from in vitro and in vivo investigations, fisetin inhibits cancer growth by altering the cell cycle, inducing cell death, metastasis, oxidative stress, angiogenesis, and inflammation, without harming normal cells. In various human cell lines, it has been shown to cause premature chromosomal segregation and escape from mitotic division before cytokinesis. Upon administration of fisetin to the culture media, Aurora B, BubR1, Cenp-F, and Bub1 lacked their kinetochore localization (Salmela et al. 2009). In the human leukemia cell line HL-60, it reduced intracellular peroxide levels, activated endonuclease, and suppressed Mcl-1 proteins.

11.3.19 Gossypol

Gossypol is a plant polyphenolic aldehyde product from cotton seed (*Gossypium hirsutum* L.). Consumption of cotton seed oil has generally been regarded as causing male infertility due to its toxicity (Coutinho 2002). Thus, gossypol is considered unfit for consumption by humans and animals.

Gossypol has been attributed to having positive effects in the treatment of breast cancer, pancreatic cancer, prostate cancer, and colon cancer. It affects energy-producing cytoplasmic and mitochondrial enzymes, resulting in cellular ATP depletion. It also has antifungal and anti-inflammatory properties. Gossypol and its analogs interact with and inactivate antiapoptotic proteins having BH3 domains (Wang et al. 2016). In human colon cancer cells, it stimulates the antitumor effect of 5-FU by inhibiting thymidylate synthase.

11.3.20 Vitamin E

Vitamin E is a lipid-soluble micronutrient first extracted from green leafy vegetables. It is found in nature as eight isomers: four tocopherols and four tocotrienols, all of which have high antioxidant properties. Tocopherols are generally sourced from nuts and vegetable oils, while tocotrienols are obtained from palm oil, cereal grains, and barley (Kamal-Eldin and Appelqvist 1996).

Multiple data sources back the beneficial effects of vitamin E isoforms in tumor prevention. Due to the activity of phenolic hydrogen (present in the chromanol ring), all vitamin E isoforms detoxify reactive oxygen species and prevent lipid peroxidation. Since the C-5 position on their chromanol ring is unmethylated, δ-tocopherol and γ-tocopherol are highly active isoforms that can scavenge reactive nitrogen species (Smolarek and Suh 2011). Expression of cytokines involved in inflammation (IL1, IL6, TNF) is regulated by NF-κB and JAK-STAT signaling pathways. These pathways are negatively regulated by tocotrienols.

11.4 Mode of Action of Phytocompounds in Cancer Therapy

Phytocompounds play their part in cancer therapy by multiple pathways. Some bioactive compounds act by bringing about cell cycle arrest in cancerous cells via downregulation of cyclin D1, CDK 4, or upregulation of ROS (Fig. 11.2).

Environmental stresses (such as UV radiation), tobacco, drugs, alcohol, and so on lead to peroxide generation, which causes DNA damage. This damage leads to tumor development, which is inhibited by certain phytocompounds. A few biocompounds also enhance DNA hypermethylation by promoting oct4, sox 2, klf 4 transcription factors in tumorous cells.

A number of bioactive agents alter signaling mechanisms such as WNT and RTK pathways and promote autophagy in cancerous cells. Thus, abnormal cells show higher generation of p53, leading to the association of proapoptotic proteins that bring about programmed cell death.

11.5 Advantages of Plant-Derived Therapeutics

For ages, plants have been a source of natural nutrition and treatments. Their nature of propagation makes it easy to scale up these medicinal herbs to allow for economic feasibility. In fact, the WHO estimated that at least 3.5 million people in developing economies use plant-based medicines for their basic healthcare. Plants are wealthy in an assortment of mixtures that have been utilized in medicines for millennia. From phenols to antibiotics to new arising cancer treatments, natural plant-based compounds are a superb raw material for modern medicine. Till date, large number of information available related to bioactive anti-cancerous phytocompounds that were tested under both in vitro (cell lines) and in vivo conditions using various animal model systems (Table 11.2.). Almost 25% of medications recommended based on dynamic ingredients derived from plants. Indeed, of the 177

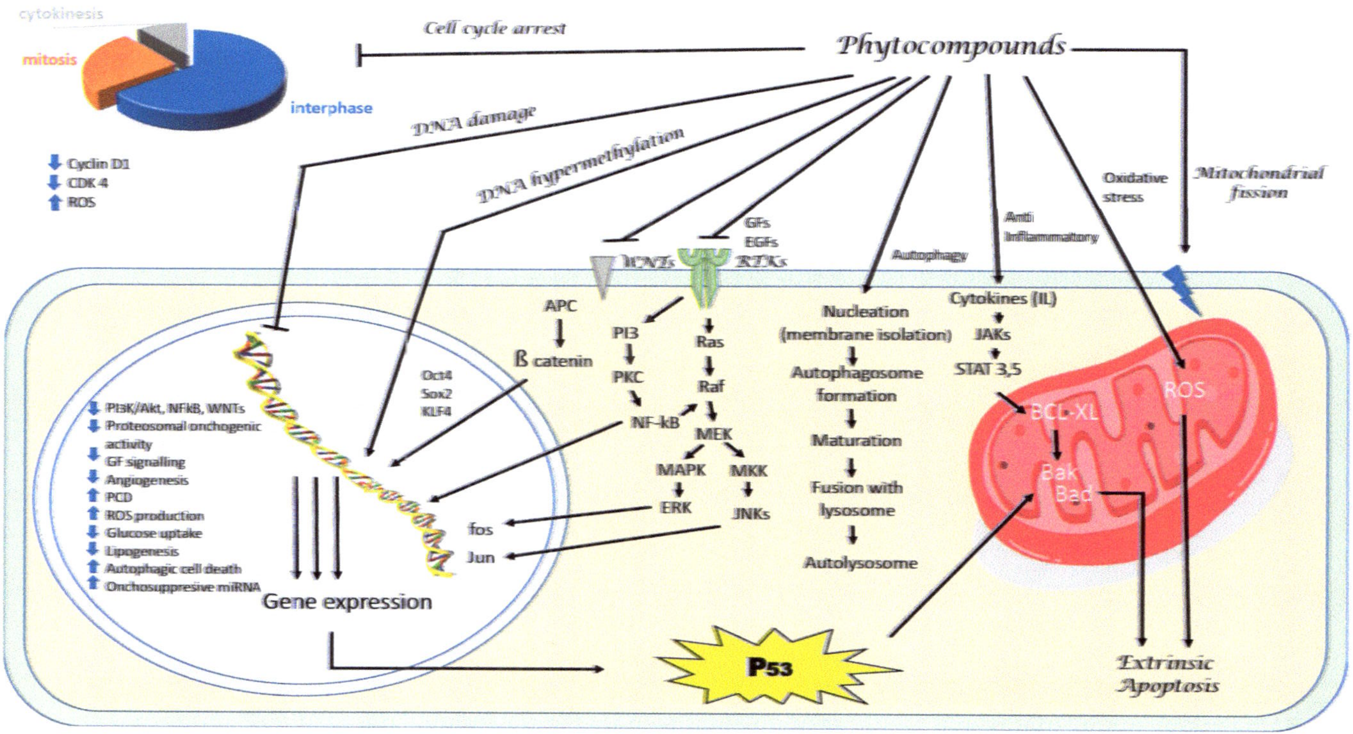

Fig. 11.2 The signal transduction route and molecular mechanism by which herbal pharmaceuticals affect at various cellular levels. *CDKs* cyclin dependent kinases, *ROS* reactive oxygen species, *PI3K/Akt* phosphatidylinositol 3-kinase/protein kinase B, *Sox2* SRY (sex-determining region Y)-box 2, *NF-kB* nuclear factor kappa B, *WNTs* wingless-related integrated site, *PCD* programmed cell death, *Oct4* octamer-binding transcription factor 4, *ERK* extracellular signal-regulated kinase, *KLF4* Kruppel like factor 4, *APC* adenomatous polyposis coli, *Ras* rat sarcoma virus, *Raf* rapidly accelerated fibrosarcoma, *MAPK* mitogen-activated protein kinase, *Fos and Jun* proto-oncogene, *Jak/Stat* Janus kinase/signal transducers and activators of transcription, *JNKs* Jun N-terminal kinases

Table 11.2 Phytochemicals, their nature, treatment for cancer, and cell lines studied

Sr. no.	Names of anti-cancerous phytochemicals	Nature of phytochemicals	Types of cancer	Cell lines	References
1.	Vinca alkaloids	Alkaloid	Testicular, breast, lung, head and neck, and bladder cancers	A2780, L1210 NSCLC-N6L2	Zhou and Rahmani (1992)
2.	Taxanes	Diterpene	Breast, prostate, and lung cancers	A121, A549, HT-29, MCF7, and MCF7-MDR	Das et al. (2021)
3.	Camptothecin	Monoterpene	Colorectal, ovarian, cervical cancers	HT29 and C26	Zahiri et al. (2021)
4.	Podophyllotoxin	Phenolic compounds	Testicular, breast, pancreatic, lung, stomach, and ovarian cancers	HepG2, HeLa, and K562/A02	Zhao et al. (2021)
5.	Colchicine	Alkaloid	Prostate and colorectal cancers	COLO 320 and Calu-3	Barreca et al. (2021)
6.	Curcumin	Polyphenol	Breast, bowel, stomach, and skin cancer cells	MCF-7, SKBR3, HT29, HCT116, HeLa, K562, and A549	Rodrigues et al. (2021)
7.	Ginsenoside	Triterpene	Breast and prostate cancers	PC3 and MDA-MB-231	Peng et al. (2019) Nakhjavani et al. (2019)
8.	Berberine	Alkaloid	Breast, gastric, melanoma, and hepatic cancers	HL-60, HepG2, HT-29, BFTC905, PC12, MCF7, and HeLa	Gaba et al. (2021)
9.	Ellipticine	Alkaloid	Breast cancer and chronic myeloid leukemia.	HeLa, Caco2, HGF-1, MDA-MB-231, and HepG2	Xu et al. (2021)
10.	Gingerol	Phenolic	Gastrointestinal tract cancers	Int407, HepG2, and H9c2	Yahyazadeh et al. (2021)
11.	Cannabinoids	Terpenophenolics	Breast, prostate, pancreatic cancer, and leukemia	M14, MIA, PaCa-2, and PANC-1.	Mangal et al. (2021)
12.	Lycopene	Terpene	Ovarian, pancreatic, prostate, and lung cancers	DU145, PC3, and LNCaP	Marzocco et al. (2021)
13.	Crocetin	Diterpene	Colon cancer	S-180, EAC, DLA, Tca8113, and HCT116	Bakshi et al. (2010)

(continued)

Table 11.2 (continued)

Sr. no.	Names of anti-cancerous phytochemicals	Nature of phytochemicals	Types of cancer	Cell lines	References
14.	Combretastatin A-4	Phenolic compounds	Thyroid cancer	518A2, HL-60, HT-29, KB–V1/Vbl, and MCF-7	Barreca et al. (2021)
15.	Apigenin	Flavonoid	Colorectal, breast, hepatic, lung, melanoma, prostate, and osteosarcoma	A549, HeLa, HepG2, and MCF-7	Nozhat et al. (2021)
16.	Geniposide	Glucoside	Colorectal, pancreatic, adenocarcinoma, human gastric carcinoma, lung, prostate cancers	H1299, DU145, PC3K562, HL-60, U266, and U93	Habtemariam and Lentini (2018)
17.	Silvestrol	Flavonoid derivative	Chronic lymphocytic leukemia and acute lymphoblastic leukemia	MDA-MB231 and PC3	Cencic et al. (2009)
18.	Fisetin	Flavonoid	Lung, breast, and colon cancers	Caco-2 and HT-29	Khan et al. (2013)
19.	Gossypol	Phenolic compound	Uterine cancer, ovarian cancer, adrenocortical cancer, and prostate cancer	SK-mel-28, SK-mel-19, H69, and K562	Shelley et al. (2000)
20.	Vitamin E	Terpenes	Colon and prostate cancers	SW480, HCT-15, HCT-116, and HT-29	Constantinou et al. (2008)

medications supported for treating disease, 70% depend on normal items (Thomas et al. 2011).

Plants use various synthesizing processes to make their own complex proteins, which would otherwise be difficult to create chemically. These natural powerhouses can be engineered to aid humanity in making effective therapeutics. Conventional protein-fabricating processes are confronting a significant worldwide capacity deficiency for the creation of biologics. Conventional techniques cannot create adequate amounts of medicinal proteins to meet patient populace needs. Plants that can fill in as protein-creating facilities can assist with mitigating this deficiency. Allopathic cell culture techniques require critical capital and work venture. On the other hand, there are altogether lower setup, production, and costs related to plant-made drugs. in addition, it is more straightforward to scale creation to fulfill ever-growing needs. Plant-based protein production can also be the savviest strategy to give patients with the advantages of more prominent and quicker admittance to new prescriptions.

Most of the time, it also costs fundamentally less to develop plants with the capacity to efficiently manufacture drug proteins since plant-based procedures do not need the similar exorbitant capital speculations as other strategies. The creation and cost–benefits of plant-based drugs can permit more cash-flow to put resources into innovative work on new therapeutics.

11.6 Disadvantages of Plant-Derived Therapeutics

As the global use of plant-based medications grows and additional products are introduced to the market, common health issues are becoming more widely acknowledged. Even while some natural remedies have a lot of promise and are widely used, a considerable lot of them still have not been approved or documented. As a result, knowledge of their probable negative effects is extremely limited, making their use and marketing more difficult.

One of the biggest disadvantages of plant-based medicines is their unrestricted use around the world. The ease of self-administering their medications without any consultation or prescriptions can be harmful at unmitigated levels. These medicines are available on the market to be bought and sold commercially without any supervisory or regulatory body overseeing the safety and effectiveness of these medicines. The overall insight that these naturally derived remedies are exceptionally protected and do not have any unfavorable repercussions is not only incorrect but also misleading. Plants have been shown to be capable of generating several adverse responses, and some of these are capable of inflicting real wounds, serious consequences, and sometimes even fatality (Ekor 2014).

11.7 Conclusion

Cancer is among the most common factors causing global morbidity and mortality; therefore, finding new ways to treat it is of paramount importance. Conventional cancer treatments frequently generate severe toxicities and side effects. The current chapter demonstrated that phytochemicals offer a promising avenue for developing new approaches for cancer treatment. Natural bioactive compounds as well as their synthetic derivatives have been discussed. They have a substantial effect as cancer treatment agents, both alone and in combination with conventional medications. Using traditional herbal medicine knowledge could lead to the development of new pharmaceuticals at a lower cost and faster pace. Nevertheless, the absence of global standardization in terms of methodologies for analyzing their constitution, toxicity, efficiency, and quality control is the main drawback associated with herbal medicine. Merging the benefits of traditional and modern medicine has proven to be a potential means for discovering and commercializing new plant-derived medicines. The efficacy and toxicity of the potential anticancer phytocompounds reviewed in this chapter ought to be further investigated and documented via preclinical studies, clinical trials (either in vitro or in vivo), and post-marketing monitoring.

References

Bakshi H, Sam S, Rozati R, Sultan P, Islam T, Rathore B, Lone Z, Sharma M, Triphati J, Saxena RC (2010) DNA fragmentation and cell cycle arrest: a hallmark of apoptosis induced by crocin from Kashmiri saffron in a human pancreatic cancer cell line. Asian Pac J Cancer Prev 11(3):675–679

Balachandran P, Govindarajan R (2005) Cancer—an ayurvedic perspective. Pharmacol Res 51(1):19–30

Barreca M, Spanò V, Raimondi MV, Tarantelli C, Spriano F, Bertoni F, Barraja P, Montalbano A (2021) Recurrence of the oxazole motif in tubulin colchicine site inhibitors with anti-tumor activity. Eur J Med Chem Rep 1:100004

Bennett JW, Bentley R (1989) What's in a name?—microbial secondary metabolism. Adv Appl Microbiol 34:1–28

Bernard MM, McConnery JR, Hoskin DW (2017) [10]-Gingerol, a major phenolic constituent of ginger root, induces cell cycle arrest and apoptosis in triple-negative breast cancer cells. Exp Mol Pathol 102(2):370–376

Bhandari M, Ravipati AS, Reddy N, Koyyalamudi SR (2015) Traditional ayurvedic medicines: pathway to develop anti-cancer drugs. J Mol Pharm Organ Process Res

Cencic R, Carrier M, Galicia-Vázquez G, Bordeleau ME, Sukarieh R, Bourdeau A, Brem B, Teodoro JG, Greger H, Tremblay ML, Porco JA Jr (2009) Antitumor activity and mechanism of action of the cyclopenta [b] benzofuran, silvestrol. PLoS One 4(4):e5223

Chaudhary A, Pandeya SN, Kumar P, Sharma PP, Gupta S, Soni N, Verma KK, Bhardwaj G (2007) Combretastatin A-4 analogs as anticancer agents. Mini Rev Med Chem 7(12):1186–1205

Constantinou C, Papas A, Constantinou AI (2008) Vitamin E and cancer: an insight into the anticancer activities of vitamin E isomers and analogs. Int J Cancer 123(4):739–752

Coutinho EM (2002) Gossypol: a contraceptive for men. Contraception 65(4):259–263

Creemers GJ, Bolis G, Gore M, Scarfone G, Lacave AJ, Guastalla JP, Despax R, Favalli G, Kreinberg R, Van Belle S, Hudson I (1996) Topotecan, an active drug in the second-line treatment of epithelial ovarian cancer: results of a large European phase II study. J Clin Oncol 14(12):3056–3061

Dai SX, Li WX, Han FF, Guo YC, Zheng JJ, Liu JQ, Wang Q, Gao YD, Li GH, Huang JF (2016) In silico identification of anti-cancer compounds and plants from traditional Chinese medicine database. Sci Rep 6(1):1–11

Das T, Anand U, Pandey SK, Ashby CR Jr, Assaraf YG, Chen ZS, Dey A (2021) Therapeutic strategies to overcome taxane resistance in cancer. Drug Resist Updat 55:100754

Ekor M (2014) The growing use of herbal medicines: issues relating to adverse reactions and challenges in monitoring safety. Front Pharmacol 4:177

Enzell CR (1969) Mass spectrometric studies of carotenoids. Pure Appl Chem 20(4):497–516

Fridlender M, Kapulnik Y, Koltai H (2015) Plant derived substances with anti-cancer activity: from folklore to practice. Front Plant Sci 6:799

Gaba S, Saini A, Singh G, Monga V (2021) An insight into the medicinal attributes of berberine derivatives: a review. Bioorg Med Chem 38:116143

Gordaliza M, Garcıa PA, Del Corral JM, Castro MA, Gómez-Zurita MA (2004) Podophyllotoxin: distribution, sources, applications and new cytotoxic derivatives. Toxicon 44(4):441–459

Gutheil WG, Reed G, Ray A, Anant S, Dhar A (2012) Crocetin: an agent derived from saffron for prevention and therapy for cancer. Curr Pharm Biotechnol 13(1):173–179

Habtemariam S, Lentini G (2018) Plant-derived anticancer agents: lessons from the pharmacology of geniposide and its aglycone, genipin. Biomedicine 6(2):39

Harborne JB (1999) Classes and functions of secondary products from plants. Chem Plants 26:1–25

Kamal-Eldin A, Appelqvist LÅ (1996) The chemistry and antioxidant properties of tocopherols and tocotrienols. Lipids 31(7):671–701

Khan N, Syed DN, Ahmad N, Mukhtar H (2013) Fisetin: a dietary antioxidant for health promotion. Antioxid Redox Signal 19(2):151–162

Kim MJ, Kim H (2015) Anticancer effect of lycopene in gastric carcinogenesis. J Cancer Prev 20(2):92

Kim S, Hwang BY, Su BN, Chai H, Mi Q, Kinghorn AD, Wild R, Swanson SM (2007) Silvestrol, a potential anticancer rocaglate derivative from *Aglaia foveolata*, induces apoptosis in LNCaP cells through the mitochondrial/apoptosome pathway without activation of executioner caspase-3 or-7. Anticancer Res 27(4B):2175–2183

Kim SH, Kaplan JA, Sun Y, Shieh A, Sun HL, Croce CM, Grinstaff MW, Parquette JR (2015) The self-assembly of anticancer camptothecin-dipeptide nanotubes: a minimalistic and high drug loading approach to increased efficacy. Chemistry 21(1):101–105

Kimira M, Arai Y, Shimoi K, Watanabe S (1998) Japanese intake of flavonoids and isoflavonoids from foods. J Epidemiol 8(3):168–175

Kovalchuk O, Kovalchuk I (2020) Cannabinoids as anticancer therapeutic agents. Cell Cycle 19(9):961–989

Kumar A, Chopra K, Mukherjee M, Pottabathini R, Dhull DK (2015) Current knowledge and pharmacological profile of berberine: an update. Eur J Pharmacol 761:288–297

Kumar A, Sharma PR, Mondhe DM (2017) Potential anticancer role of colchicine-based derivatives: an overview. Anti-Cancer Drugs 28(3):250–262

Lo CY, Hsu LC, Chen MS, Lin YJ, Chen LG, Kuo CD, Wu JY (2013) Synthesis and anticancer activity of a novel series of 9-O-substituted berberine derivatives: a lipophilic substitute role. Bioorg Med Chem Lett 23(1):305–309

Mangal N, Erridge S, Habib N, Sadanandam A, Reebye V, Sodergren MH (2021) Cannabinoids in the landscape of cancer. J Cancer Res Clin Oncol 147(9):2507–2534

Marzocco S, Singla RK, Capasso A (2021) Multifaceted effects of lycopene: a boulevard to the multitarget-based treatment for cancer. Molecules 26(17):5333

Nafees S, Zafaryab M, Mehdi SH, Zia B, Rizvi MA, Khan MA (2021) Anti-cancer effect of gingerol in cancer prevention and treatment. Anticancer Agents Med Chem 21(4):428–432

Nakhjavani M, Palethorpe HM, Tomita Y, Smith E, Price TJ, Yool AJ et al (2019) Stereoselective anti-cancer activities of ginsenoside Rg3 on triple negative breast cancer cell models. Pharmaceuticals 12(3):117

Nirmala MJ, Samundeeswari A, Sankar PD (2011) Natural plant resources in anti-cancer therapy—a review. Res Plant Biol 1(3):01–14

Nozhat Z, Heydarzadeh S, Memariani Z, Ahmadi A (2021) Chemoprotective and chemosensitizing effects of apigenin on cancer therapy. Cancer Cell Int 21(1):1–26

Ozbey U, Attar R, Romero MA, Alhewairini SS, Afshar B, Sabitaliyevich UY, Hanna-Wakim L, Ozcelik B, Farooqi AA (2019) Apigenin as an effective anticancer natural product: spotlight on TRAIL, WNT/β-catenin, JAK-STAT pathways, and microRNAs. J Cell Biochem 120(2):1060–1067

Pathania S, Ramakrishnan SM, Bagler G (2015) Phytochemica: a platform to explore phytochemicals of medicinal plants. Database 2015

Peng Y, Zhang R, Yang X, Zhang Z, Kang N, Bao L et al (2019) Ginsenoside Rg3 suppresses the proliferation of prostate cancer cell line PC3 through ROS-induced cell cycle arrest. Oncol Lett 17(1):1139–1145

Petzke TL, Shi QW, Sauriol F, Mamer O, Zamir LO (2004) Taxanes from rooted cuttings of Taxus canadensis. J Nat Prod 67(11):1864–1869

Pfander H, Rychener M (1982) Separation of crocetin glycosyl esters by high-performance liquid chromatography. J Chromatogr A 234(2):443–447

Rodrigues FC, Kumar NA, Thakur G (2021) The potency of heterocyclic curcumin analogues: an evidence-based review. Pharmacol Res 166:105489

Roja G, Rao PS (2000) Anticancer compounds from tissue cultures of medicinal plants. J Herbs Spices Med Plants 7(2):71–102

Salim AA, Garson MJ, Craik DJ (2004) New indole alkaloids from the roots of Ochrosia acuminata. J Nat Prod 67(10):1719–1721

Salmela AL, Pouwels J, Varis A, Kukkonen AM, Toivonen P, Halonen PK, Perälä M, Kallioniemi O, Gorbsky GJ, Kallio MJ (2009) Dietary flavonoid fisetin induces a forced exit from mitosis by targeting the mitotic spindle checkpoint. Carcinogenesis 30(6):1032–1040

Shankar E, Goel A, Gupta K, Gupta S (2017) Plant flavone apigenin: an emerging anticancer agent. Curr Pharmacol Rep 3(6):423–446

Sharifi-Rad J, Ozleyen A, Boyunegmez Tumer T, Oluwaseun Adetunji C, El Omari N, Balahbib A, Taheri Y, Bouyahya A, Martorell M, Martins N, Cho WC (2019) Natural products and synthetic analogs as a source of antitumor drugs. Biomol Ther 9(11):679

Shelley MD, Hartley L, Groundwater PW, Fish RG (2000) Structure-activity studies on gossypol in tumor cell lines. Anti-Cancer Drugs 11(3):209–216

Smolarek AK, Suh N (2011) Chemopreventive activity of vitamin E in breast cancer: a focus on γ-and δ-tocopherol. Nutrients 3(11):962–986

Spencer CM, Faulds D (1994) Paclitaxel. Drugs 48(5):794–847

Stiborova M, Rupertova M, Schmeiser HH, Frei E (2006) Molecular mechanisms of antineoplastic action of an anticancer drug ellipticine. Biomed Pap Palacky Univ Olomouc 150(1):13

Sun S, Qi LW, Du GJ, Mehendale SR, Wang CZ, Yuan CS (2011) Red notoginseng: higher ginsenoside content and stronger anticancer potential than Asian and American ginseng. Food Chem 125(4):1299–1305

Sung H, Ferlay J, Siegel RL, Laversanne M, Soerjomataram I, Jemal A, Bray F (2021) Global cancer statistics 2020: GLOBOCAN estimates of incidence and mortality worldwide for 36 cancers in 185 countries. CA Cancer J Clin 71(3):209–249

Thomas DR, Penney CA, Majumder A, Walmsley AM (2011) Evolution of plant-made pharmaceuticals. Int J Mol Sci 12(5):3220–3236

Uesato S, Ueda S, Kobayashi K, Miyauchi M, Itoh H, Inouye H (1986) Intermediacy of 8-epiiridodial in the biosynthesis of iridoid glucosides by *Gardenia jasminoides* cell cultures. Phytochemistry 25(10):2309–2314

Van Wyk BE, Wink M (2018) Medicinal plants of the world. CABI

Wang CY, Bai XY, Wang CH (2014) Traditional Chinese medicine: a treasured natural resource of anticancer drug research and development. Am J Chin Med 42(03):543–559

Wang X, Zhang C, Yan X, Lan B, Wang J, Wei C, Cao X, Wang R, Yao J, Zhou T, Zhou M (2016) A novel bioavailable BH3 mimetic efficiently inhibits colon cancer via cascade effects of mitochondria. Clin Cancer Res 22(6):1445–1458

Xiao J, Gao M, Sun Z, Diao Q, Wang P, Gao F (2020) Recent advances of podophyllotoxin/epipodophyllotoxin hybrids in anticancer activity, mode of action, and structure-activity relationship: an update (2010-2020). Eur J Med Chem 208:112830

Xu J, Qi G, Wang W, Sun XS (2021) Advances in 3D peptide hydrogel models in cancer research. NPJ Sci Food 5(1):1–10

Yahyazadeh R, Baradaran Rahimi V, Yahyazadeh A, Mohajeri SA, Askari VR (2021) Promising effects of gingerol against toxins: a review article. Biofactors 47:885

Yan Z, Lai Z, Lin J (2017) Anticancer properties of traditional Chinese medicine. Comb Chem High Throughput Screen 20(5):423–429

Zahiri M, Taghdisi SM, Abnous K, Ramezani M, Alibolandi M (2021) Fabrication of versatile targeted lipopolymersomes for improved camptothecin efficacy against colon adenocarcinoma in vitro and in vivo. Expert Opin Drug Deliv 18(9):1309–1322

Zhao W, Cong Y, Li HM, Li S, Shen Y, Qi Q, Zhang Y, Li YZ, Tang YJ (2021) Challenges and potential for improving the druggability of podophyllotoxin-derived drugs in cancer chemotherapy. Nat Prod Rep 38(3):470–488

Zhao Y, Feng JH, Ding HX, Xiong Y, Cheng CH, Hao XJ et al (2006) Synthesis and cytotoxicity of racemic isodeoxypodophyllotoxin analogues with isoprene-derived side chains. J Nat Prod 69(8):1145–1152

Zhou XJ, Rahmani R (1992) Preclinical and clinical pharmacology of vinca alkaloids. Drugs 44(4):1–16

Zhu H, He YS, Ma J, Zhou J, Kong M, Wu CY, Mao Q, Lin G, Li SL (2021) The dual roles of ginsenosides in improving the anti-tumor efficiency of cyclophosphamide in mammary carcinoma mice. J Ethnopharmacol 265:113271

MIX
Papier aus verantwortungsvollen Quellen
Paper from responsible sources
FSC® C105338

If you have any concerns about our products,
you can contact us on
ProductSafety@springernature.com

In case Publisher is established outside the EU,
the EU authorized representative is:
Springer Nature Customer Service Center GmbH
Europaplatz 3, 69115 Heidelberg, Germany

Printed by Libri Plureos GmbH
in Hamburg, Germany